# THE MATH BOOK

# Books by Clifford A. Pickover

*The Alien IQ Test*
*Archimedes to Hawking*
*A Beginner's Guide to Immortality*
*Black Holes: A Traveler's Guide*
*Calculus and Pizza*
*Chaos and Fractals*
*Chaos in Wonderland*
*Computers, Pattern, Chaos, and Beauty*
*Computers and the Imagination*
*Cryptorunes: Codes and Secret Writing*
*Dreaming the Future*
*Egg Drop Soup*
*Future Health*
*Fractal Horizons: The Future Use of Fractals*
*Frontiers of Scientific Visualization*
*The Girl Who Gave Birth to Rabbits*
*The Heaven Virus*
*Jews in Hyperspace*
*Keys to Infinity*
*Liquid Earth*
*The Lobotomy Club*
*The Loom of God*
*The Mathematics of Oz*
*The Math Book*
*Mazes for the Mind: Computers and the Unexpected*
*Mind-Bending Visual Puzzles* (calendars and card sets)
*The Möbius Strip*
*The Paradox of God and the Science of Omniscience*
*A Passion for Mathematics*
*The Pattern Book: Fractals, Art, and Nature*
*The Science of Aliens*
*Sex, Drugs, Einstein, and Elves*
*Spider Legs* (with Piers Anthony)
*Spiral Symmetry* (with Istvan Hargittai)
*Strange Brains and Genius*
*Sushi Never Sleeps*
*The Stars of Heaven*
*Surfing Through Hyperspace*
*Time: A Traveler's Guide*
*Visions of the Future*
*Visualizing Biological Information*
*Wonders of Numbers*

# THE MATH BOOK

CLIFFORD A. PICKOVER

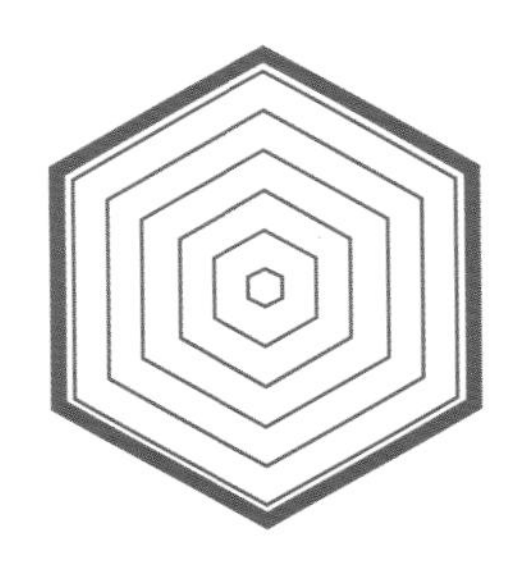

## 250 MILESTONES IN THE HISTORY OF MATHEMATICS

BARNES & NOBLE
NEW YORK

*For Martin Gardner*



This 2013 edition published by Barnes & Noble, Inc.



Cover design: Steve Attardo
Endpapers: Shutterstock.com

Barnes & Noble, Inc.
122 Fifth Avenue
New York, NY 10011

ISBN: 978-1-4351-4803-1

Printed and bound in China

3 5 7 9 10 8 6 4 2

*"Mathematics, rightly viewed, possesses not only truth, but supreme beauty—a beauty cold and austere, like that of sculpture."*

—Bertrand Russell, *Mysticism and Logic*, 1918

*"Mathematics is a wonderful, mad subject, full of imagination, fantasy, and creativity that is not limited by the petty details of the physical world, but only by the strength of our inner light."*

—Gregory Chaitin, "Less Proof, More Truth," *New Scientist*, July 28, 2007

*"Perhaps an angel of the Lord surveyed an endless sea of chaos, then troubled it gently with his finger. In this tiny and temporary swirl of equations, our cosmos took shape."*

—Martin Gardner, *Order and Surprise*, 1950

*"The great equations of modern physics are a permanent part of scientific knowledge, which may outlast even the beautiful cathedrals of earlier ages."*

—Steven Weinberg, in Graham Farmelo's *It Must Be Beautiful*, 2002

# Contents

# Introduction

## The Beauty and Utility of Mathematics

> *"An intelligent observer seeing mathematicians at work might conclude that they are devotees of exotic sects, pursuers of esoteric keys to the universe."*
>
> —Philip Davis and Reuben Hersh, *The Mathematical Experience*

Mathematics has permeated every field of scientific endeavor and plays an invaluable role in biology, physics, chemistry, economics, sociology, and engineering. Mathematics can be used to help explain the colors of a sunset or the architecture of our brains. Mathematics helps us build supersonic aircraft and roller coasters, simulate the flow of Earth's natural resources, explore subatomic quantum realities, and image faraway galaxies. Mathematics has changed the way we look at the cosmos.

In this book, I hope to give readers a taste for mathematics using few formulas, while stretching and exercising the imagination. However, the topics in this book are not mere curiosities with little value to the average reader. In fact, reports from the U.S. Department of Education suggest that successfully completing a mathematics class in high school results in better performance at college *whatever major* the student chooses to pursue.

The *usefulness* of mathematics allows us to build spaceships and investigate the geometry of our universe. Numbers may be our first means of communication with intelligent alien races. Some physicists have even speculated that an understanding of higher dimensions and of *topology*—the study of shapes and their interrelationships—may someday allow us to escape our universe, when it ends in either great heat or cold, and then we could call all of space-time our home.

Simultaneous discovery has often occurred in the history of mathematics. As I mention in my book *The Möbius Strip*, in 1858 the German mathematician August Möbius (1790–1868) simultaneously and independently discovered the Möbius strip (a wonderful twisted object with just one side) along with a contemporary scholar, the German mathematician Johann Benedict Listing (1808–1882). This simultaneous discovery of the Möbius band by Möbius and Listing, just like that of calculus by English polymath Isaac Newton (1643–1727) and German mathematician Gottfried Wilhelm Leibniz (1646–1716), makes me wonder why so many discoveries in science were made at the same time by people working independently. For another example, British naturalists Charles Darwin (1809–1882) and Alfred Wallace (1823–1913) both developed the theory of evolution independently and simultaneously. Similarly, Hungarian mathematician János Bolyai (1802–1860) and Russian mathematician Nikolai Lobachevsky (1793–1856) seemed to have developed hyperbolic geometry independently, and at the same time.

Most likely, such simultaneous discoveries have occurred because the time was ripe for such discoveries, given humanity's accumulated knowledge at the time the discoveries were made. Sometimes, two scientists are stimulated by reading the same preliminary research of one of their contemporaries. On the other hand, mystics have suggested that a deeper meaning exists to such coincidences. Austrian biologist Paul Kammerer (1880–1926) wrote, "We thus arrive at the image of a world-mosaic or cosmic kaleidoscope, which, in spite of constant shuffling and rearrangements, also takes care of bringing like and like together." He compared events in our world to the tops of ocean waves that seem isolated and unrelated. According to his controversial theory, we notice the tops of the waves, but beneath the surface some kind of synchronistic mechanism may exist that mysteriously connects events in our world and causes them to cluster.

Georges Ifrah in *The Universal History of Numbers* discusses simultaneity when writing about Mayan mathematics:

> We therefore see yet again how people who have been widely separated in time or space have... been led to very similar if not identical results.... In some cases, the explanation for this may be found in contacts and influences between different groups of people.... The true explanation lies in what we have previously referred to as the profound unity of culture: the intelligence of *Homo sapiens* is universal and its potential is remarkably uniform in all parts of the world.

Ancient people, like the Greeks, had a deep fascination with numbers. Could it be that in difficult times numbers were the only constant thing in an ever-shifting world? To the Pythagoreans, an ancient Greek sect, numbers were tangible, immutable, comfortable, eternal—more reliable than friends, less threatening than Apollo and Zeus.

Many entries in this book deal with whole numbers, or integers. The brilliant mathematician Paul Erdös (1913–1996) was fascinated by number theory—the study of integers—and he had no trouble posing problems, using integers, that were often simple to state but notoriously difficult to solve. Erdös believed that if one can state a problem in mathematics that is unsolved for more than a century, then it is a problem in number theory.

Many aspects of the universe can be expressed by whole numbers. Numerical patterns describe the arrangement of florets in a daisy, the reproduction of rabbits, the orbit of the planets, the harmonies of music, and the relationships between elements in the periodic table. Leopold Kronecker (1823–1891), a German algebraist and number theorist, once said, "The integers came from God and all else was man-made." His implication was that the primary source of all mathematics is the integers.

Since the time of Pythagoras, the role of integer ratios in musical scales has been widely appreciated. More important, integers have been crucial in the evolution of humanity's scientific understanding. For example, French chemist Antoine Lavoisier (1743–1794) discovered that chemical compounds are composed of fixed proportions of elements corresponding to the ratios of small integers. This was very strong evidence for the existence of atoms. In 1925, certain integer relations between the wavelengths of spectral lines emitted by excited atoms gave early clues to the structure of atoms. The near-integer ratios of atomic weights were evidence that the atomic nucleus is made up of an integer number of similar nucleons (protons and neutrons). The deviations from integer ratios led to the discovery of elemental isotopes (variants with nearly identical chemical behavior but with different numbers of neutrons).

Small divergences in the atomic masses of pure isotopes from exact integers confirmed Einstein's famous equation $E = mc^2$ and also the possibility of atomic bombs. Integers are everywhere in atomic physics. Integer relations are fundamental strands in the mathematical weave—or as German mathematician Carl Friedrich Gauss (1777–1855) said, "Mathematics is the queen of sciences—and number theory is the queen of mathematics."

Our mathematical description of the universe grows forever, but our brains and language skills remain entrenched. New kinds of mathematics are being discovered or created all the time, but we need fresh ways to think and to understand. For example, in the last few years, mathematical proofs have been offered for famous problems in the history of mathematics, but the arguments have been far too long and complicated for experts to be certain they are correct. Mathematician Thomas Hales had to wait *five years* before expert reviewers of his geometry paper—submitted to the journal *Annals of Mathematics*—finally decided that they could find no errors and that the journal should publish Hales's proof, but only with the disclaimer saying they were not certain it was right! Moreover, mathematicians like Keith Devlin have admitted in the *New York Times* that "the story of mathematics has reached a stage of such abstraction that many of its frontier problems cannot even be understood by the experts." If experts have such trouble, one can easily see the challenge of conveying this kind of information to a general audience. We do the best we can. Mathematicians can construct theories and perform computations, but they may not be sufficiently able to fully comprehend, explain, or communicate these ideas.

A physics analogy is relevant here. When Werner Heisenberg worried that human beings might never truly understand atoms, Niels Bohr was a bit more optimistic. He replied in the early 1920s, "I think we may yet be able to do so, but in the process we may have to learn what the word *understanding* really means." Today, we use computers to help us reason beyond the limitations of our own intuition. In fact, experiments with computers are leading mathematicians to discoveries and insights never dreamed of before the ubiquity of these devices. Computers and computer graphics allow mathematicians to discover results long before they can prove them formally and open entirely new fields of mathematics. Even simple computer tools like spreadsheets give modern mathematicians power that Gauss, Leonhard Euler, and Newton would have lusted after. As just one example, in the late 1990s, computer programs designed by David Bailey and Helaman Ferguson helped produce new formulas that related pi to log 5 and two other constants. As Erica Klarreich reports in *Science News*, once the computer had produced the formula, proving that it was correct was extremely easy. Often, simply *knowing* the answer is the largest hurdle to overcome when formulating a proof.

Mathematical theories have sometimes been used to predict phenomena that were not confirmed until years later. For example, Maxwell's equations, named after physicist James Clerk Maxwell, predicted radio waves. Einstein's field equations suggested that gravity would bend light and that the universe is expanding. Physicist Paul Dirac once noted that the abstract mathematics we study now gives us a glimpse of physics in the future. In fact, his equations predicted the existence of antimatter, which was subsequently discovered. Similarly, mathematician Nikolai Lobachevsky said that "there is no branch of mathematics, however abstract, which may not someday be applied to the phenomena of the real world."

In this book, you will encounter various interesting geometries that have been thought to hold the keys to the universe. Galileo Galilei (1564–1642) suggested that "Nature's great book is written in mathematical symbols." Johannes Kepler (1571–1630) modeled the solar system with Platonic solids such as the dodecahedron. In the 1960s, physicist Eugene Wigner (1902–1995) was impressed with the "unreasonable effectiveness of mathematics in the natural sciences." Large Lie groups, like $E_8$—which is discussed in the entry "The Quest for Lie Group $E_8$ (2007)"—may someday help us create a unified theory of physics. In 2007, Swedish American cosmologist Max Tegmark published both scientific and popular articles on the mathematical universe hypothesis, which states that our physical reality is a mathematical structure—in other words, our universe is not just *described* by mathematics—it *is* mathematics.

## Book Organization and Purpose

*"At every major step, physics has required, and frequently stimulated, the introduction of new mathematical tools and concepts. Our present understanding of the laws of physics, with their extreme precision and universality, is only possible in mathematical terms."*

—Sir Michael Atiyah, "Pulling the Strings," *Nature*

One common characteristic of mathematicians is a passion for completeness—an urge to return to first principles to explain their works. As a result, readers of mathematical texts must often wade through pages of background before getting to the essential findings. To avoid this problem, each entry in this book is short, at most only a few paragraphs in length. This format allows readers to jump right in to ponder a subject, without having to sort through a lot of verbiage. Want to know about infinity? Turn to the entries "Cantor's Transfinite Numbers" (1874) or "Hilbert's Grand Hotel" (1925), and you'll have a quick mental workout. Interested in the first commercially successful portable mechanical calculator, developed by a prisoner in a Nazi concentration camp? Turn to "Curta Calculator" (1948) for a brief introduction.

Wonder how an amusing-sounding theorem may one day help scientists form nanowires for electronics devices? Then browse through the book and read the "Hairy Ball Theorem" (1912) entry. Why did the Nazis compel the president of the Polish Mathematical Society to feed his own blood to lice? Why was the first female mathematician murdered? Is it really possible to turn a sphere inside out? Who was the "Number Pope"? When did humans tie their first knots? Why don't we use Roman numerals anymore? Who was the earliest named individual in the history of mathematics? Can a surface have only one side? We'll tackle these and other thought-provoking questions in the pages that follow.

Of course, my approach has some disadvantages. In just a few paragraphs, I can't go into any depth on a subject. However, I provide suggestions for further reading in the "Notes and Further Reading" section. While I sometimes list primary sources, I have often explicitly listed excellent secondary references that readers can frequently obtain more easily than older primary sources. Readers interested in pursuing any subject can use the references as a useful starting point.

My goal in writing *The Math Book* is to provide a wide audience with a brief guide to important mathematical ideas and thinkers, with entries short enough to digest in a few minutes. Most entries are ones that interested me personally. Alas, not all of the great mathematical milestones are included in this book in order to prevent the book from growing too large. Thus, in celebrating the wonders of mathematics in this short volume, I have been forced to omit many important mathematical marvels. Nevertheless, I believe that I have included a majority of those with historical significance and that have had a strong influence on mathematics, society, or human thought. Some entries are eminently practical, involving topics that range from slide rules and other calculating devices to geodesic domes and the invention of zero. Occasionally, I include several lighter moments, which were nonetheless significant, such as the rise of the Rubik's Cube puzzle or the solving of the Bed Sheet Problem. Sometimes, snippets of information are repeated so that each entry can be read on its own. Occasional text in boldface type points the reader to related entries. Additionally, a small "See also" section at the bottom of each entry helps weave entries together in a web of interconnectedness and may help the reader traverse the book in a playful quest for discovery.

*The Math Book* reflects my own intellectual shortcomings, and while I try to study as many areas of science and mathematics as I can, it is difficult to become fluent in all aspects, and this book clearly indicates my own personal interests, strengths, and weaknesses. I am responsible for the choice of pivotal entries included in this book and, of course, for any errors and infelicities. This is not a comprehensive or scholarly dissertation, but rather it is intended as recreational reading for students of science and mathematics and interested laypeople. I welcome feedback and suggestions for improvement from readers, as I consider this an ongoing project and a labor of love.

This book is organized chronologically, according to the year of a mathematical milestone or finding. In some cases, the literature may report slightly different dates for the milestone because some sources give the publication date as the discovery date of a finding, while other sources give the actual date that a mathematical principle was discovered, regardless of the fact that the publication date is sometimes a year or more later. If I was uncertain of a precise earlier date of discovery, I often used the publication date.

Dating of entries can also be a question of judgment when more than one individual made a contribution. Often, I have used the earliest date where appropriate, but sometimes I have surveyed colleagues and decided to use the date when a concept gained particular prominence. For example, consider the Gray code, which is used to facilitate error correction in digital communications, such as in TV signal transmission, and to make transmission systems less susceptible to noise. This code was named after Frank Gray, a physicist at Bell Telephone Laboratories in the 1950s and 1960s. During this time, these kinds of codes gained particular prominence, partly due to his patent filed in 1947 and the rise of modern communications. The Gray code entry is thus dated as 1947, although it might also have been dated much earlier, because the roots of the idea go back to Émile Baudot (1845–1903), the French pioneer of the telegraph. In any case, I have attempted to give readers a feel for the span of possible dates in each entry or in the "Notes and Further Reading" section.

Scholars sometimes have disputes with respect to the person to whom a discovery is traditionally attributed. For example, author Heinrich Dörrie cites four scholars who do not believe that a particular version of Archimedes' cattle problem is due to Archimedes, but he also cites four authors who believe that the problem *should* be attributed to Archimedes. Scholars also dispute the authorship of Aristotle's wheel paradox. Where possible, I mention such disputes either in the main text or the "Notes and Further Reading" section.

You will notice that a significant number of milestones have been achieved in just the last few decades. As just one example, in 2007, researchers finally "solved" the game of checkers, proving that if an opponent plays perfectly, the game ends in a draw. As already mentioned, part of the rapid recent progress in mathematics is due to the use of the computer as a tool for mathematical experiments. For the checkers solution, the analysis actually began in 1989 and required dozens of computers for the complete solution. The game has roughly 500 billion billion possible positions.

Sometimes, science reporters or famous researchers are quoted in the main entries, but purely for brevity I don't list the source of the quote or the author's full credentials in the entry. I apologize in advance for this occasional compact approach; however, references in the back of the book should help to make the author's identity clearer.

Even the naming of a theorem can be a tricky business. For example, mathematician Keith Devlin writes in his 2005 column for the Mathematical Association of America:

> Most mathematicians prove many theorems in their lives, and the process whereby their name gets attached to one of them is very haphazard. For instance, Euler, Gauss, and Fermat each proved hundreds of theorems, many of them important ones, and yet their names are attached to just a few of them. Sometimes theorems acquire names that are incorrect. Most famously, perhaps, Fermat almost certainly did not prove "Fermat's Last Theorem"; rather, that name was attached by someone else, after his death, to a conjecture the French mathematician had scribbled in the margin of a textbook. And Pythagoras's theorem was known long before Pythagoras came onto the scene.

In closing, let us note that mathematical discoveries provide a framework in which to explore the nature of reality, and mathematical tools allow scientists to make predictions about the universe; thus, the discoveries in this book are among humanity's greatest achievements.

At first glance, this book may seem like a long catalogue of isolated concepts and people with little connection between them. But as you read, I think you'll begin to see many linkages. Obviously, the final goal of scientists and mathematicians is not simply the accumulation of facts and lists of formulas, but rather they seek to understand the patterns, organizing principles, and relationships between these facts to form theorems and entirely new branches of human thought. For me, mathematics cultivates a perpetual state of wonder about the nature of mind, the limits of thoughts, and our place in this vast cosmos.

Our brains, which evolved to make us run from lions on the African savanna, may not be constructed to penetrate the infinite veil of reality. We may need mathematics, science, computers, brain augmentation, and even literature, art, and poetry to help us tear away the veils. For those of you who are about to embark on reading the *The Math Book* from cover to cover, look for the connections, gaze in awe at the evolution of ideas, and sail on the shoreless sea of imagination.


## Acknowledgments

I thank Teja Krašek, Dennis Gordon, Nick Hobson, Pete Barnes, and Mark Nandor for their comments and suggestions. I would also like to especially acknowledge Meredith Hale, my editor for this book, as well as Jos Leys, Teja Krašek, and Paul Nylander for allowing me to include their mathematically inspired artworks.

While researching the milestones and pivotal moments presented in this book, I studied a wide array of wonderful reference works and Web sites, many of which are listed in the "Notes and Further Reading" section toward the end of the book. These references include "The MacTutor History of Mathematics Archive" (*www-history.mcs.st-and.ac.uk*), "Wikipedia: The Free Encyclopedia" (*en.wikipedia.org*), "MathWorld" (*mathworld.wolfram.com*), Jan Gullberg's *Mathematics: From the Birth of Numbers*, David Darling's *The Universal Book of Mathematics*, Ivars Peterson's "Math Trek Archives" (*www.maa.org/mathland/mathland_archives.html*), Martin Gardner's *Mathematical Games* (a CD-ROM made available from The Mathematical Association of America), and some of my own books such as *A Passion for Mathematics*.

# Ant Odometer

Ants are social insects that evolved from vespoid wasps in the mid-Cretaceous period, about 150 million years ago. After the rise of flowering plants, about 100 million years ago, ants diversified into numerous species.

The Saharan desert ant, *Cataglyphis fortis*, travels immense distances over sandy terrain, often completely devoid of landmarks, as it searches for food. These creatures are able to return to their nest using a direct route rather than by retracing their outbound path. Not only do they judge directions, using light from the sky for orientation, but they also appear to have a built-in "computer" that functions like a pedometer that counts their steps and allows them to measure exact distances. An ant may travel as far as 160 feet (about 50 meters) until it encounters a dead insect, whereupon it tears a piece to carry directly back to its nest, accessed via a hole often less than a millimeter in diameter.

By manipulating the leg lengths of ants to give them longer and shorter strides, a research team of German and Swiss scientists discovered that the ants "count" steps to judge distance. For example, after ants had reached their destination, the legs were lengthened by adding stilts or shortened by partial amputation. The researchers then returned the ants so that the ants could start on their journey back to the nest. Ants with the stilts traveled too far and passed the nest entrance, while those with the amputated legs did not reach it. However, if the ants *started* their journey from their nest with the modified legs, they were able to compute the appropriate distances. This suggests that stride length is the crucial factor. Moreover, the highly sophisticated computer in the ant's brain enables the ant to compute a quantity related to the horizontal projection of its path so that it does not become lost even if the sandy landscape develops hills and valleys during its journey.

**SEE ALSO** Primates Count (c. 30 Million B.C.) and Cicada-Generated Prime Numbers (c. 1 Million B.C.).

*Saharan desert ants may have built-in "pedometers" that count steps and allow the ants to measure exact distances. Ants with stilts glued to their legs (shown in red) travel too far and pass their nest entrance, suggesting that stride length is important for distance determination.*

# Primates Count

Around 60 million years ago, small, lemur-like primates had evolved in many areas of the world, and 30 million years ago, primates with monkeylike characteristics existed. Could such creatures count? The meaning of *counting* by animals is a highly contentious issue among animal behavior experts. However, many scholars suggest that animals have some sense of number. H. Kalmus writes in his *Nature* article "Animals as Mathematicians":

> There is now little doubt that some animals such as squirrels or parrots can be trained to count.…Counting faculties have been reported in squirrels, rats, and for pollinating insects. Some of these animals and others can distinguish numbers in otherwise similar visual patterns, while others can be trained to recognize and even to reproduce sequences of acoustic signals. A few can even be trained to tap out the numbers of elements (dots) in a visual pattern.…The lack of the spoken numeral and the written symbol makes many people reluctant to accept animals as mathematicians.

Rats have been shown to "count" by performing an activity the correct number of times in exchange for a reward. Chimpanzees can press numbers on a computer that match numbers of bananas in a box. Testsuro Matsuzawa of the Primate Research Institute at Kyoto University in Japan taught a chimpanzee to identify numbers from 1 to 6 by pressing the appropriate computer key when she was shown a certain number of objects on the computer screen.

Michael Beran, a research scientist at Georgia State University in Atlanta, Georgia, trained chimps to use a computer screen and joystick. The screen flashed a numeral and then a series of dots, and the chimps had to match the two. One chimp learned numerals 1 to 7, while another managed to count to 6. When the chimps were tested again after a gap of three years, both chimps were able to match numbers, but with double the error rate.

**SEE ALSO** Ant Odometer (c. 150 Million B.C.) and Ishango Bone (c. 18,000 B.C.).

*Primates appear to have some sense of number, and the higher primates can be taught to identify numbers from 1 to 6 by pressing the appropriate computer key when shown a certain number of objects.*

# Cicada-Generated Prime Numbers

Cicadas are winged insects that evolved around 1.8 million years ago during the Pleistocene epoch, when glaciers advanced and retreated across North America. Cicadas of the genus *Magicicada* spend most of their lives below the ground, feeding on the juices of plant roots, and then emerge, mate, and die quickly. These creatures display a startling behavior: Their emergence is synchronized with periods of years that are usually the prime numbers 13 and 17. (A prime number is an integer such as 11, 13, and 17 that has only two integer divisors: 1 and itself.) During the spring of their 13th or 17th year, these periodical cicadas construct an exit tunnel. Sometimes more than 1.5 million individuals emerge in a single acre; this abundance of bodies may have survival value as they overwhelm predators such as birds that cannot possibly eat them all at once.

Some researchers have speculated that the evolution of prime-number life cycles occurred so that the creatures increased their chances of evading shorter-lived predators and parasites. For example, if these cicadas had 12-year life cycles, all predators with life cycles of 2, 3, 4, or 6 years might more easily find the insects. Mario Markus of the Max Planck Institute for Molecular Physiology in Dortmund, Germany, and his coworkers discovered that these kinds of prime-number cycles arise naturally from evolutionary mathematical models of interactions between predator and prey. In order to experiment, they first assigned random life-cycle durations to their computer-simulated populations. After some time, a sequence of mutations always locked the synthetic cicadas into a stable prime-number cycle.

Of course, this research is still in its infancy and many questions remain. What is special about 13 and 17? What predators or parasites have actually existed to drive the cicadas to these periods? Also, a mystery remains as to why, of the 1,500 cicada species worldwide, only a small number of the genus *Magicicada* are known to be periodical.

**SEE ALSO** Ant Odometer (c. 150 Million B.C.), Ishango Bone (c. 18,000 B.C.), Sieve of Eratosthenes (240 B.C.), Goldbach Conjecture (1742), Constructing a Regular Heptadecagon (1796), Gauss's *Disquisitiones Arithmeticae* (1801), Proof of the Prime Number Theorem (1896), Brun's Constant (1919), Gilbreath's Conjecture (1958), Sierpiński Numbers (1960), Ulam Spiral (1963), Erdös and Extreme Collaboration (1971), and Andrica's Conjecture (1985).

*Certain cicadas display a startling behavior: Their emergence from the soil is synchronized with periods that are usually the prime numbers 13 and 17. Sometimes more than 1.5 million individuals emerge in a single acre within a short interval of time.*

c.100,000 B.C.

# Knots

The use of knots may predate modern humans (*Homo sapiens*). For example, seashells colored with ocher, pierced with holes, and dated to 82,000 years ago have been discovered in a Moroccan cave. Other archeological evidence suggests much older bead use in humans. The piercing implies the use of cords and the use of a knot to hold the objects to a loop, such as a necklace.

The quintessence of ornamental knots is exemplified by *The Book of Kells*, an ornately illustrated Gospel Bible, produced by Celtic monks in about A.D. 800. In modern times, the study of knots, such as the trefoil knot with three crossings, is part of a vast branch of mathematics dealing with closed twisted loops. In 1914, German mathematician Max Dehn (1878–1952) showed that the trefoil knot's mirror images are not equivalent.

For centuries, mathematicians have tried to develop ways to distinguish tangles that *look* like knots (called *unknots*) from true knots and to distinguish true knots from one another. Over the years, mathematicians have created seemingly endless tables of distinct knots. So far, more than 1.7 million nonequivalent knots with pictures containing 16 or fewer crossings have been identified.

Entire conferences are devoted to knots today. Scientists study knots in fields ranging from molecular genetics—to help us understand how to unravel a loop of DNA—to particle physics, in an attempt to represent the fundamental nature of elementary particles.

Knots have been crucial to the development of civilization, where they have been used to tie clothing, to secure weapons to the body, to create shelters, and to permit the sailing of ships and world exploration. Today, knot theory in mathematics has become so advanced that mere mortals find it challenging to understand its most profound applications. In a few millennia, humans have transformed knots from simple necklace ties to models of the very fabric of reality.

**SEE ALSO** Quipu (c. 3000 B.C.), Borromean Rings (834), Perko Knots (1974), Jones Polynomial (1984), and Murphy's Law and Knots (1988).

*The quintessence of ornamental knots is exemplified by* The Book of Kells, *an ornately illustrated Gospel Bible, produced by Celtic monks in about A.D. 800. Various knot-like forms can be seen in the details of this illustration.*

c. 18,000 B.C.

# Ishango Bone

In 1960, Belgian geologist and explorer Jean de Heinzelin de Braucourt (1920–1998) discovered a baboon bone with markings in what is today the Democratic Republic of the Congo. The Ishango bone, with its sequence of notches, was first thought to be a simple tally stick used by a Stone Age African. However, according to some scientists, the marks suggest a mathematical prowess that goes beyond counting of objects.

The bone was found in Ishango, near the headwaters of the Nile River, the home of a large population of upper Paleolithic people prior to a volcanic eruption that buried the area. One column of marks on the bone begins with three notches that double to six notches. Four notches double to eight. Ten notches halve to five. This may suggest a simple understanding of doubling or halving. Even more striking is the fact that numbers in other columns are all odd (9, 11, 13, 17, 19, and 21). One column contains the prime numbers between 10 and 20, and the numbers in each column sum to 60 or 48, both multiples of 12.

A number of tally sticks have been discovered that predate the Ishango bone. For example, the Swaziland Lebombo bone is a 37,000-year-old baboon fibula with 29 notches. A 32,000-year-old wolf tibia with 57 notches, grouped in fives, was found in Czechoslovakia. Although quite speculative, some have hypothesized that the markings on the Ishango bone form a kind of lunar calendar for a Stone Age woman who kept track of her menstrual cycles, giving rise to the slogan "menstruation created mathematics." Even if the Ishango was a simple bookkeeping device, these tallies seem to set us apart from the animals and represent the first steps to symbolic mathematics. The full mystery of the Ishango bone can't be solved until other similar bones are discovered.

**SEE ALSO** Primates Count (c. 30 Million B.C.), Cicada-Generated Prime Numbers (c. 1 Million B.C.), and Sieve of Eratosthenes (240 B.C.).

*The Ishango baboon bone, with its sequence of notches, was first thought to be a simple tally stick used by a Stone Age African. However, some scientists believe that the marks suggest a mathematical prowess that goes beyond counting of objects.*

c. 3000 B.C.

# Quipu

The ancient Incas used *quipus* (pronounced "key-poos"), memory banks made of strings and knots, for storing numbers. Until recently, the oldest-known quipus dated from about A.D. 650. However, in 2005, a quipu from the Peruvian coastal city of Caral was dated to about 5,000 years ago.

The Incas of South America had a complex civilization with a common state religion and a common language. Although they did not have writing, they kept extensive records encoded by a logical-numerical system on the quipus, which varied in complexity from three to around a thousand cords. Unfortunately, when the Spanish came to South America, they saw the strange quipus and thought they were the works of the Devil. The Spanish destroyed thousands of them in the name of God, and today only about 600 quipus remain.

Knot types and positions, cord directions, cord levels, and color and spacing represent numbers mapped to real-world objects. Different knot groups were used for different powers of 10. The knots were probably used to record human and material resources and calendar information. The quipus may have contained more information such as construction plans, dance patterns, and even aspects of Inca history. The quipu is significant because it dispels the notion that mathematics flourishes only after a civilization has developed writing; however, societies can reach advanced states without ever having developed written records. Interestingly, today there are computer systems whose file managers are called quipus, in honor of this very useful ancient device.

One sinister application of the quipu by the Incas was as a death calculator. Yearly quotas of adults and children were ritually slaughtered, and this enterprise was planned using a quipu. Some quipus represented the empire, and the cords referred to roads and the knots to sacrificial victims.

**SEE ALSO** Knots (c. 100,000 B.C.) and Abacus (c. 1200).

*The ancient Incas used quipus made of knotted strings to store numbers. Knot types and positions, cord directions, cord levels, and colors often represented dates and counts of people and objects.*

c. 3000 B.C.

# Dice

Imagine a world without random numbers. In the 1940s, the generation of statistically random numbers was important to physicists simulating thermonuclear explosions, and today, many computer networks employ random numbers to help route Internet traffic to avoid congestion. Political poll-takers use random numbers to select unbiased samples of potential voters.

Dice, originally made from the anklebones of hoofed animals, were one of the earliest means for producing random numbers. In ancient civilizations, the gods were believed to control the outcome of dice tosses; thus, dice were relied upon to make crucial decisions, ranging from the selection of rulers to the division of property in an inheritance. Even today, the metaphor of God controlling dice is common, as evidenced by astrophysicist Stephen Hawking's quote, "Not only does God play dice, but He sometimes confuses us by throwing them where they can't be seen."

The oldest-known dice were excavated together with a 5,000-year-old backgammon set from the legendary Burnt City in southeastern Iran. The city represents four stages of civilization that were destroyed by fires before being abandoned in 2100 B.C. At this same site, archeologists also discovered the earliest-known artificial eye, which once stared out hypnotically from the face of an ancient female priestess or soothsayer.

For centuries, dice rolls have been used to teach probability. For a single roll of an $n$-sided die with a different number on each face, the probability of rolling any value is $1/n$. The probability of rolling a particular sequence of $i$ numbers is $1/n^i$. For example, the chance of rolling a 1 followed by a 4 on a traditional die is $1/6^2 = 1/36$. Using two traditional dice, the probability of throwing any given sum is the number of ways to throw that sum divided by the total number of combinations, which is why a sum of 7 is much more likely than a sum of 2.

**SEE ALSO** Law of Large Numbers (1713), Buffon's Needle (1777), Least Squares (1795), Laplace's *Théorie Analytique des Probabilités* (1812), Chi-Square (1900), Lost in Hyperspace (1921), The Rise of Randomizing Machines (1938), Pig Game Strategy (1945), and Von Neumann's Middle-Square Randomizer (1946).

*Dice were originally made from the anklebones of animals and were among the earliest means for producing random numbers. In ancient civilizations, people used dice to predict the future, believing that the gods influenced dice outcomes.*

c. 2200 B.C.

# Magic Squares

**Bernard Frénicle de Bessy** (1602–1675)

Legends suggest that magic squares originated in China and were first mentioned in a manuscript from the time of Emperor Yu, around 2200 B.C. A *magic square* consists of $N^2$ boxes, called *cells*, filled with integers that are all different. The sums of the numbers in the horizontal rows, vertical columns, and main diagonals are all equal.

If the integers in a magic square are the consecutive numbers from 1 to $N^2$, the square is said to be of the Nth order, and the *magic number*, or sum of each row, is a constant equal to $N(N^2 + 1)/2$. Renaissance artist Albrecht Dürer created this wonderful $4 \times 4$ magic square below in 1514.

| | | | |
|---|---|---|---|
| 16 | 3 | 2 | 13 |
| 5 | 10 | 11 | 8 |
| 9 | 6 | 7 | 12 |
| 4 | 15 | 14 | 1 |

Note the two central numbers in the bottom row read "1514," the year of its construction. The rows, columns, and main diagonals sum to 34. In addition, 34 is the sum of the numbers of the corner squares ($16 + 13 + 4 + 1$) and of the central $2 \times 2$ square ($10 + 11 + 6 + 7$).

As far back as 1693, the 880 different fourth-order magic squares were published posthumously in *Des quassez ou tables magiques* by Bernard Frénicle de Bessy, an eminent amateur French mathematician and one of the leading magic square researchers of all time.

We've come a long way from the simplest $3 \times 3$ magic squares venerated by civilizations of almost every period and continent, from the Mayan Indians to the Hasua people of Africa. Today, mathematicians study these magic objects in high dimensions—for example, in the form of four-dimensional hypercubes that have magic sums within all appropriate directions.

**SEE ALSO** Franklin Magic Square (1769) and Perfect Magic Tesseract (1999).

*The Sagrada Família church in Barcelona, Spain, features a 4 × 4 magic square with a magic constant of 33, the age at which Jesus died according to many biblical interpretations. Note that this is not a traditional magic square because some numbers are repeated.*

1 14 14 4
11 7 6 9
8 10 10 5
13 2 3 15

c. 1800 B.C.

# Plimpton 322

**George Arthur Plimpton** (1855–1936)

Plimpton 322 refers to a mysterious Babylonian clay tablet featuring numbers in cuneiform script in a table of 4 columns and 15 rows. Eleanor Robson, a historian of science, refers to it as "one of the world's most famous mathematical artifacts." Written around 1800 B.C., the table lists Pythagorean triples—that is, whole numbers that specify the side lengths of right triangles that are solutions to the Pythagorean theorem $a^2 + b^2 = c^2$. For example, the numbers 3, 4, and 5 are a Pythagorean triple. The fourth column in the table simply contains the row number. Interpretations vary as to the precise meaning of the numbers in the table, with some scholars suggesting that the numbers were solutions for students studying algebra or trigonometry-like problems.

Plimpton 322 is named after New York publisher George Plimpton who, in 1922, bought the tablet for $10 from a dealer and then donated the tablet to Columbia University. The tablet can be traced to the Old Babylonian civilization that flourished in Mesopotamia, the fertile valley of the Tigris and Euphrates rivers, which is now located in Iraq. To put the era into perspective, the unknown scribe who generated Plimpton 322 lived within about a century of King Hammurabi, famous for his set of laws that included "an eye for an eye, a tooth for a tooth." According to biblical history, Abraham, who is said to have led his people west from the city of Ur on the bank of the Euphrates into Canaan, would have been another near contemporary of the scribe.

The Babylonians wrote on wet clay by pressing a stylus or wedge into the clay. In the Babylonian number system, the number 1 was written with a single stroke and the numbers 2 through 9 were written by combining multiples of a single stroke.

**SEE ALSO** Pythagorean Theorem and Triangles (c. 600 B.C.).

*Plimpton 322 refers to a Babylonian clay tablet featuring numbers in cuneiform script. These whole numbers specify the side lengths of right triangles that are solutions to the Pythagorean theorem* $a^2 + b^2 = c^2$.

c.1650 B.C.

# Rhind Papyrus

**Ahmes** (c. 1680 B.C.–c. 1620 B.C.), **Alexander Henry Rhind** (1833–1863)

The Rhind Papyrus is considered to be the most important known source of information concerning ancient Egyptian mathematics. This scroll, about a foot (30 centimeters) high and 18 feet (5.5 meters) long, was found in a tomb in Thebes on the east bank of the river Nile. Ahmes, the scribe, wrote it in hieratic, a script related to the hieroglyphic system. Given that the writing occurred in around 1650 B.C., this makes Ahmes the earliest-named individual in the history of mathematics! The scroll also contains the earliest-known symbols for mathematical operations—*plus* is denoted by a pair of legs walking toward the number to be added.

In 1858, Scottish lawyer and Egyptologist Alexander Henry Rhind had been visiting Egypt for health reasons when he bought the scroll in a market in Luxor. The British Museum in London acquired the scroll in 1864.

Ahmes wrote that the scroll gives an "accurate reckoning for inquiring into things, and the knowledge of all things, mysteries…all secrets." The content of the scroll concerns mathematical problems involving fractions, arithmetic progressions, algebra, and pyramid geometry, as well as practical mathematics useful for surveying, building, and accounting. The problem that intrigues me the most is Problem 79, the interpretation of which was initially baffling.

Today, many interpret Problem 79 as a puzzle, which may be translated as "Seven houses contain seven cats. Each cat kills seven mice. Each mouse had eaten seven ears of grain. Each ear of grain would have produced seven hekats (measures) of wheat. What is the total of all of these?" Interestingly, this indestructible puzzle meme, involving the number 7 and animals, seems to have persisted through thousands of years! We observe something quite similar in Fibonacci's *Liber Abaci* (*Book of Calculation*), published in 1202, and later in the St. Ives puzzle, an Old English children's rhyme involving 7 cats.

**SEE ALSO** *Ganita Sara Samgraha* (850), Fibonacci's *Liber Abaci* (1202), and *Treviso Arithmetic* (1478).

*The Rhind Papyrus is the most important source of information concerning ancient Egyptian mathematics. The scroll, a portion of which is shown here, includes mathematical problems involving fractions, arithmetic progressions, algebra, geometry, and accounting.*

c. 1300 B.C.

# Tic Tac Toe

The game of Tic Tac Toe (TTT) is a among humanity's best-known and most ancient games. Although the precise date of TTT with its modern rules may be relatively recent, archeologists can trace what appear to be "three-in-a-row games" to ancient Egypt around 1300 B.C., and I suspect that similar kinds of games originated at the very dawn of human societies. For TTT, two players, O and X, take turns marking their symbols in the spaces of a $3 \times 3$ grid. The player who first places three of his own marks in a horizontal, vertical, or diagonal row wins. A draw can always be obtained for the $3 \times 3$ board.

In ancient Egypt, during the time of the great pharaohs, board games played an important role in everyday life, and TTT-like games are known to have been played during these ancient days. TTT may be considered an "atom" upon which the molecules of more advanced games of position were built through the centuries. With the slightest of variations and extensions, the simple game of TTT becomes a fantastic challenge requiring significant time to master.

Mathematicians and puzzle aficionados have extended TTT to larger boards, higher dimensions, and strange playing surfaces such as rectangular or square boards that are connected at their edges to form a torus (doughnut shape) or Klein bottle (a surface with just one side).

Consider some TTT curiosities. Players can place their Xs and Os on the TTT board in $9! = 362,880$ ways. There are 255,168 possible games in TTT when considering all possible games that end in 5, 6, 7, 8, and 9 moves. In the early 1980s, computer geniuses Danny Hillis, Brian Silverman, and friends built a Tinkertoy® computer that played TTT. The device was made from 10,000 Tinkertoy parts. In 1998, researchers and students at the University of Toronto created a robot to play three-dimensional ($4 \times 4 \times 4$) TTT with a human.

**SEE ALSO** Go (548 B.C.), Icosian Game (1857), Solving the Game of Awari (2002), and Checkers Is Solved (2007).

*Philosophers Patrick Grim and Paul St. Denis offer an analytic presentation of all possible Tic-Tac-Toe games. Each cell in the Tic-Tac-Toe board is divided into smaller boards to show various possible choices.*

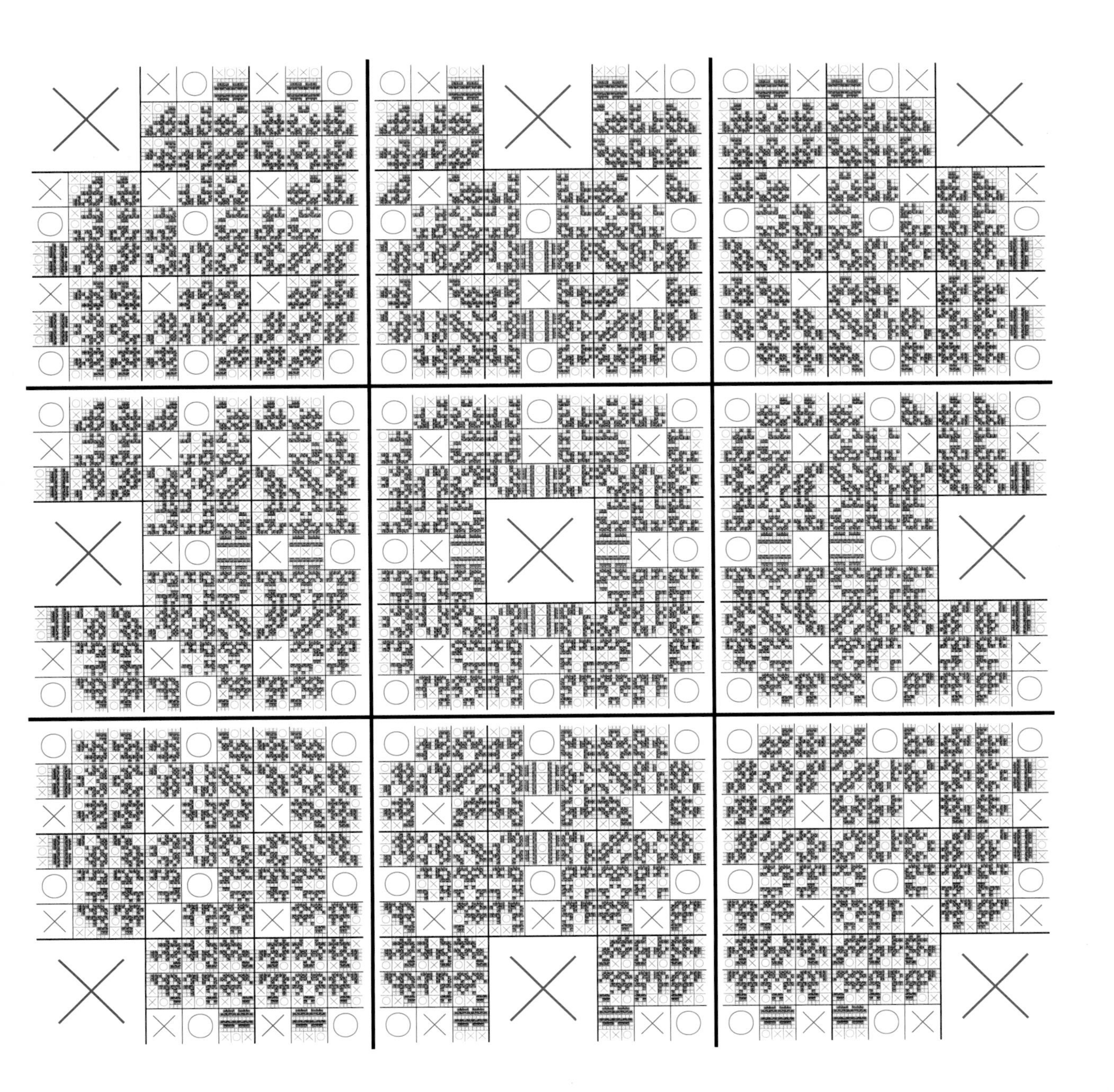

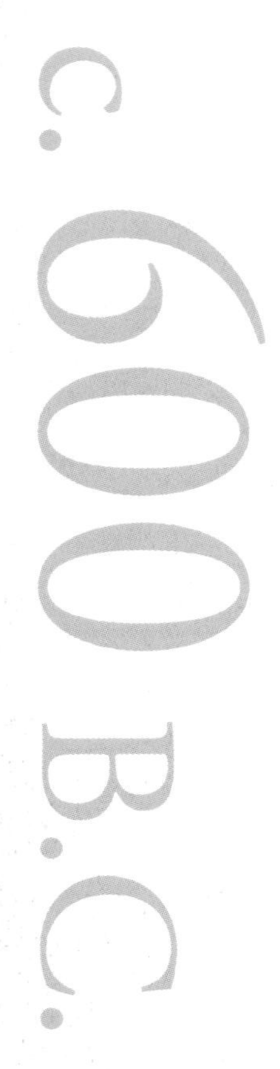


# Pythagorean Theorem and Triangles

**Baudhayana** (c. 800 B.C.), **Pythagoras of Samos** (c. 580 B.C.–c. 500 B.C.)

Today, young children sometimes first hear of the famous Pythagorean theorem from the mouth of the Scarecrow, when he finally gets a brain in MGM's 1939 film version of *The Wizard of Oz*. Alas, the Scarecrow's recitation of the famous theorem is completely wrong!

The Pythagorean theorem states that for any right triangle, the square of the hypotenuse length $c$ is equal to the sum of the squares on the two (shorter) "leg" lengths $a$ and $b$—which is written as $a^2 + b^2 = c^2$. The theorem has more published proofs than any other, and Elisha Scott Loomis's book *Pythagorean Proposition* contains 367 proofs.

Pythagorean triangles (PTs) are right triangles with integer sides. The "3-4-5" PT—with legs of lengths 3 and 4, and a hypotenuse of length 5—is the only PT with three sides as consecutive numbers and the only triangle with integer sides, the sum of whose sides (12) is equal to double its area (6). After the 3-4-5 PT, the next triangle with consecutive leg lengths is 21-20-29. The tenth such triangle is much larger: 27304197-27304196-38613965.

In 1643, French mathematician Pierre de Fermat (1601–1665) asked for a PT, such that both the hypotenuse $c$ and the sum $(a + b)$ had values that were square numbers. It was startling to find that the *smallest* three numbers satisfying these conditions are 4,565,486,027,761, 1,061,652,293,520, and 4,687,298,610,289. It turns out that the second such triangle would be so "large" that if its numbers were represented as feet, the triangle's legs would project from Earth to beyond the sun!

Although Pythagoras is often credited with the formulation of the Pythagorean theorem, evidence suggests that the theorem was developed by the Hindu mathematician Baudhayana centuries earlier around 800 B.C. in his book *Baudhayana Sulba Sutra*. Pythagorean triangles were probably known even earlier to the Babylonians.

**SEE ALSO** Plimpton 322 (c. 1800 B.C.), Pythagoras Founds Mathematical Brotherhood (c. 530 B.C.), Quadrature of the Lune (c. 440 B.C.), Law of Cosines (c. 1427), and Viviani's Theorem (1659).

*Persian mathematician Nasr al-Din al-Tusi (1201–1274) presented a version of Euclid's proof of the Pythagorean theorem. Al-Tusi was a prolific mathematician, astronomer, biologist, chemist, philosopher, physician, and theologian.*

[illegible]

ر آ اط كح فنصل ر آ آح خطا واحدا لكون زاويتي
ح س آح قائمتين وكذلك س اط ونخرج من آ آل موازيا
لـ د ويقع داخل المثلث لان زاوية د ... آكبر من قائمة فيكون
زاوية آ ل اصغر من زاوية ... آح القائمة ويقطع لا محالة ... ح
على م ويقسم به مربع ... الى سطحي ... ل ... ونصل
آد ... فلان في مثلثي ح ... آد ضلعي ... ح
وزاوية ح ... مساوية لضلعي ... وزاوية ...
يكون المثلثان متساويين ومثلث ح ... يساوي نصف مربع
... لكونهما على قاعدة
... بين متوازيين ...
... وكذلك مثلث ...
يساوي نصف سطح ...
لكونهما على قاعدة ...
بين متوازيين ...
فمربع ... يساوي
سطح ... المساوي نصفيهما وبمثل ذلك تبين ان مربع ... يساوي
سطح ... فاذن مربع ... يساوي مربعي ... وذلك ما اردناه

548 B.C.

# Go

Go is a two-player board game that originated in China around 2000 B.C. The earliest written references to the game are from the earliest Chinese work of narrative history, *Zuo Zhuan* (*Chronicle of Zuo*), which describes a man in 548 B.C. who played the game. The game spread to Japan, where it became popular in the thirteenth century. Two players alternately place black and white stones on intersections of a 19 × 19 playing board. A stone or a group of stones is captured and removed if it is tightly surrounded by stones of the opposing color. The objective is to control a larger territory than one's opponent.

Go is complex for many reasons, including its large game board, multifaceted strategies, and huge numbers of variations in possible games. Simply having more stones than an opponent does not ensure victory. After taking symmetry into account, there are 32,940 opening moves, of which 992 are considered to be strong ones. The number of possible board configurations is usually estimated to be on the order of $10^{172}$, with about $10^{768}$ possible games. Typical games between talented players consist of about 150 moves, with an average of about 250 choices per move. While powerful chess software is capable of defeating top chess players, the best Go programs often lose to skillful children.

Go-playing computers find it difficult to "look ahead" in the game to judge outcomes because many more reasonable moves must be considered in Go than in chess. The process of evaluating the favorability of a position is also quite difficult because a difference of a single unoccupied grid point can affect large groups of stones.

In 2006, two Hungarian researchers reported that an algorithm called UCT (for Upper Confidence bounds applied to Trees) could compete with professional Go players, but only on 9 × 9 boards. UCT helps the computer focus its search on the most promising moves.

**SEE ALSO** Tic Tac Toe (c. 1300 B.C.), Solving the Game of Awari (2002), and Checkers Is Solved (2007).

*The game of Go is complex, due in part to the large game board, complicated strategies, and huge numbers of variations in possible games. While powerful chess software is capable of defeating top chess players, the best Go programs often lose to skillful children.*

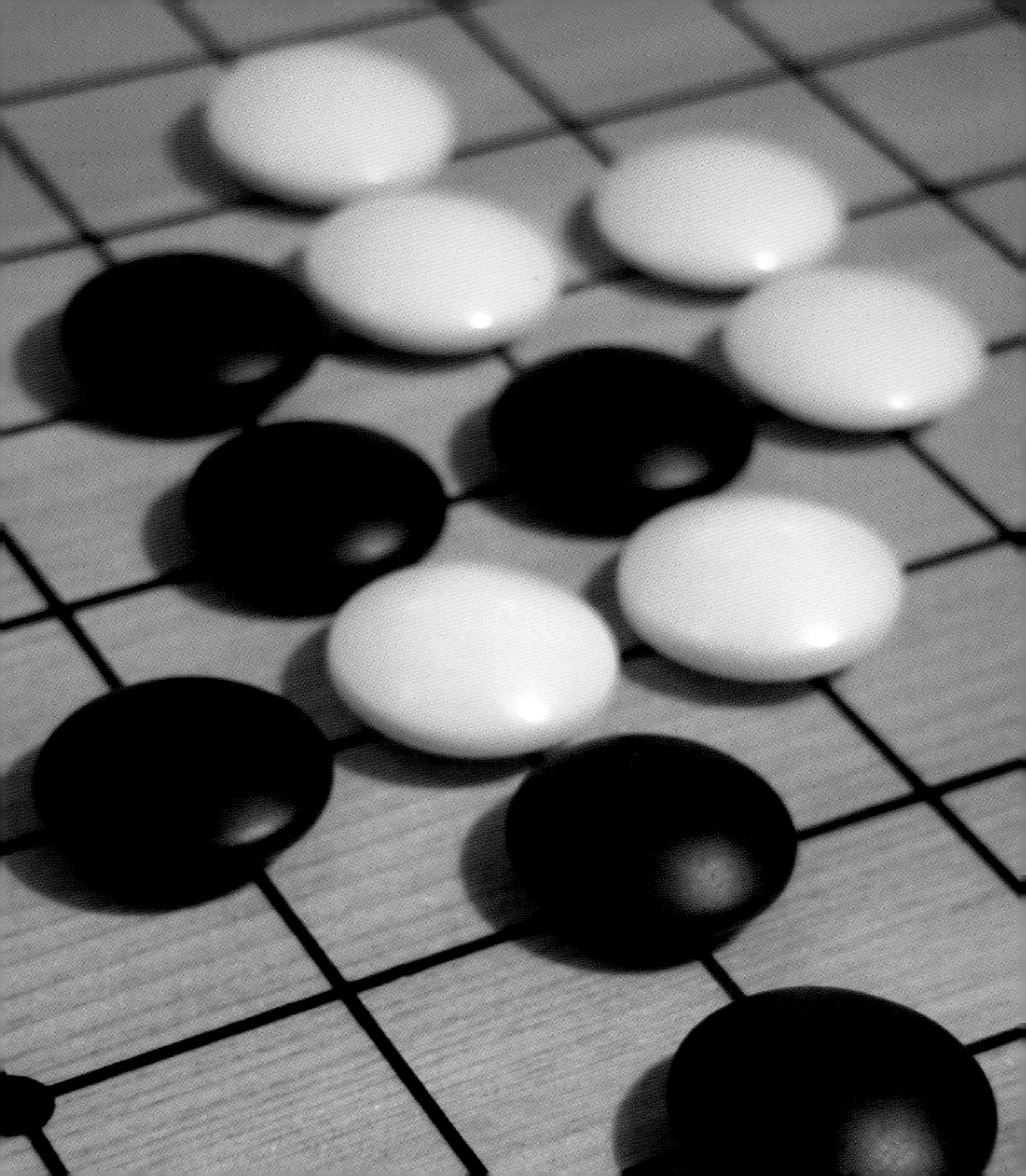

c. 530 B.C.

# Pythagoras Founds Mathematical Brotherhood

**Pythagoras of Samos** (c. 580 B.C.–c. 500 B.C.)

Around 530 B.C., the Greek mathematician Pythagoras moved to Croton, Italy, to teach mathematics, music, and reincarnation. Although many of Pythagoras's accomplishments may actually have been due to his disciples, the ideas of his brotherhood influenced both numerology and mathematics for centuries. Pythagoras is usually credited with discovering mathematical relationships relevant to musical harmonies. For example, he observed that vibrating strings produce harmonious sounds when the ratios of the lengths of the strings are whole numbers. He also studied triangular numbers (based on patterns of dots in a triangular shape) and perfect numbers (integers that are the sum of their proper positive divisors). Although the famous theorem that bears his name, $a^2 + b^2 = c^2$ for a right triangle with legs $a$ and $b$ and hypotenuse $c$, may have been known to the Indians and Babylonians much earlier, some scholars have suggested that Pythagoras or his students were among the first Greeks to prove it.

To Pythagoras and his followers, numbers were like gods, pure and free from material change. Worship of the numbers 1 through 10 was a kind of polytheism for the Pythagoreans. They believed that numbers were alive, with a telepathic form of consciousness. Humans could relinquish their three-dimensional lives and telepathize with these number beings by using various forms of meditation.

Some of these seemingly odd ideas are not foreign to modern mathematicians who often debate whether mathematics is a creation of the human mind or if it is simply a part of the universe, independent of human thought. To the Pythagoreans, mathematics was an ecstatic revelation. Mathematical and theological blending flourished under the Pythagoreans and eventually affected much of the religious philosophy in Greece, played a role in religion of the Middle Ages, and extended to philosopher Immanuel Kant in modern times. Bertrand Russell mused that if it were not for Pythagoras, theologians would not have so frequently sought logical proofs of God and immortality.

**SEE ALSO** Plimpton 322 (c. 1800 B.C.) and Pythagorean Theorem and Triangles (c. 600 B.C.).

*Pythagoras (the bearded man at bottom left with a book) is teaching music to a youth in* The School of Athens *by Raphael (1483–1520), the famous Renaissance Italian painter and architect.*

c. 445 B.C.

# Zeno's Paradoxes

**Zeno of Elea** (c. 490 B.C.–c. 430 B.C.)

For more than a thousand years, philosophers and mathematicians have tried to understand Zeno's paradoxes, a set of riddles that suggest that motion should be impossible or that it is an illusion. Zeno was a pre-Socratic Greek philosopher from southern Italy. His most famous paradox involves the Greek hero Achilles and a slow tortoise that Achilles can never overtake during a race once the tortoise is given a head start. In fact, the paradox seems to imply that you can never leave the room you are in. In order to reach the door, you must first travel half the distance there. You'll also need to continue to half the remaining distance, and half again, and so on. You won't reach the door in a finite number of jumps! Mathematically one can represent this limit of an infinite sequence of actions as the sum of the series (1/2 + 1/4 + 1/8 + ...). One modern tendency is to attempt to resolve Zeno's paradox by insisting that the sum of this infinite series 1/2 + 1/4 + 1/8 is *equal* to 1. If each step is done in half as much time, the actual time to complete the infinite series is no different than the real time required to leave the room.

However, this approach may not provide a satisfying resolution because it does not explain how one is able to *finish* going through an *infinite* number of points, one after the other. Today, mathematicians make use of infinitesimals (unimaginably tiny quantities that are almost but not quite zero) to provide a microscopic analysis of the paradox. Coupled with a branch of mathematics called nonstandard analysis and, in particular, internal set theory, we may have resolved the paradox, but debate continues. Some have also argued that if space and time are *discrete*, the total number of jumps in going from one point to another *must* be finite.

**SEE ALSO** Aristotle's Wheel Paradox (c. 320 B.C.), Harmonic Series Diverges (c. 1350), Discovery of Series Formula for π (c. 1500), Discovery of Calculus (c. 1665), St. Petersburg Paradox (1738), Barber Paradox (1901), Banach-Tarski Paradox (1924), Hilbert's Grand Hotel (1925), Birthday Paradox (1939), Coastline Paradox (c. 1950), Newcomb's Paradox (1960), and Parrondo's Paradox (1999).

*According to Zeno's most famous paradox, the rabbit can never overtake the tortoise once the tortoise is given a head start. In fact, the paradox seems to imply that neither can ever cross the finish line.*

FINISH

c. 440 B.C.

# Quadrature of the Lune

**Hippocrates of Chios** (c. 470 B.C.–c. 400 B.C.)

Ancient Greek mathematicians were enchanted by the beauty, symmetry, and order of geometry. Succumbing to this passion, Greek mathematician Hippocrates of Chios demonstrated how to construct a square with an area equal to a particular lune. A lune is a crescent-shaped area, bounded by two concave circular arcs, and this Quadrature of the Lune is one of the earliest-known proofs in mathematics. In other words, Hippocrates demonstrated that the area of these lunes could be expressed exactly as a rectilinear area, or "quadrature." In the example depicted here, two yellow lunes associated with the sides of a right triangle have a combined area equal to that of the triangle.

For the ancient Greeks, finding the quadrature meant using a straightedge and compass to construct a square equal in area to a given shape. If such a construction is possible, the shape is said to be "quadrable" (or "squareable"). The Greeks had accomplished the quadrature of polygons, but curved forms were more difficult. In fact, it must have seemed unlikely, at first, that curved objects could be quadrable at all.

Hippocrates is also famous for compiling the first-known organized work on geometry, nearly a century before Euclid. Euclid may have used some of Hippocrates' ideas in his own work, *Elements*. Hippocrates' writings were significant because they provided a common framework upon which other mathematicians could build.

Hippocrates' lune quest was actually part of a research effort to achieve the "quadrature of the circle"—that is, to construct a square with the same area as a circle. Mathematicians had tried to solve the problem of "squaring the circle" for more than 2,000 years, until Ferdinand von Lindemann in 1882 proved that it is impossible. Today, we know that only five types of lune exist that are quadrable. Three of these were discovered by Hippocrates, and two more kinds were found in the mid-1770s.

**SEE ALSO** Pythagorean Theorem and Triangles (c. 600 B.C.), Euclid's *Elements* (300 B.C.), Descartes' *La Géométrie* (1637), and Transcendental Numbers (1844).

*The two lunes (the yellow crescent-shaped areas) associated with the sides of a right triangle have a combined area equal to that of the triangle. Ancient Greek mathematicians were enchanted by the elegance of these kinds of geometrical findings.*

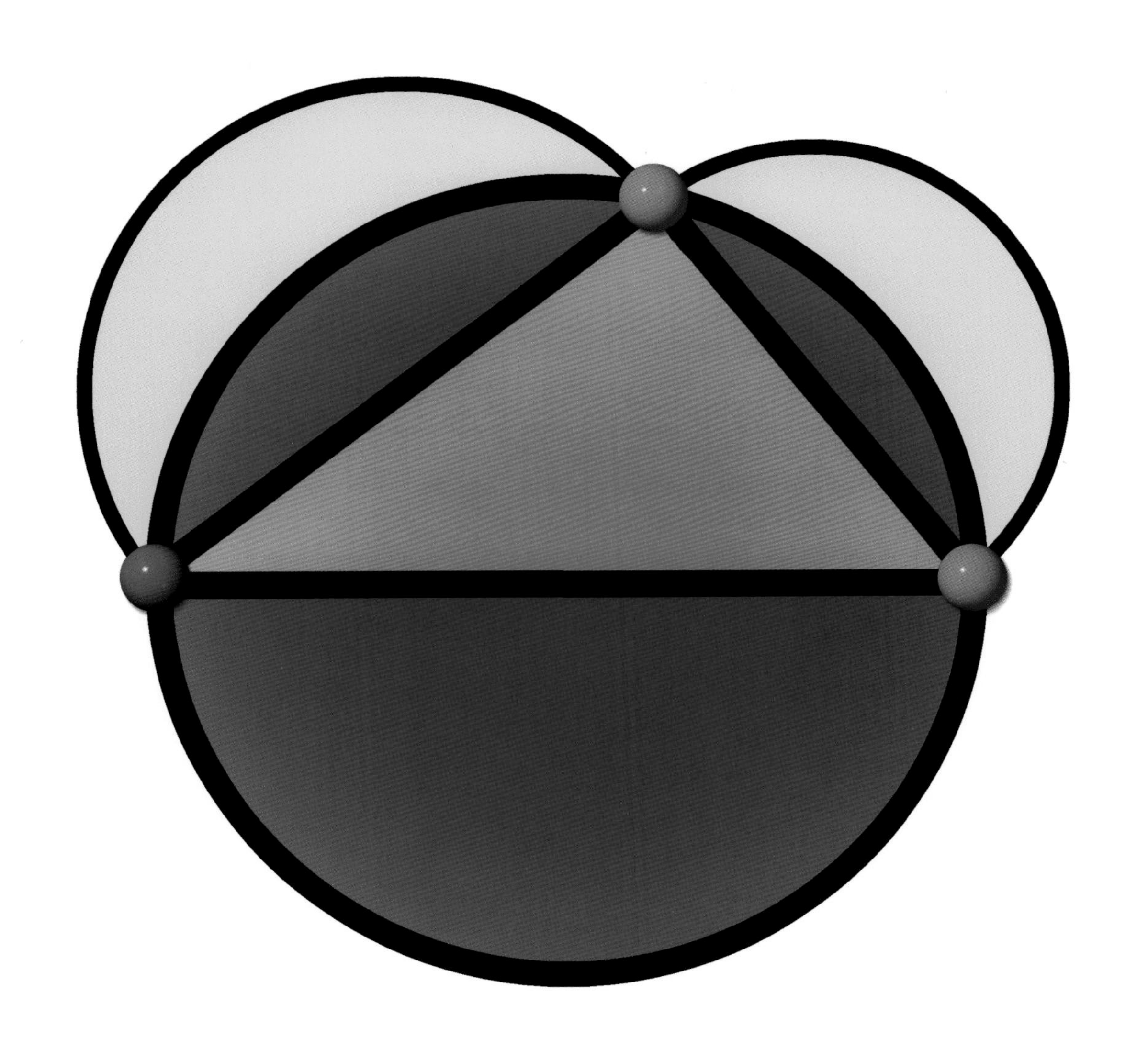

c. 350 B.C.

# Platonic Solids

**Plato** (c. 428 B.C.–c. 348 B.C.)

A *Platonic solid* is a convex multifaceted 3-D object whose faces are all identical polygons, with sides of equal length and angles of equal degrees. A Platonic solid also has the same number of faces meeting at every vertex. The best-known example of a Platonic solid is the cube, whose faces are six identical squares.

The ancient Greeks recognized and proved that only five Platonic solids can be constructed: the tetrahedron, cube, octahedron, dodecahedron, and icosahedron. For example, the icosahedron has 20 faces, all in the shape of equilateral triangles.

Plato described the five Platonic solids in *Timaeus* in around 350 B.C. He was not only awestruck by their beauty and symmetry, but he also believed that the shapes described the structures of the four basic elements thought to compose the cosmos. In particular, the tetrahedron was the shape that represented fire, perhaps because of the polyhedron's sharp edges. The octahedron was air. Water was made up of icosahedra, which are smoother than the other Platonic solids. Earth consisted of cubes, which look sturdy and solid. Plato decided that God used the dodecahedron for arranging the constellations in the heavens.

Pythagoras of Samos—the famous mathematician and mystic who lived in the time of Buddha and Confucius, around 550 B.C.—probably knew of three of the five Platonic solids (the cube, tetrahedron, and dodecahedron). Slightly rounded versions of the Platonic solids made of stone have been discovered in areas inhabited by the late Neolithic people of Scotland at least 1,000 years before Plato. The German astronomer Johannes Kepler (1571–1630) constructed models of Platonic solids nested within one another in an attempt to describe the orbits of the planets about the sun. Although Kepler's theories were wrong, he was one of the first scientists to insist on a geometrical explanation for celestial phenomena.

**SEE ALSO** Pythagoras Founds Mathematical Brotherhood (c. 530 B.C.), Archimedean Semi-Regular Polyhedra (c. 240 B.C.), Euler's Formula for Polyhedra (1751), Icosian Game (1857), Pick's Theorem (1899), Geodesic Dome (1922), Császár Polyhedron (1949), Szilassi Polyhedron (1977), Spidrons (1979), and Solving of the Holyhedron (1999).

*A traditional dodecahedron is a polyhedron with 12 pentagonal faces. Shown here is Paul Nylander's graphical approximation of a hyperbolic dodecahedron, which uses a portion of a sphere for each face.*

c. 350 B.C.

# Aristotle's *Organon*

**Aristotle** (384 B.C.–322 B.C.)

Aristotle was a Greek philosopher and scientist, a pupil of Plato, and a teacher of Alexander the Great. The *Organon* (*Instrument*) refers to the collection of six of Aristotle's works on logic: *Categories*, *Prior Analytics*, *De Interpretatione*, *Posterior Analytics*, *Sophistical Refutations*, and *Topics*. Andronicus of Rhodes determined the ordering of the six works around 40 B.C. Although Plato (c. 428–348 B.C.) and Socrates (c. 470–399 B.C.) delved into logical themes, Aristotle actually systematized the study of logic, which dominated scientific reasoning in the Western world for 2,000 years.

The goal of the *Organon* is not to tell readers what is true, but rather to give approaches for how to investigate truth and how to make sense of the world. The primary tool in Aristotle's tool kit is the syllogism, a three-step argument, such as "All women are mortal; Cleopatra is a woman; therefore, Cleopatra is mortal." If the two premises are true, we know that the conclusion must be true. Aristotle also made a distinction between particulars and universals (general categories). *Cleopatra* is a particular term. *Woman* and *mortal* are universal terms. When universals are used, they are preceded by "all," "some," or "no." Aristotle analyzed many possible kinds of syllogisms and showed which of them are valid.

Aristotle also extended his analysis to syllogisms that involved modal logic—that is, statements containing the words "possibly" or "necessarily." Modern mathematical logic can depart from Aristotle's methodologies or extend his work into other kinds of sentence structures, including ones that express more complex relationships or ones that involve more than one quantifier, such as "No women like all women who dislike some women." Nevertheless, Aristotle's systematic attempt at developing logic is considered to be one of humankind's greatest achievements, providing an early impetus for fields of mathematics that are in close partnership with logic and even influencing theologians in their quest to understand reality.

**SEE ALSO** Euclid's *Elements* (300 B.C.), Boolean Algebra (1854), Venn Diagrams (1880), *Principia Mathematica* (1910–1913), Gödel's Theorem (1931), and Fuzzy Logic (1965).

*Italian Renaissance artist Raphael depicts Aristotle (right), holding his* Ethics, *next to Plato. This Vatican fresco,* The School of Athens, *was painted between 1510 and 1511.*

ETI

c. 320 B.C.

# Aristotle's Wheel Paradox

**Aristotle** (384 B.C.–322 B.C.)

The paradox of Aristotle's wheel is mentioned in the ancient Greek text *Mechanica*. The problem has haunted some of the greatest mathematicians for centuries. Consider a small wheel mounted on a large wheel, diagrammed as two concentric circles. A one-to-one correspondence exists between points on the larger circle and those on the smaller circle; that is, for each point in the large circle, there is exactly one point on the small circle, and vice versa. Thus, the wheel assembly might be expected to travel the same horizontal distance regardless of whether it is rolled on a rod that touches the smaller wheel or rolled along the bottom wheel that touches the road. But how can this be? After all, we know that the two circumferences of the circles are different.

Today, mathematicians know that a one-to-one correspondence of points doesn't mean that two curves must have the same length. Georg Cantor (1845–1918) showed that the number, or cardinality, of points on line segments of any length is the same. He called this **Transfinite Number** of points the "continuum." For example, all the points in a segment from zero to one can even be put in one-to-one correspondence with all points of an infinite line. Of course, before the work of Cantor, mathematicians had quite a difficult time with this problem. Also note that, from a physical standpoint, if the large wheel did roll along the road, the smaller wheel would skip and be dragged along the line that touches its surface.

The precise date and authorship of *Mechanica* may forever be shrouded in mystery. Although often attributed as the work of Aristotle, many scholars doubt that *Mechanica*, the oldest-known textbook on engineering, was actually written by Aristotle. Another possible candidate for authorship is Aristotle's student Straton of Lampsacus (also known as Strato Physicus), who died around 270 B.C.

**SEE ALSO** Zeno's Paradoxes (c. 445 B.C.), St. Petersburg Paradox (1738), Cantor's Transfinite Numbers (1874), Barber Paradox (1901), Banach-Tarski Paradox (1924), Hilbert's Grand Hotel (1925), Birthday Paradox (1939), Coastline Paradox (c. 1950), Newcomb's Paradox (1960), Continuum Hypothesis Undecidability (1963), and Parrondo's Paradox (1999).

*Consider a small wheel glued to a large wheel. Describe the motion of the wheel assembly as it moves from right to left along a rod that touches the smaller wheel and a road that touches the bottom wheel.*

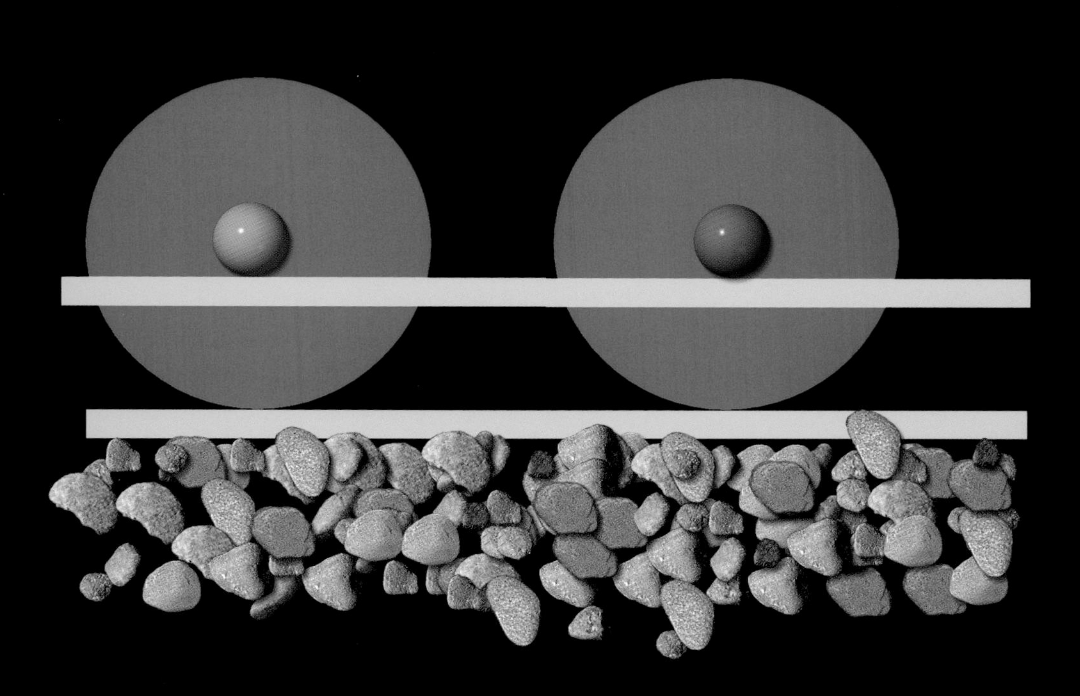

300 B.C.

# Euclid's *Elements*

**Euclid of Alexandria** (c. 325 B.C.–c. 270 B.C.)

The geometer Euclid of Alexandria lived in Hellenistic Egypt, and his book *Elements* is one of the most successful textbooks in the history of mathematics. His presentation of plane geometry is based on theorems that can all be derived from just five simple axioms, or postulates, one of which is that only one straight line can be drawn between any two points. Given a point and a line, another famous postulate suggests that only one line through the point is parallel to the first line. In the 1800s, mathematicians finally explored **Non-Euclidean Geometries**, in which the parallel postulate was no longer always required. Euclid's methodical approach of proving mathematical theorems by logical reasoning not only laid the foundations of geometry but also shaped countless other areas concerning logic and mathematical proofs.

*Elements* consists of 13 books that cover two- and three-dimensional geometries, proportions, and the theory of numbers. *Elements* was one of the first books to be printed after the invention of the printing press and was used for centuries as part of university curricula. More than 1,000 editions of *Elements* have been published since its original printing in 1482. Although Euclid was probably not the first to prove the various results in *Elements*, his clear organization and style made the work of lasting significance. Mathematical historian Thomas Heath called *Elements* "the greatest mathematical textbook of all time." Scientists like Galileo Galilei and Isaac Newton were strongly influenced by *Elements*. Philosopher and logician Bertrand Russell wrote, "At the age of eleven, I began Euclid, with my brother as my tutor. This was one of the great events of my life, as dazzling as first love. I had not imagined that there was anything so delicious in the world." The poet Edna St. Vincent Millay wrote, "Euclid alone has looked on Beauty bare."

**SEE ALSO** Pythagorean Theorem and Triangles (c. 600 B.C.), Quadrature of the Lune (c. 440 B.C.), Aristotle's *Organon* (c. 350 B.C.), Descartes' *La Géométrie* (1637), Non-Euclidean Geometry (1829), and Weeks Manifold (1985).

*This is the frontispiece of Adelard of Bath's translation of Euclid's* Elements, *c. 1310. This translation from Arabic to Latin is the oldest surviving Latin translation of* Elements.

c. 250 B.C.

# Archimedes: Sand, Cattle & Stomachion

**Archimedes of Syracuse** (c. 287 B.C.–c. 212 B.C.)

In 1941, mathematician G. H. Hardy wrote, "Archimedes will be remembered when [playwright] Aeschylus is forgotten, because languages die and mathematical ideas do not. 'Immortality' may be a silly word, but probably a mathematician has the best chance of whatever it may mean." Indeed, Archimedes, the ancient Greek geometer, is often regarded as the greatest mathematician and scientist of antiquity and one of the four greatest mathematicians to have walked the Earth—together with Isaac Newton, Carl Friedrich Gauss, and Leonhard Euler. Interestingly, Archimedes sometimes sent his colleagues false theorems in order to trap them when they stole his ideas.

In addition to many other mathematical ideas, he is famous for his contemplation of tremendously large numbers. In his book *The Sand Reckoner*, Archimedes estimated that $8 \times 10^{63}$ grains of sand would fill the universe.

More amazingly, the number $7.760271406486818269530232833213\ldots \times 10^{202544}$ is the solution to one version of Archimedes' famous "cattle problem," which involves computing the total number of cattle in a puzzle concerning four hypothetical herds of different colors. Archimedes wrote that anyone who could solve the problem would be "crowned with glory" and would be "adjudged perfect in this species of wisdom." Not until 1880 did mathematicians have an approximate answer. A more precise number was first calculated in 1965 by Canadian mathematicians Hugh C. Williams, R. A. German, and C. Robert Zarnke using an IBM 7040 computer.

In 2003, math historians discovered long lost information on the *Stomachion of Archimedes*. In particular, an ancient parchment, overwritten by monks nearly a thousand years ago, describes Archimedes' Stomachion, a puzzle involving combinatorics. *Combinatorics* is a field of math dealing with the number of ways a given problem can be solved. The goal of the Stomachion is to determine in how many ways the 14 pieces shown here can be put together to make a square. In 2003, four mathematicians determined that the number is 17,152.

**SEE ALSO** π (c. 250 B.C.), Euler's Polygon Division Problem (1751), Googol (c. 1920), and Ramsey Theory (1928).

*For Archimedes' Stomachion puzzle, one goal is to determine in how many ways the 14 pieces shown here can be put together to make a square. In 2003, four mathematicians determined that the number is 17,152. (Rendering by Teja Krašek.)*

c. 250 B.C.

# π

**Archimedes of Syracuse** (c. 287 B.C.–c. 212 B.C.)

Pi, symbolized by the Greek letter π, is the ratio of a circle's circumference to its diameter and is approximately equal to 3.14159. Perhaps ancient peoples observed that for every revolution of a cartwheel, a cart moves forward about three times the diameter of the wheel—an early recognition that the circumference is about three times the diameter. An ancient Babylonian tablet states that the ratio of the circumference of a circle to the perimeter of an inscribed hexagon is 1 to 0.96, implying a value of pi of 3.125. Greek mathematician Archimedes (c. 250 B.C.) was the first to give us a mathematically rigorous range for π—a value between 223/71 and 22/7. The Welsh mathematician William Jones (1675–1749) introduced the symbol π in 1706, most likely after the Greek word for periphery, which starts with the letter π.

The most famous ratio in mathematics is π, on Earth and probably for any advanced civilization in the universe. The digits of π never end, nor has anyone detected an orderly pattern in their arrangement. The speed with which a computer can compute π is an interesting measure of a computer's computational ability, and today we know more than a trillion digits of π.

We usually associate π with a circle, and so did pre-seventeenth-century humanity. However, in the seventeenth century, π was freed from the circle. Many curves were invented and studied (for example, various arches, hypocycloids, and curves known as *witches*), and it was found that their areas could be expressed in terms of π. Finally, π appeared to flee geometry altogether, and today π relates to unaccountably many areas in number theory, probability, complex numbers, and series of simple fractions, such as $\pi/4 = 1 - 1/3 + 1/5 - 1/7\ldots$In 2006, Akira Haraguchi, a retired Japanese engineer, set a world record for memorizing and reciting 100,000 digits of π.

**SEE ALSO** Archimedes: Sand, Cattle & Stomachion (c. 250 B.C.), Discovery of Series Formula for π (c. 1500), Rope around the Earth Puzzle (1702), Euler's Number, *e* (1727), Euler-Mascheroni Constant (1735), Buffon's Needle (1777), Transcendental Numbers (1844), Holditch's Theorem (1858), and Normal Number (1909).

*Pi is approximately equal to 3.14 and is the ratio of a circle's circumference to its diameter. Ancient peoples may have noticed that for every revolution of a cart wheel, the cart moves forward about three times the diameter of the wheel.*

c. 240 B.C.

# Sieve of Eratosthenes

**Eratosthenes** (c. 276 B.C.–c. 194 B.C.)

A prime number is a number larger than 1, such as 5 or 13, that is divisible only by itself or 1. The number 14 is not prime because $14 = 7 \times 2$. Prime numbers have fascinated mathematicians for more than two thousand years. Around 300 B.C., Euclid showed that there is no "largest prime" and that an infinitude of prime numbers exists. But how can we determine if a number is prime? Around 240 B.C, the Greek mathematician Eratosthenes developed the first-known test for primality, which we today call the Sieve of Eratosthenes. In particular, the Sieve can be used to find all prime numbers up to a specified integer. (The ever-versatile Eratosthenes served as the director of the famous library in Alexandria and was also the first person to provide a reasonable estimation of the diameter of the Earth.)

The French theologian and mathematician Marin Mersenne (1588–1648) was also fascinated by prime numbers, and he tried to find a formula that he could use to find all primes. Although he did not find such a formula, his work on Mersenne numbers of the form $2^p - 1$, where $p$ is an integer, continues to be of interest to us today. Mersenne numbers, with $p$ a prime number, are the easiest type of number to prove prime, so they are usually the largest primes of which humanity is aware. The forty-fifth known Mersenne prime ($2^{43,112,609} - 1$) was discovered in 2008, and it contains 12,978,189 digits!

Today, prime numbers play an important role in public-key cryptography algorithms that may be used for sending secure messages. More important, for pure mathematicians, prime numbers have been at the heart of many intriguing unsolved conjectures through history, including the **Riemann Hypothesis**, which concerns the distribution of prime numbers, and the strong **Goldbach Conjecture**, which states that every even integer greater than 2 can be written as a sum of two primes.

**SEE ALSO** Cicada-Generated Prime Numbers (c. 1 Million B.C.), Ishango Bone (c. 18,000 B.C.), Goldbach Conjecture (1742), Constructing a Regular Heptadecagon (1796), Gauss's *Disquisitiones Arithmeticae* (1801), Riemann Hypothesis (1859), Proof of the Prime Number Theorem (1896), Brun's Constant (1919), Gilbreath's Conjecture (1958), Sierpiński Numbers (1960), Ulam Spiral (1963), Erdös and Extreme Collaboration (1971), Public-Key Cryptography (1977), and Andrica's Conjecture (1985).

*Polish artist Andreas Guskos creates contemporary art by concatenating thousands of prime numbers and using them as textures on various surfaces. This work is called* Eratosthenes, *after the Greek mathematician who developed the first-known test for primality.*

c. 240 B.C.

# Archimedean Semi-Regular Polyhedra

**Archimedes of Syracuse** (c. 287 B.C.–c. 212 B.C.)

Like **Platonic Solids**, Archimedean semi-regular polyhedra (ASRP) are convex, multifaceted 3-D objects whose faces are all regular polygons that have sides of equal length and angles of equal degrees. However, for the ASRP, the faces are of different kinds. For example, the polyhedron formed by 12 pentagons and 20 hexagons, which resembles a modern soccer ball, was described by Archimedes along with 12 other such polyhedra. Around every vertex (corner) of these kinds of solids, the same polygons appear in the same sequence—for example, hexagon-hexagon-triangle.

Archimedes' original writings that described the 13 ASRP are lost and known only from other sources. During the Renaissance, artists discovered all but one ASRP. In 1619, Kepler presented the entire set in his book *Harmonices Mundi (The Harmonies of the World)*. ASRPs may be specified using a numerical notation that indicates the shapes around a vertex. For example, 3,5,3,5 means that a triangle, pentagon, triangle, and pentagon appear in that order. Using this notation, we have the following ASRPs: 3,4,3,4 (a cuboctahedron); 3,5,3,5 (an icosidodecahedron); 3,6,6 (a truncated tetrahedron); 4,6,6 (a truncated octahedron); 3,8,8 (a truncated cube); 5,6,6 (a truncated icosahedron, or soccer ball); 3,10,10 (a truncated dodecahedron); 3,4,4,4 (a rhombicuboctahedron); 4,6,8 (a truncated cuboctahedron); 3,4,5,4 (a rhombicosidodecahedron); 4,6,10 (a truncated icosidodecahedron); 3,3,3,3,4 (a snub cube, or snub cuboctahedron); and 3,3,3,3,5 (a snub dodecahedron, or snub icosidodecahedron).

The 32-faced truncated icosahedron is particularly fascinating. Soccer ball shapes are based on this Archimedean solid, and this was also the configuration used for arranging lenses that focused the explosive shock waves of the detonators in the "Fat Man" atomic bomb detonated over Nagasaki, Japan, in World War II. In the 1980s, chemists succeeded in creating the world's tiniest soccer ball, a carbon molecule with 60 atoms at the vertices of a truncated icosahedron. These so-called Buckyballs have fascinating chemical and physical properties that are being explored in applications ranging from lubrication to AIDS treatments.

**SEE ALSO** Platonic Solids (c. 350 B.C.), Archimedes: Sand, Cattle & Stomachion (c. 250 B.C.), Euler's Formula for Polyhedra (1751), Icosian Game (1857), Pick's Theorem (1899), Geodesic Dome (1922), Császár Polyhedron (1949), Szilassi Polyhedron (1977), Spidrons (1979), and Solving of the Holyhedron (1999).

*Slovenian artist Teja Krašek explores the 13 Archimedean semi-regular polyhedra in her artwork titled* Harmonices Mundi II, *in honor of Johannes Kepler's presentation of these objects in his 1619 book* Harmonices Mundi.

225 B.C.

# Archimedes' Spiral

**Archimedes of Syracuse** (c. 287 B.C.–c. 212 B.C.)

The term *spiral* is often used generically to describe any geometrically smooth curve that winds about a central point or axis while also receding from it. When we think of examples of spirals, both the mundane and the exotic come to mind—the gentle curl of a fern tendril, the shape of an octopus's retracted arm, the death form assumed by a centipede, the spiral intestine of a giraffe, the shape of a butterfly's tongue, and the spiral cross section of a scroll. Spirals possess a simple beauty that humans have copied in their arts and tools, and that nature has used in the creation of many structures of life.

The mathematics of the simplest spiral form, the spiral of Archimedes, was first discussed by Archimedes in 225 B.C. in his book *On Spirals*. This spiral can be expressed by the equation $r = a + b\theta$. The parameter $a$ rotates the entire spiral, and $b$ controls the distance between the successive turnings. The most commonly observed spirals are of the Archimedean type: tightly wound springs, edges of rolled-up rugs, and decorative spirals on jewelry. Practical uses of the Archimedes spiral have included the transformation of rotary to linear motion in sewing machines. The Archimedean spiral spring is particularly interesting in its ability to respond to both torsional and translational force.

Ancient examples of Archimedean spirals include prehistoric spiral mazes, terra-cotta pot spiral designs from the sixth century B.C., decorations from ancient Altaic works (middle of the first millennium B.C.), engravings on threshold stones of initiation chambers in the Bronze Age in Ireland, scrollworks for Irish manuscripts, and Tibetan Tanka artworks, the latter of which are painted or embroidered Buddhist designs sometimes hung in monasteries. In fact, the spiral is a ubiquitous symbol throughout the ancient world. Its frequent appearance at burial sites suggests that this symbol may have represented a cycle of life, death, and rebirth, as with the continual rising and falling of the sun.

**SEE ALSO** Golden Ratio (1509), Loxodrome (1537), Fermat's Spiral (1636), Logarithmic Spiral (1638), Voderberg Tilings (1936), Ulam Spiral (1963), and Spidrons (1979).

*The fiddlehead fern exhibits the spiral of Archimedes, a shape discussed by Archimedes in 225 B.C. in his book* On Spirals.

c. 180 B.C.

# Cissoid of Diocles

**Diocles** (c. 240 B.C.–c. 180 B.C.)

The cissoid of Diocles was discovered by Greek mathematician Diocles, around 180 B.C., during his attempts to use its remarkable properties to double a cube. "Doubling the cube" refers to a famous and ancient challenge of constructing a cube with a volume twice the volume of a given smaller cube, which means that the larger cube has an edge that is $\sqrt[3]{2}$ times larger than the first cube. Diocles' use of the cissoid, and its intersection with a straight line, was theoretically correct, but did not rigorously follow the rules of Euclidean construction that allowed the use of only a compass and a straightedge.

The name *cissoid* comes from the Greek term meaning "ivy-shaped." The graph of the curve extends to infinity along both directions of the *y*-axis and has a single cusp. Both branches of the curve that extend away from the cusp approach the same vertical asymptote. If we draw a circle that passes through the cusp at point O and that is tangent to the asymptote, then *any* line joining the cusp and a point M on the cissoid can be extended so that it intersects the asymptote at B. The length of the linear extension from C to B is always equal to the length between O and M. The curve may be expressed in polar coordinates as $r = 2a(\sec\theta - \cos\theta)$ or in rectangular coordinates as $y^2 = x^3/(2a - x)$. Interestingly, the cissoid can be produced by tracing the vertex of a parabola as it rolls, without slipping, on a second parabola of the same size.

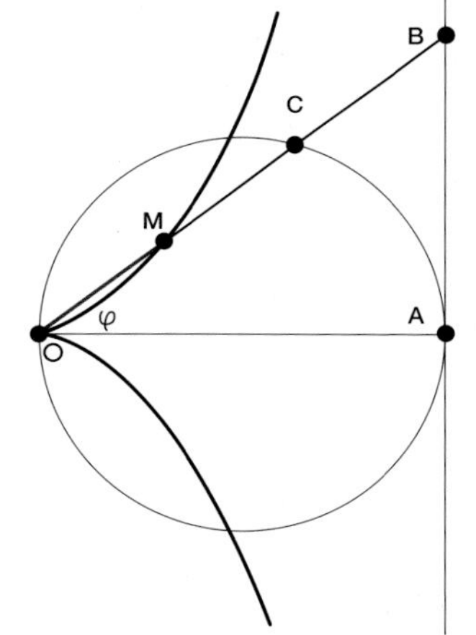


Diocles was fascinated by curves known as *conic sections*, and in his work *On Burning Mirrors* he discussed the focal point of a parabola. One of his goals was to find a mirror surface that focuses the maximum amount of heat when it is placed in sunlight.

**SEE ALSO** Cardioid (1637), The Length of Neile's Semicubical Parabola (1657), and Astroid (1674).

*Parabolic telecommunication antenna. The Greek mathematician Diocles was fascinated by curves such as these, and in his work* On Burning Mirrors, *he discussed the focal point of a parabola. Diocles sought to find a mirror surface that focused the maximum amount of heat when illuminated.*

# Ptolemy's *Almagest*

**Claudius Ptolemaeus** (c. 90–c. 168)

Mathematician and astronomer Ptolemy of Alexandria wrote the 13-book *Almagest*, a comprehensive treatise on virtually all aspects of astronomy known at his time. In the *Almagest*, Ptolemy deals with the apparent movement of the planets and stars. His geocentric model, in which the Earth is at the center of the universe and the Sun and planets orbit around the Earth, was accepted as correct for more than a thousand years in Europe and in the Arabic world.

*Almagest* is the Latin form of the Arabic name *al-kitabu-l-mijisti* (*The Great Book*), and the work is of particular interest to mathematicians due to its trigonometric content, which includes the equivalent of a table of sine values for angles from 0° to 90° at 15-minute intervals and an introduction to spherical trigonometry. The *Almagest* also contains theorems that correspond to our modern "law of sines" and compound-angle and half-angle identities. Jan Gullberg writes, "That so much of early Greek work on astronomy has been lost could be a result of the completeness and elegance of presentation of Ptolemy's *Almagest*, making all earlier works appear superfluous." Gerd Grasshoff writes, "Ptolemy's *Almagest* shares with Euclid's *Elements* the glory of being the scientific text longest in use. From its conception in the second century up to the late Renaissance, this work determined astronomy as a science."

The *Almagest* was translated into Arabic around the year 827 and then from Arabic to Latin in the twelfth century. The Persian mathematician and astronomer Abu al-Wafa (940–998) built upon the *Almagest* and systemized trigonometric theorems and proofs.

Interestingly, Ptolemy attempted to compute the size of the universe based on his models of planetary motions in which a planet moves in a small circle, called an epicycle, which in turn moves along a larger circle. He estimated that a sphere containing the faraway "fixed stars" was 20,000 times the radius of the Earth.

**SEE ALSO** Euclid's *Elements* (300 B.C.) and Law of Cosines (c. 1427).

*Ptolemy's* Almagest *describes a geocentric model of the universe, which places the Earth at the center of the universe, and the Sun and planets orbit around the Earth. This model was accepted for more than a thousand years in Europe and in the Arabic world.*

# Diophantus's *Arithmetica*

**Diophantus of Alexandria** (c. 200–c. 284)

Greek mathematician Diophantus of Alexandria, sometimes called the "father of algebra," was the author of *Arithmetica* (c. 250), a series of mathematical texts that has influenced mathematics for centuries. *Arithmetica*, the most famous work on algebra in all of Greek mathematics, contains various problems along with numerical solutions to equations. Diophantus is also important due to his advances in mathematical notation and his treatment of fractions as numbers. In the dedication to *Arithmetica*, Diophantus writes to Dionysus (most likely the bishop of Alexandria) that although the material in the book may be difficult, "it will be easy to grasp, with your enthusiasm and my teaching."

Diophantus's various works were preserved by the Arabs and translated into Latin in the sixteenth century. Diophantine equations, with their integer solutions, are named in his honor. Pierre de Fermat scribbled his famous **Fermat's Last Theorem** involving integer solutions of $a^n + b^n = c^n$ in a French translation of *Arithmetica*, published in 1681.

In *Arithmetica*, Diophantus was often interested in finding integer solutions for equations like $ax^2 + bx = c$. Although the Babylonians were aware of some methods for solving linear and quadratic equations of the kind that fascinated Diophantus, Diophantus is special, according to J. D. Swift, for being "the first to introduce extensive and consistent algebraic notation representing a tremendous improvement over the purely verbal style of his predecessors (and many successors)....The rediscovery of *Arithmetica* through Byzantine sources greatly aided the renaissance of mathematics in Western Europe and stimulated many mathematicians, of whom the greatest was Fermat."

Note that Persian mathematician al-Khwarizmi (780–850) also shares the title "father of algebra" for his own book ***Algebra***, which contained a systematic solution of linear and quadratic equations. Al-Khwarizmi introduced Hindu-Arabic numerals and algebraic concepts into European mathematics, and the words *algorithm* and *algebra* are derived from his name and *al-jabr*, respectively. *Al-jabr* is an Arabic word for a mathematical operation used to solve quadratic equations.

**SEE ALSO** The Death of Hypatia (415), Al-Khwarizmi's *Algebra* (830), *Sumario Compendioso* (1556), and Fermat's Last Theorem (1637).

*Title page of the 1621 edition of Diophantus's* Arithmetica, *translated into Latin by French mathematician Claude Gaspard Bachet de Méziriac. The European rediscovery of* Arithmetica *stimulated the renaissance of mathematics in Western Europe.*

# DIOPHANTI ALEXANDRINI ARITHMETICORVM LIBRI SEX,

*ET DE NVMERIS MVLTANGVLIS LIBER VNVS.*

*Nunc primùm Græcè & Latinè editi, atque absolutissimis Commentariis illustrati.*

AVCTORE CLAVDIO GASPARE BACHETO MEZIRIACO SEBVSIANO, V.C.

LVTETIAE PARISIORVM,
Sumptibus SEBASTIANI CRAMOISY, via Iacobæa, sub Ciconiis.

M. DC. XXI.

*CVM PRIVILEGIO REGIS.*

c. 340

# Pappus's Hexagon Theorem

**Pappus of Alexandria** (c. 290–c. 350)

A farmer wishes to plant nine maple trees so that they form ten straight rows with three trees in each row. One curious way to achieve this goal makes use of Pappus's theorem. If three points A, B, C are located *anywhere* along one line, and three points D, E, F are located *anywhere* on a second line, Pappus's theorem guarantees that the intersections X, Y, Z of opposite sides of a crossed hexagon A, F, B, D, C, E lie on a straight line. The farmer can solve his problem and form a tenth row by sliding tree B to bring B, Y, and E into alignment.

Pappus was one of the most important Hellenistic mathematicians of his age and famous for *Synagoge* (*Collection*) written in c. 340. The work focuses on topics in geometry that include polygons, polyhedra, circles, spirals, and honeycomb construction by bees. The *Synagoge* is also valuable because it includes results based on ancient works that have subsequently been lost. Thomas Heath writes of *Synagoge*, "Obviously written with the object of reviving the classical Greek geometry, it covers practically the whole field."

About the famous theorem of Pappus, Max Dehn writes that it "marks an event in the history of geometry. From the beginning, geometry was concerned with measures: lengths of lines, areas of plane figures, and volumes of bodies. Here, we have for the first time a theorem which is established by the ordinary theory of measures but is itself free of all elements of measurements." In other words, the theorem demonstrates the existence of a figure that is determined only through the incidence of lines and points. Dehn also says that this figure is the "first configuration of **projective geometry**."

*Synagoge* became widely known in Europe after 1588, when a Latin translation by Federico Commandino was printed. Pappus's figure intrigued Isaac Newton and René Descartes. About 1,300 years after Pappus wrote *Synagoge*, French mathematician Blaise Pascal provided an interesting generalization of Pappus's theorem.

**SEE ALSO** Descartes' *La Géométrie* (1637), Projective Geometry (1639), and Sylvester's Line Problem (1893).

*If three points A, B, C are located* anywhere *along one line, and three points D, E, F are located* anywhere *on a second line, Pappus's theorem guarantees that the intersections X, Y, Z lie on a straight line.*

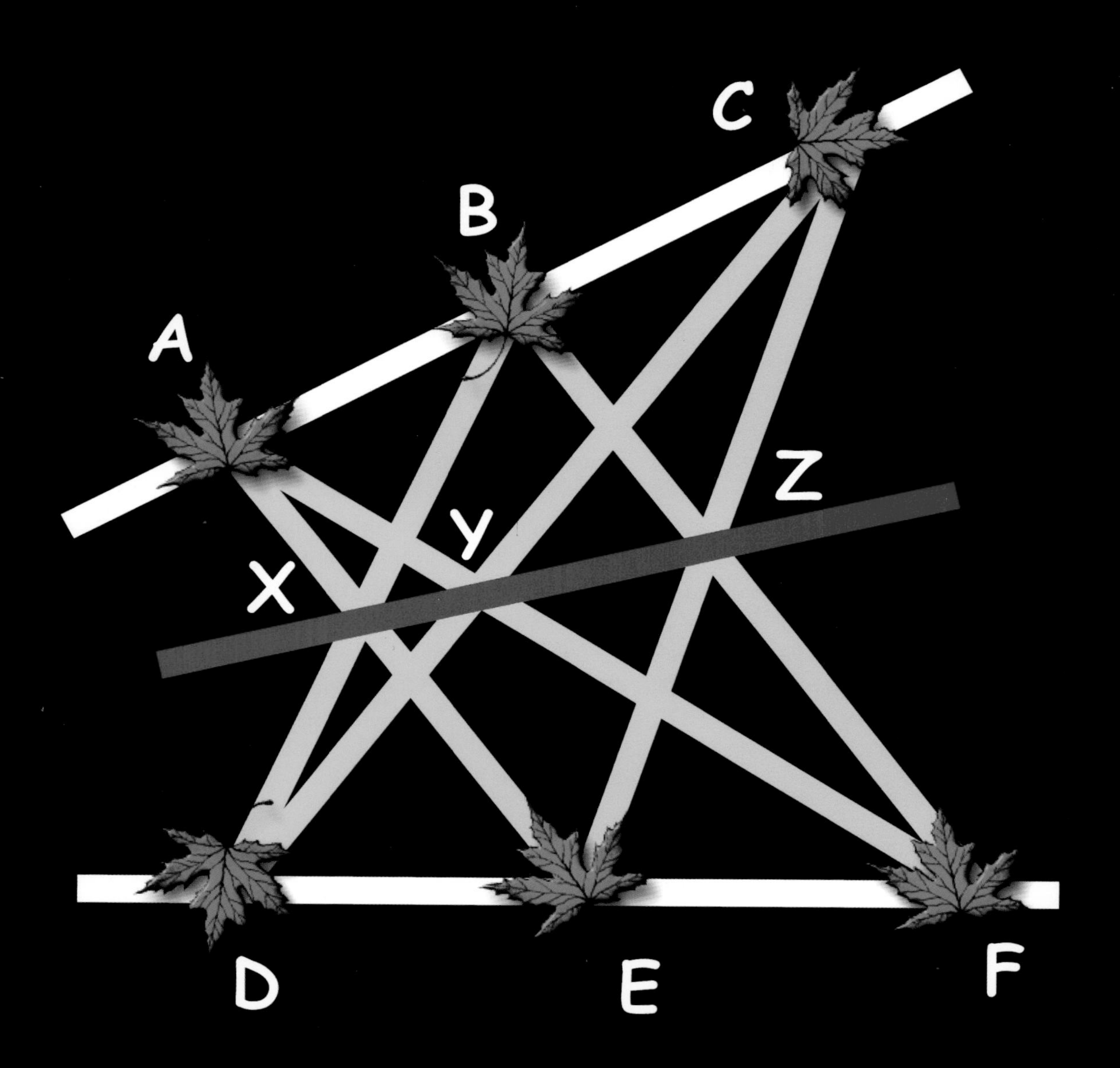
C
B
A
Z
Y
X
D
E
F

c. 350

# Bakhshali Manuscript

The Bakhshali manuscript is a famous mathematical collection, discovered in 1881 within a stone enclosure in northwest India, and it may even date back to the third century. When it was discovered, a large part of the manuscript had been destroyed, and only about 70 leaves of birch bark survived to the time of its discovery. The Bakhshali manuscript provides techniques and rules for solving arithmetic, algebra, and geometry problems, and it provides a formula for calculating the square root.

Here is one problem from the manuscript: "Before you is a group of 20 people comprising men, women, and children. They earn 20 coins between them. Each man earns 3 coins, each woman 1.5 coins, and each child half a coin. How many men, women, and children are there?" Can you solve this? The solution turns out to be 2 men, 5 women, and 13 children. We can let the number of men, women, and children be $m$, $w$, and $c$, respectively. Two formulas describe our situation: $m + w + c = 20$ and $3m + (3/2)w + (1/2)c = 20$. The solution given is the only valid one.

The manuscript was found near the village of Bakhshali in the Yusufzai subdivision of the Peshawar district (now in Pakistan). The date of the manuscript is subject to much debate; however, a number of scholars believe it to be a commentary on an older work that may have existed around A.D. 200 to 400. One unusual feature of the Bakhshali notation is the use of a "+" sign placed after a number to indicate a negative. Equations are given with a large dot representing the unknown value that is being sought. A similar dot is used to represent zero. Dick Teresi writes, "Most important is that the Bakhshali manuscript is the first document depicting a form of Indian mathematics devoid of religious association."

**SEE ALSO** Diophantus's *Arithmetica* (250), Zero (c. 650), and *Ganita Sara Samgraha* (850).

*A fragment of the Bakhshali manuscript, discovered in 1881 in northwest India.*

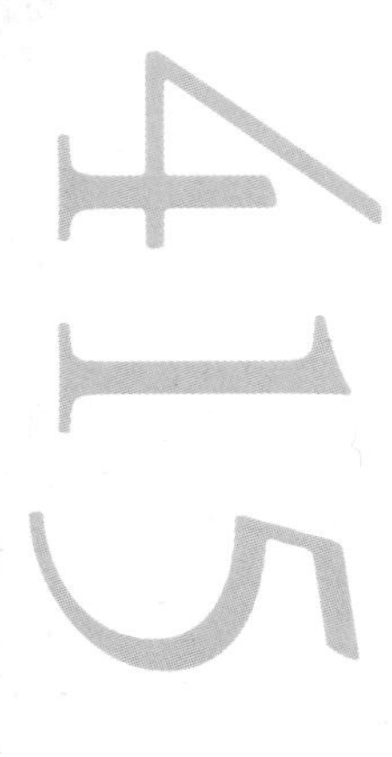

# The Death of Hypatia

**Hypatia of Alexandria** (c. 370–c. 415)

Hypatia of Alexandria was martyred by being torn to shreds by a Christian mob, partly because she did not adhere to strict Christian principles. She considered herself a neo-Platonist, a pagan, and a follower of Pythagorean ideas. Interestingly, Hypatia is the first woman mathematician in the history of humanity of whom we have reasonably secure and detailed knowledge. She was said to be physically attractive and determinedly celibate. When asked why she was obsessed with mathematics and would not marry, she replied that she was wedded to the truth.

Hypatia's works include commentaries on **Diophantus's *Arithmetica***. In one of her mathematical problems for her students, she asked them for the integer solution of the pair of simultaneous equations: $x - y = a$ and $x^2 - y^2 = (x - y) + b$, where $a$ and $b$ are known. Can you find any integer values for $x$, $y$, $a$, and $b$ that make both of these formulas true?

The Christians were her strongest philosophical rivals, and they officially discouraged her Platonic assertions about the nature of God and the afterlife. On a warm March day in A.D. 414, a crowd of Christian zealots seized her, stripped her, and proceeded to scrape her flesh from her bones using sharp shells. Next, they cut up her body and burned the pieces. Like some victims of religious terrorism today, she may have been seized merely because she was a famous person on the other side of the religious divide. It was not until after the Renaissance that another woman, Maria Agnesi, made her name as a famous mathematician.

Hypatia's death triggered the departure of many scholars from Alexandria and, in many ways, marked the end of centuries of Greek progress in mathematics. During the European Dark Ages, Arabs and Hindus were the ones to play the leading roles in fostering the progress of mathematics.

**SEE ALSO** Pythagoras Founds Mathematical Brotherhood (c. 530 B.C.), Diophantus's *Arithmetica* (250), Agnesi's *Instituzioni Analitiche* (1748), and The Doctorate of Kovalevskaya (1874).

*In 1885, British painter Charles William Mitchell depicted Hypatia moments before her death at the hands of a Christian mob that stripped her and slaughtered her in a church. According to some reports, she was flayed with sharp objects and then the pieces of her dismembered body were burned.*

c. 650

# Zero

**Brahmagupta** (c. 598–c. 668), **Bhaskara** (c. 600–c. 680),
**Mahavira** (c. 800–c. 870)

The ancient Babylonians originally had no symbol for zero, which caused uncertainty in their notation, just as today we would be confused if numbers like 12, 102, and 1,002 had no zero to distinguish them. The Babylonian scribes only left a space where a zero should be, and it was not easy to distinguish the number of spaces in the middle or at the ends of numbers. Eventually, the Babylonians did invent a symbol to mark the gap between their digits, but they probably had no concept of zero as an actual number.

Around A.D. 650, the use of the number was common in Indian mathematics, and a stone tablet was found in Gwalior, south of Delhi, with the numbers 270 and 50. The numbers on the tablet, dated to A.D. 876, look very similar to modern numbers, except that the zeros are smaller and raised. Indian mathematicians such as Brahmagupta, Mahavira, and Bhaskara used zero in mathematical operations. For example, Brahmagupta explained that a number subtracted from itself gives zero, and he noted that any number when multiplied by zero is zero. The **Bakhshali Manuscript** may be the first documented evidence of zero used for mathematical purposes, but its date is unclear.

Around A.D. 665, the Mayan civilization in Central America also developed the number zero, but its achievement did not seem to influence other peoples. On the other hand, the Indian concept of zero spread to the Arabs, Europeans, and Chinese, and changed the world.

Mathematician Hossein Arsham writes, "The introduction of zero into the decimal system in the thirteenth century was the most significant achievement in the development of a number system, in which calculation with large numbers became feasible. Without the notion of zero, the... modeling processes in commerce, astronomy, physics, chemistry, and industry would have been unthinkable. The lack of such a symbol is one of the serious drawbacks in the Roman numeral system."

**SEE ALSO** Bakhshali Manuscript (c. 350), *Ganita Sara Samgraha* (850), *Chapters in Indian Mathematics* (c. 953), Al-Samawal's *The Dazzling* (c. 1150), and Fibonacci's *Liber Abaci* (1202).

*The notion of zero ignited a fire that eventually allowed humanity to more easily work with large numbers and to become efficient in calculations in fields ranging from commerce to physics.*

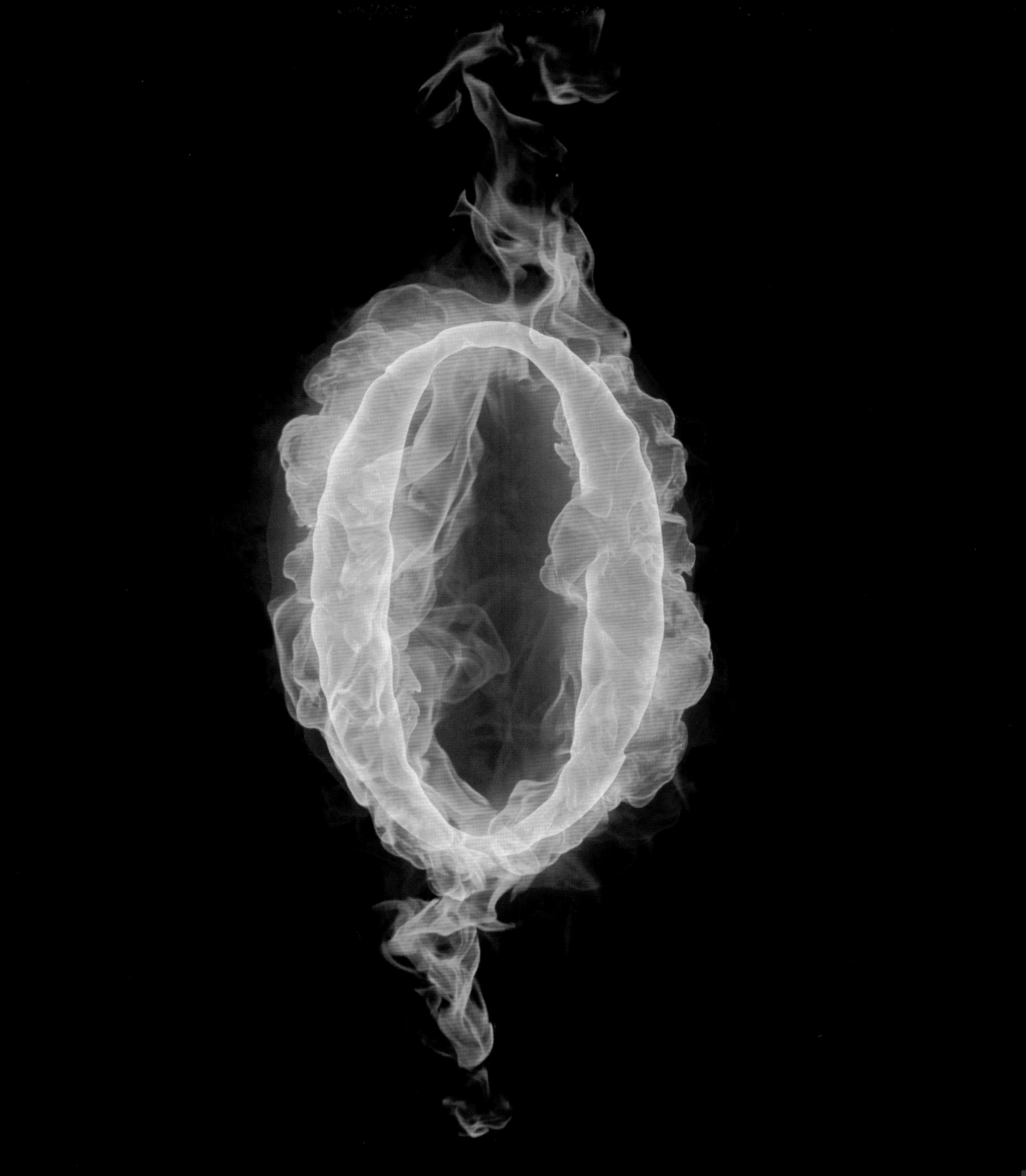

# Alcuin's *Propositiones ad Acuendos Juvenes*

**Alcuin of York** (c. 735–c. 804), **Gerbert of Aurillac** (c. 946–c. 1003)

Flaccus Albinus Alcuinus, also known as Alcuin of York, was a scholar from York, England. At the invitation of King Charlemagne, he became a leading teacher at the Carolingian court, where he wrote theological treatises and poems. He was abbot of the Abbey of Saint Martin at Tours in 796 and foremost scholar of the revival of learning known as the Carolingian Renaissance.

Scholars speculate that his mathematics book *Propositiones ad acuendos juvenes* (*Problems to Sharpen the Young*) contributed to the education of the last Pope Mathematician, Gerbert of Aurillac, who was fascinated by mathematics and elected as Pope Sylvester II in 999. This pope's advanced knowledge of mathematics convinced some of his enemies that he was an evil magician.

In the city of Reims, France, the "Number Pope" transformed the floor of the cathedral into a giant **abacus**. He also adopted Arabic numerals (1, 2, 3, 4, 5, 6, 7, 8, and 9) as a replacement for Roman numerals. He contributed to the invention of the pendulum clock, invented devices that tracked planetary orbits, and wrote on geometry. When he realized that he lacked knowledge of formal logic, he studied under German logicians. The Number Pope said, "The just man lives by faith, but it is good that he should combine science with his faith."

Alcuinus's *Propositiones* contained roughly 50 word problems with solutions, the most famous of which involve river crossings, counting doves on a ladder, a dying father leaving vessels of wine to his sons, and three jealous husbands, each of whom can't let another man be alone with his wife. Several major classes of problems appear for the first time in *Propositiones*. Mathematics writer Ivars Peterson notes that "Browsing the problems (and solutions) in *Propositiones* provides fascinating glimpses of various aspects of life in medieval times. And it testifies to the enduring power of puzzles in mathematical education."

**SEE ALSO** Rhind Papyrus (1650 B.C.), Al-Khwarizmi's *Algebra* (830), and Abacus (c. 1200).

*Alcuin's mathematical work very likely contributed to the education of the last Pope Mathematician, Gerbert of Aurillac, who was fascinated by mathematics and elected as Pope Sylvester II in 999. Shown here is a statue of the Number Pope, located in Aurillac, Auvergne, France.*

A GERBERT SYLVESTRE II
PREMIER PAPE FRANÇAIS
MORT A ROME EN 1003
L'AUVERGNE SA PATRIE
ET LA MUNICIPALITÉ D'AURILLAC
GENESTE

# Al-Khwarizmi's *Algebra*

**Abu Ja'far Muhammad ibn Musa al-Khwarizmi** (c. 780–c. 850)

Al-Khwarizmi was a Persian mathematician and astronomer who spent most of his life in Baghdad. His book on algebra, *Kitab al-mukhtasar fi hisab al-jabr wa'l-muqabala* (*The Compendious Book on Calculation by Completion and Balancing*) was the first book on the systematic solution of linear and quadratic equations and is sometimes referred to by the shortened title *Algebra*. Along with **Diophantus**, he is considered the "father of algebra." The Latin translation of his works introduced the decimal positional number system to Europe. Interestingly, the word *algebra* comes from *al-jabr*, one of the two operations used in his book to solve quadratic equations.

For al-Khwarizmi, al-jabr is a method in which we can eliminate negative quantities in an equation by adding the same quantity to each side. For example, we can reduce $x^2 = 50x - 5x^2$ to $6x^2 = 50x$ by adding $5x^2$ to both sides. *Al-muqabala* is a method whereby we gather quantities of the same type to the same side of the equation. For example, $x^2 + 15 = x + 5$ is reduced to $x^2 + 10 = x$.

The book helped readers to solve equations such as those of the forms $x^2 + 10x = 39$, $x^2 + 21 = 10x$, and $3x + 4 = x^2$, but more generally, al-Khwarizmi believed that the difficult mathematical problems could be solved if broken down into a series of smaller steps. Al-Khwarizmi intended his book to be practical, helping people to make calculations that deal with money, property inheritance, lawsuits, trade, and the digging of canals. His book also contained example problems and solutions.

Al-Khwarizmi worked most of his life in the Baghdad House of Wisdom, a library, translation institute, and place of learning that was a major intellectual center of the Islamic Golden Age. Alas, the Mongols destroyed the House of Wisdom in 1258, and legend says that the waters of the Tigris ran black with ink from the books tossed into its waters.

**SEE ALSO** Diophantus's *Arithmetica* (250) and Al-Samawal's *The Dazzling* (c. 1150).

*A stamp from the Soviet Union, issued in 1983 in honor of al-Khwarizmi, the Persian mathematician and astronomer whose book on algebra offered a systematic solution to a wide variety of equations.*

ПОЧТА СССР

4

1983

1200
ЛЕТ

Мухаммед
аль-Хорезми

# Borromean Rings

**Peter Guthrie Tait** (1831–1901)

A simple yet intriguing set of interlocking objects of interest to mathematicians and chemists is formed by Borromean rings—three mutually interlocked rings named after the Italian Renaissance family who used them on its coat of arms in the fifteenth century.

Notice that Borromean rings have no two rings that are linked, so if we cut any one of the rings, all three rings come apart. Some historians speculate that the ancient ring configurations once represented the three families of Visconti, Sforza, and Borromeo, who formed a tenuous union through intermarriages. The rings also appear in 1467 in the Church of San Pancrazio in Florence. Even older, triangular versions were used by the Vikings, one famous example of which was found on a bedpost of a prominent woman who died in 834.

The rings appear in a mathematical context in the 1876 paper on **knots** by Scottish mathematical physicist Peter Tait. Because two choices (over or under) are possible for each ring crossing, $2^6 = 64$ possible interlaced patterns exist. If we take symmetry into account, only 10 of these patterns are geometrically distinct.

Mathematicians now know that we cannot actually construct a true set of Borromean rings with *flat* circles, and, in fact, you can see this for yourself if you try to create the interlocked rings out of wire, which requires some deformation or kinks in the wires. In 1987, Michael Freedman and Richard Skora proved the theorem stating that Borromean rings are impossible to construct with flat circles.

In 2004, UCLA chemists created a molecular Borromean ring compound that was 2.5 nanometers across and that included six metal ions. Researchers are currently contemplating ways in which they may use molecular Borromean rings in such diverse fields as spintronics (a technology that exploits electron spin and charge) and medical imaging.

**SEE ALSO** Knots (c. 100,000 B.C.), Johnson's Theorem (1916), and Murphy's Law and Knots (1988).

*This Borromean ring motif was found in a thirteenth-century French manuscript where it symbolized the Christian Trinity. The original contains* trinitas *(Latin for "Trinity" or "three in one") broken into its three syllables—tri, ni, and tas—that were written in the three circles.*

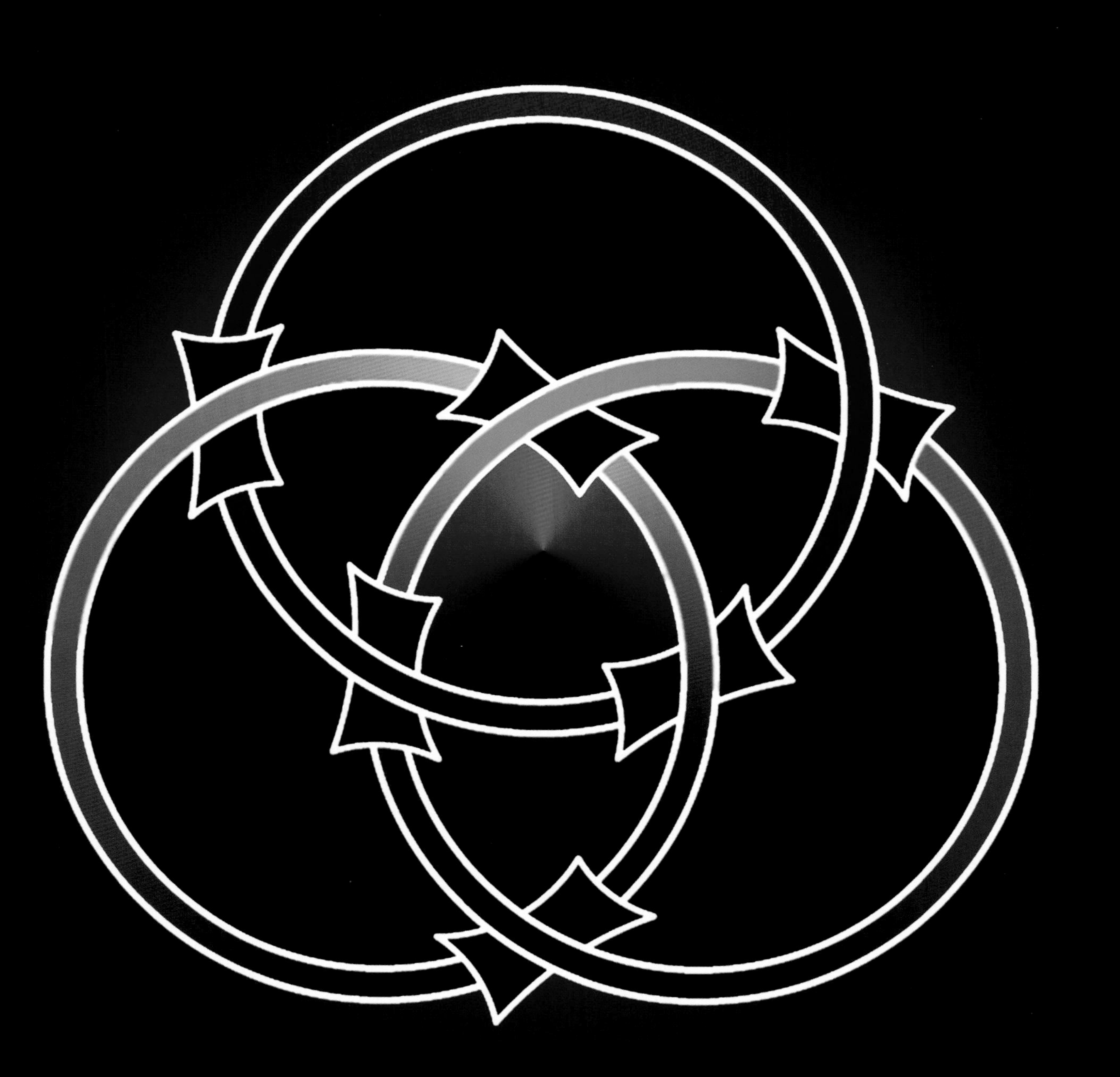

# *Ganita Sara Samgraha*

**Mahavira** (c. 800–c. 870)

The *Ganita Sara Samgraha* (*Compendium of the Essence of Mathematics*), dated A.D. 850, is special for several reasons. First, it is the only existing treatise on arithmetic by a Jaina scholar. Second, it included essentially all mathematical knowledge of mid-ninth-century India. It is the earliest extant Indian text devoted entirely to mathematics.

*Ganita* was written by Mahavira (or Mahaviracharya, meaning "Mahavira the Teacher"), who lived in southern India. One particular problem in the book has delighted scholars for centuries and was worded as follows: A young lady has a quarrel with her husband and damages her necklace. One-third of the necklace's pearls scatter toward the lady. One-sixth fall on the bed. One-half of what remains (and one-half of what remains thereafter and again one-half of what remains after that, and so on, counting six times in all) fall everywhere else. A total of 1,161 pearls were found to remain unscattered. How many pearls did the girl originally have in total?

The astounding answer is that the girl originally had 148,608 pearls on her necklace! Let's reflect on the problem. One-sixth fell on the bed. One-third scattered toward her. This means the remaining pearls that are neither on the bed nor near her are half of all the pearls. The remaining pearls are halved six times, so $((1/2)^7)x = 1{,}161$, where $x$ is the total number of pearls; thus, $x$ is 148,608. The Indian woman's huge necklace was well worth quarreling over!

*Ganita* was notable for its explicit assertion that the square root of a negative number did not exist. In *Ganit*a, Mahavira also discussed the properties of the number zero and provided a naming scheme for numbers from 10 up to $10^{24}$, methods for obtaining the sum of a series whose terms are squares of an arithmetical progression, rules for determining the area and perimeter of an ellipse, and methods for solving linear and quadratic equations.

**SEE ALSO** Bakhshali Manuscript (c. 350), Zero (c. 650), and *Treviso Arithmetic* (1478).

*The* Ganita Sara Samgraha *discusses a mathematical problem involving a woman who has a quarrel with her husband and damages her necklace. The pearls scatter according to a particular set of rules, and we must determine how many pearls the necklace originally contained.*

# Thabit Formula for Amicable Numbers

**Thabit ibn Qurra** (826–901)

The Pythagoreans of ancient Greece were fascinated by amicable numbers for which each such number is the sum of the proper divisors of the other. (A proper divisor of a number does not include the number itself.) The smallest such pair is 220 and 284. The number 220 is evenly divisible by 1, 2, 4, 5, 10, 11, 20, 22, 44, 55, and 110, which sum to 284, and 284 is evenly divisible by 1, 2, 4, 71, and 142, which sum to 220.

In 850, Thabit ibn Qurra, an Arab astronomer and mathematician, presented a formula that could be used to generate amicable numbers. Compute $p = 3 \times 2^{n-1} - 1$, $q = 3 \times 2^n - 1$, and $r = 9 \times 2^{2n-1} - 1$ for an integer $n > 1$. If $p$, $q$, and $r$ are prime numbers, then $2^n pq$ and $2^n r$ are a pair of amicable numbers. When $n = 2$, this gives the numbers 220 and 284, but the formula does not generate every amicable number that exists. In every known case, the numbers of a pair are either both even or both odd. Will we ever discover an even-odd amicable pair? Amicable numbers are quite difficult to find. For example, by 1747, Leonhard Euler, a Swiss mathematician and physicist, had found only 30 pairs. Today, we know of more than 11 million pairs, but only 5,001 of such pairs have both numbers less than $3.06 \times 10^{11}$.

In Genesis 32:14, Jacob gives a present of 220 goats to his brother. According to mystics, this was a "hidden secret arrangement" because 220 is one of a pair of amicable numbers, and Jacob sought to secure friendship with Esau. Martin Gardner, a popular mathematics and science writer, stated, "One poor Arab of the eleventh century recorded that he once tested the erotic effect of *eating* something labeled with 284, at the same time having someone else swallow something labeled 220, but he failed to add how the experiment worked out."

**SEE ALSO** Pythagoras Founds Mathematical Brotherhood (c. 530 B.C.).

*In Genesis, Jacob gives a present of 220 goats to his brother. According to mystics, this was a "hidden secret arrangement" because 220 is one of a pair of amicable numbers, and Jacob sought to secure friendship with Esau.*

c. 953

# *Chapters in Indian Mathematics*

**Abu'l Hasan Ahmad ibn Ibrahim al-Uqlidisi** (c. 920–c. 980)

Al-Uqlidisi ("the Euclidian") was an Arab mathematician whose *Kitab al-fusul fi al-hisab al-Hindi* (*Chapters in Indian Mathematics*) is the earliest-known Arabic work discussing the positional use of the Hindu-Arabic numerals, meaning the use of digits corresponding to 0 through 9 in which each position starting from the right of a multi-digit number corresponds to a power of 10 (for example, 1, 10, 100, and 1,000). Al-Uqlidisi's work also represents the earliest-known arithmetic extant in Arabic. Although al-Uqlidisi was born and died in Damascus, he was well traveled and may have learned about Hindu mathematics in India. Only one copy of this manuscript remains today.

Al-Uqlidisi also discussed the problems of previous mathematicians in terms of the new system of numerals. Dick Teresi, the author of several books about science and technology, writes, "His name was evidence of his reverence for the Greeks. He copied the works of Euclid, hence the name al-Uqlidisi. One of his legacies is paper-and-pen mathematics." During al-Uqlidisi's time, it was common in India and the Islamic world to perform mathematical calculations in the sand or in dust, erasing steps with one's hand as one proceeded. Al-Uqlidisi suggested that paper and pen be used instead. Written arithmetic preserves the process, and although his scheme did not involve erasure of ink numbers, it did permit greater flexibility in calculation. In a sense, paper drove the evolution of modern methods for performing multiplication and long division.

Régis Morelon, the editor of the *Encyclopedia of the History of Arabic Science*, writes, "One of the most remarkable ideas in the arithmetic of al-Uqlidisi is the use of decimal fractions" and the use of the decimal symbol. For example, to halve 19 successively, al-Uqlidisi gave the following: 19, 9.5, 4.75, 2.375, 1.1875, 0.59375. Eventually, the advanced calculations enabled by the decimal system led to its common use throughout the region and the world.

**SEE ALSO** Zero (c. 650).

*In India and the Islamic world during al-Uqlidisi's time, mathematical calculations were often performed in the sand or in dust, erasing steps with one's hand as one proceeded. With al-Uqlidisi's paper-and-pen approach, written arithmetic preserved the calculation process and permitted greater flexibility in calculation.*

1070

# Omar Khayyam's *Treatise*

**Omar Khayyam** (1048–1131)

Omar Khayyam, the Persian mathematician, astronomer, and philosopher, is best known for his collection of poems, the *Rubaiyat of Omar Khayyam*. However, he has also achieved great fame for his influential *Treatise on Demonstration of Problems of Algebra* (1070). Here, he derived methods for solving cubic and some higher-order equations. An example of a cubic equation that he solved is $x^3 + 200x = 20x^2 + 2000$. Although his approaches were not entirely new, his generalizations that could be used to solve all cubics were noteworthy. His *Treatise* contains a comprehensive classification of cubic equations with geometric solutions found by means of intersecting conic sections.

Khayyam also was able to show how to obtain the $n$th power of the binomial $a + b$ in powers of $a$ and $b$, when $n$ is any whole number. As background, consider the expression $(a + b)^n$, which equals $(a + b) \times (a + b) \times (a + b)\ldots$ with $n$ repetitions of $(a + b)$. According to the binomial expansion, for example, $(a + b)^5 = a^5 + 5a^4b + 10a^3b^2 + 10a^2b^3 + 5ab^4 + b^5$. The numerical coefficients (1, 5, 10, 10, 5, and 1) are referred to as binomial coefficients, which are the values in a row of **Pascal's Triangle**. Some of Khayyam's work on this subject actually appears in another book to which he refers but which is now lost.

Khayyam's 1077 work on geometry, *Sharh ma ashkala min musadarat kitab Uqlidis* (*Commentaries on the Difficulties in the Postulates of Euclid's Book*), provides an interesting look at Euclid's famous parallel postulate. In *Sharh*, Khayyam discussed properties of **Non-Euclidean Geometries** and thus stumbled into a realm of mathematics that would not flourish until the 1800s.

A literal translation of Khayyam's name is "tent maker," the possible trade of his father. Khayyam once referred to himself as one "who stitched the tents of science."

**SEE ALSO** Euclid's *Elements* (300 B.C.), Cardano's *Ars Magna* (1545), Pascal's Triangle (1654), Normal Distribution Curve (1733), and Non-Euclidean Geometry (1829).

*Tomb of Omar Khayyam in Neishapur, Iran. The open structure features inscriptions of the poet's verse.*

c.1150

# Al-Samawal's *The Dazzling*

**Ibn Yahya al-Maghribi al-Samawal** (c. 1130–c. 1180), **Abu Bakr ibn Muhammad ibn al Husayn al-Karaji** (c. 953–c. 1029)

Al-Samawal (also known as Samau'al al-Maghribi) was born in Baghdad to a Jewish family. He started to develop his passion for mathematics at the age of 13 when he began his study using Hindu methods of calculation. By the time he was 18 years old, he had read almost all the available mathematical literature that existed in his day. Al-Samawal wrote his most famous work, *al-Bahir fi'l-jabr* (translated as either *The Brilliant in Algebra* or *The Dazzling in Algebra*), when he was only 19 years old. *The Dazzling* is significant for both its original ideas as well as its information concerning the lost works of the tenth-century Persian mathematician al-Karaji.

*The Dazzling* emphasizes the principles of arithmetization of algebra, explaining how unknown arithmetic quantities, or variables, can be treated just like ordinary numbers when considering arithmetic operations. Al-Samawal goes on to define powers of numbers, polynomials, and methods for finding roots of polynomials. Many scholars consider *The Dazzling* to be the first treatise to assert that $x^0 = 1$ (in modern notation). In other words, al-Samawal realized and published the idea that any number raised to the power of 0 is 1. He was also quite comfortable with using negative numbers and zero in his work, considering such concepts (in modern notation) as $0 - a = -a$. He also understood how to handle multiplication involving negative numbers and was proud of his finding $1^2 + 2^2 + 3^2 + \ldots + n^2 = n(n + 1)(2n + 1)/6$, an expression that does not seem to appear in earlier works.

In 1163, after a great deal of study and contemplation, al-Samawal converted from Judaism to Islam. He would have converted earlier in his life but delayed because he did not wish to hurt his father's feelings. His work *Decisive Refutation of the Christians and Jews* still survives today.

**SEE ALSO** Diophantus's *Arithmetica* (250), Zero (c. 650), Al-Khwarizmi's *Algebra* (830), and Fundamental Theorem of Algebra (1797).

*Al-Samawal's* The Dazzling *is likely to be the first treatise to assert that* $x^0 = 1$ *(in modern notation). In other words, al-Samawal realized and published the idea that any number raised to the power of 0 is 1.*

$$x^0 = 1$$

c.1200

# Abacus

In 2005, Forbes.com readers, editors, and a panel of experts ranked the abacus as the second most important tool of all time in terms of its impact on human civilization. (First and third on the list were the knife and the compass, respectively.)

The modern abacus with beads and wires, used for counting, has it roots in ancient devices such as the Salamis tablet, the oldest surviving counting board used by the Babylonians around 300 B.C. These boards were usually wood, metal, or stone, and they contained lines or grooves along which beads or stones were moved. Around A.D.1000, the Aztecs invented the *nepohualtzitzin* (referred to by aficionados as the "Aztec computer"), an abacus-like device that made use of corn kernels threaded through wooden frames to help operators perform computations.

The abacus, as we know it today, which contains beads that move along wires, was used in A.D. 1200 in China, where it was called the *suan-pan*. In Japan, the abacus is called the *soroban*. In some sense, the abacus may be considered the ancestor of the computer, and like the computer, the abacus serves as a tool to allow humans to perform fast calculations in commerce and in engineering. Abaci are still used in China, Japan, parts of the Soviet Union, and Africa, and sometimes by blind people, with slight variations in design. Although the abacus is generally used for fast addition and subtraction operations, experienced users are able to quickly multiply, divide, and calculate square roots. In 1946 in Tokyo, a calculating speed contest featured a competition between a Japanese soroban operator and a person using an electric calculator of that time. The soroban operator usually beat the electric calculator.

**SEE ALSO** Quipu (c. 3000 B.C.), Alcuin's *Propositiones ad Acuendos Juvenes* (c. 800), Slide Rule (1621), Babbage Mechanical Computer (1822), and Curta Calculator (1948).

*The abacus is among the most important tools of all time in terms of its impact on human civilization. For many centuries, this device served as a tool to allow humans to perform fast calculations in commerce and in engineering.*

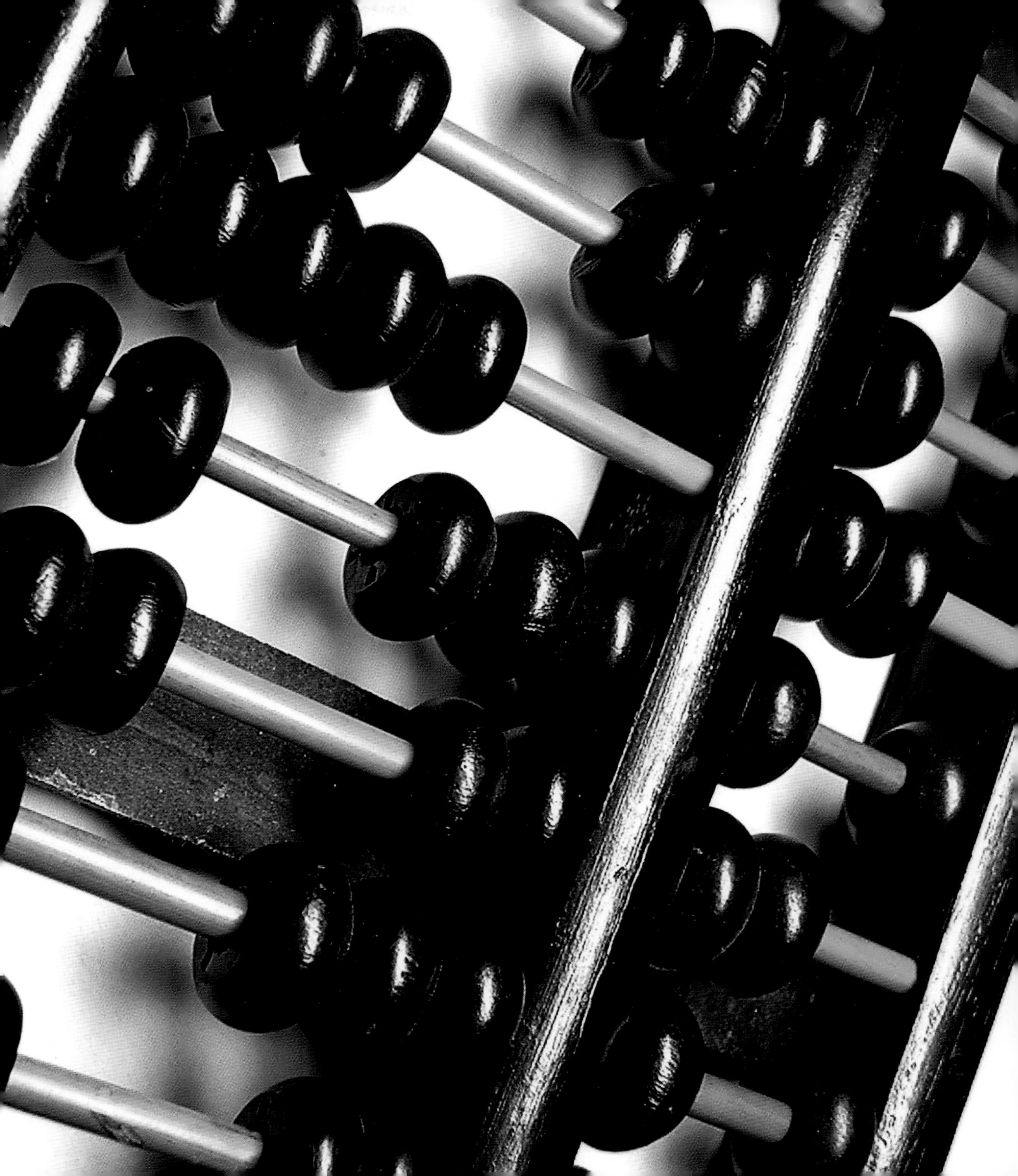

1202

# Fibonacci's *Liber Abaci*

**Leonardo of Pisa** (also known as Fibonacci, c. 1175–c. 1250)

Carl Boyer refers to Leonardo of Pisa, also known as Fibonacci, as "without a doubt, the most original and most capable mathematician of the medieval Christian world." Fibonacci, a wealthy Italian merchant, traveled through Egypt, Syria, and Barbary (Algeria), and in 1202 published the book *Liber Abaci* (*The Book of the Abacus*), which introduced the Hindu-Arabic numerals and decimal number system to Western Europe. This system is now used throughout the world, having overcome the terribly cumbersome Roman numerals common in Fibonacci's time. In *Liber Abaci*, Fibonacci notes, "These are the nine figures of the Indians: 9 8 7 6 5 4 3 2 1. With these nine figures, and with this sign 0, which in Arabic is called *zephirum*, any number can be represented, as will be demonstrated."

Although *Liber Abaci* was not the first European book to describe the Hindu-Arabic numerals—and even though decimal numerals did not gain widespread use in Europe directly after its publication—the book is nevertheless considered to have had a strong impact on European thought because it was directed to both academicians and businesspeople.

*Liber Abaci* also introduced Western Europe to the famous number sequence 1, 1, 2, 3, 5, 8, 13…, which today is called the *Fibonacci sequence*. Notice that except for the first two numbers, every successive number in the sequence equals the sum of the previous two. These numbers appear in an amazing number of mathematical disciplines and in nature.

Is God a mathematician? Certainly, the universe seems to be reliably understood using mathematics. Nature *is* mathematics. The arrangement of seeds in a sunflower can be understood using Fibonacci numbers. Sunflower heads, like those of other flowers, contain families of interlaced spirals of seeds—one spiral winding clockwise, the other counterclockwise. The number of spirals in such heads, as well as the number of petals in flowers, is very often a Fibonacci number.

**SEE ALSO** Zero (c. 650), *Treviso Arithmetic* (1478), Fermat's Spiral (1636), and Benford's Law (1881).

*Sunflower heads contain families of interlaced spirals of seeds—one spiral winding clockwise, the other counterclockwise. The number of spirals in such heads, as well as the number of petals in flowers, is very often a Fibonacci number.*

1256

# Wheat on a Chessboard

**Abu-l 'Abbas Ahmad ibn Khallikan** (1211–1282), **Dante Alighieri** (1265–1321)

The problem of Sissa's chessboard is notable in the history of mathematics because it has been used for centuries to demonstrate the nature of geometric growth or geometric progressions, and it is one of the earliest mentions of chess in puzzles. The Arabic scholar Ibn Khallikan in 1256 appears to be the first author to discuss the story of Grand Vizier Sissa ben Dahir, who, according to legend, was asked by the Indian King Shirham what reward he wanted for inventing the game of chess.

Sissa addressed the king: "Majesty, I would be happy if you were to give me a grain of wheat to place on the first square of the chessboard, and two grains of wheat to place on the second square, and four grains of wheat to place on the third, and eight grains of wheat to place on the fourth, and so on for the sixty-four squares."

"And is that all you wish, Sissa, you fool?" the astonished King shouted.

The king did not realize how many grains Sissa would be awarded! One way to determine the solution is to compute the sum of the first 64 terms of a geometrical progression, $1 + 2 + 2^2 + \ldots + 2^{63} = 2^{64} - 1$, which is a walloping 18,446,744,073,709, 551,615 grains of wheat.

It is possible that some version of this story was known to Dante, because he referred to a similar concept in the *Paradiso* to describe the abundance of Heaven's lights: "They were so many that their number piles up faster than the chessboard doubling." Jan Gullberg writes, "With about 100 grains to a cubic centimeter, the total volume of [Sissa's] wheat would be nearly…two hundred cubic kilometers, to be loaded on two thousand million railway wagons, which would make up a train reaching a thousand times around the Earth."

**SEE ALSO** Harmonic Series Diverges (c. 1350), Rope around the Earth Puzzle (1702), and Rubik's Cube (1974).

*The famous problem of Sissa's chessboard demonstrates the nature of geometric progressions. In the smaller version depicted here, how many pieces of candy will the hungry beetle get if the progression 1 + 2 + 4 + 8 + 16 . . . continues?*

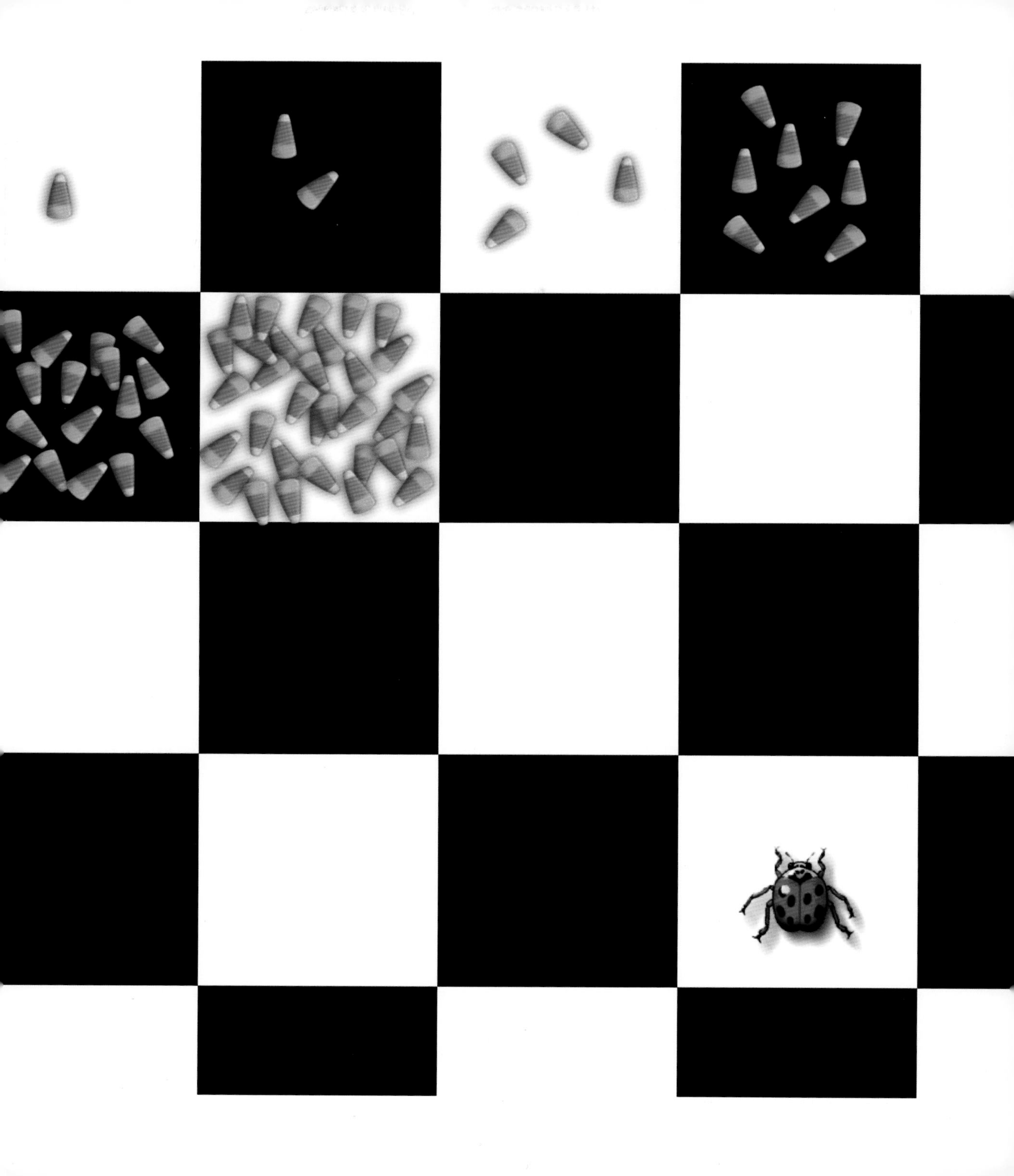

# Harmonic Series Diverges

**Nicole Oresme** (1323–1382), **Pietro Mengoli** (1626–1686), **Johann Bernoulli** (1667–1748), **Jacob Bernoulli** (1654–1705)

If God were infinity, then *divergent series* would be His angels flying higher and higher to reach Him. Given an eternity, all such angels approach their Creator. For example, consider the following infinite series: $1 + 2 + 3 + 4\ldots.$ If we add one term of the series each year, in four years the sum will be 10. Eventually, after an infinite number of years, the sum reaches infinity. Mathematicians call such series divergent because they explode to infinity, given an infinite number of terms. For this entry, we are interested in a series that diverges much more slowly. We're interested in a more magical series, an angel, perhaps, with weaker wings.

Consider the harmonic series, the first famous example of a divergent series whose terms approach zero: $1 + 1/2 + 1/3 + 1/4 + \ldots.$ Of course, this series explodes more slowly than does our previous example, but it still grows to infinity. In fact, it grows so incredibly slowly that if we added a term a year, in $10^{43}$ years we'd have a sum less than 100. William Dunham writes, "Seasoned mathematicians tend to forget how surprising this phenomenon appears to the uninitiated student—that, by adding ever more negligible terms, we nonetheless reach a sum greater than any preassigned quantity."

Nicole Oresme, the famous French philosopher of the Middle Ages, was the first to prove the divergence of the harmonic series (c. 1350). His results were lost for several centuries, and the result was proved again by Italian mathematician Pietro Mengoli in 1647 and by Swiss mathematician Johann Bernoulli in 1687. His brother Jacob Bernoulli published a proof in his 1689 work *Tractatus de Seriebus Infinitis* (*Treatise on Infinite Series*), which he closes with: "So the soul of immensity dwells in minutia. And in narrowest limits no limits inhere. What joy to discern the minute in infinity! The vast to perceive in the small, what divinity!"

**SEE ALSO** Zeno's Paradoxes (c. 445 B.C.), Wheat on a Chessboard (1256), Discovery of Series Formula for $\pi$ (c. 1500), Brun's Constant (1919), and Polygon Circumscribing (c. 1940).

*Depiction of Nicole Oresme from his* Tractatus de origine, natura, jure et mutationibus monetarum (On the Origin, Nature, Juridical Status, and Variations of Coinage), *which was published around the year 1360.*

c.1427

# Law of Cosines

**Ghiyath al-Din Jamshid Mas'ud al-Kashi** (c. 1380–1429), **François Viète** (1540–1603)

The law of cosines may be used for calculating the length of one side of a triangle when the angle opposite this side, and the length of the other two sides, are known. The law may be expressed as $c^2 = a^2 + b^2 - 2ab\cos(C)$, where $a$, $b$, and $c$ are triangle side lengths, and $C$ is an angle between sides $a$ and $b$. Because of its generality, the application of the law ranges from land surveying to calculating the flight paths of aircraft.

Notice how the law of cosines becomes the **Pythagorean Theorem** ($c^2 = a^2 + b^2$) for right triangles, when C becomes 90° and the cosine becomes zero. Also note that if all three side lengths of a triangle are known, we can use the law of cosines to compute the angles of a triangle.

**Euclid's *Elements*** (c. 300 B.C.) contains the seeds of concepts that lead to the law of cosines. In the fifteenth century, the Persian astronomer and mathematician al-Kashi provided accurate trigonometric tables and expressed the theorem in a form suitable for modern usage. French mathematician François Viète discovered the law independently of al-Kashi.

In French, the law of cosines is named *Théorème d'Al-Kashi*, after al-Kashi's unification of existing works on the subject. Al-Kashi's most important work is *The Key to Arithmetic*, completed in 1427, which discusses mathematics used in astronomy, surveying, architecture, and accounting. Al-Kashi uses decimal fractions in calculating the total surface area needed for certain *muqarnas*, decorative structures in Islamic and Persian architecture.

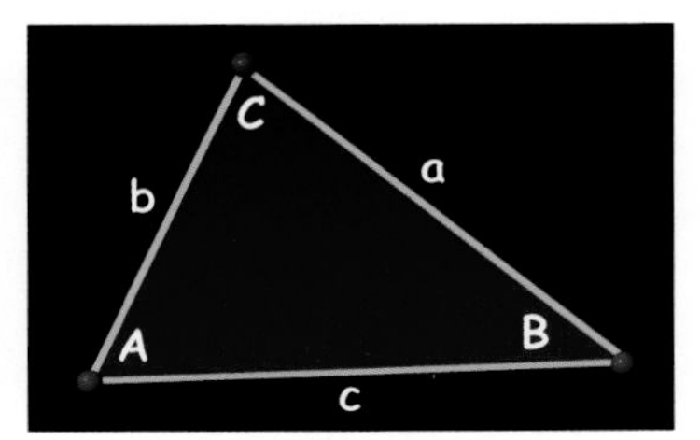


Viète had a fascinating life. At one point, he successfully broke the codes of Philip II of Spain for Henry IV of France. Philip believed that the very complex cipher could not be broken by mere men, and when he discovered that the French knew of his military plans, he complained to the Pope that black magic was being used against his country.

**SEE ALSO** Pythagorean Theorem and Triangles (c. 600 B.C.), Euclid's *Elements* (300 B.C.), Ptolemy's *Almagest* (c. 150), and *Polygraphiae Libri Sex* (1518).

*An Iranian stamp issued in 1979 commemorating al-Kashi. In French, the law of cosines is called* Théorème d'Al-Kashi, *after al-Kashi's unification of existing works on the subject.*

GHYATH-AL-DIN JAMSHID KASHANI

(14–15) A.C.

1478

# *Treviso Arithmetic*

European arithmetic texts in the fifteenth and sixteenth centuries often presented mathematical word problems related to commerce in order to teach mathematical concepts. The general idea of word problems for students dates back for centuries, and some of the oldest-known word problems were presented in ancient Egypt, China, and India.

*Treviso Arithmetic* is brimming with word problems, many involving merchants investing money and who wish to avoid being cheated. The book was written in a Venetian dialect and published in 1478 in the town of Treviso, Italy. The unknown author of the book writes, "I have often been asked by certain youths in whom I have much interest, and who look forward to mercantile pursuits, to put into writing the fundamental principles of arithmetic. Therefore, being impelled by my affection for them, and by the value of the subject, I have to the best of my small ability undertaken to satisfy them in some slight degree." He then gives numerous word problems involving merchants with names like Sebastiano and Jacomo who invest their money for gain in a partnership. The book also shows several ways for performing multiplication and includes information from the Fibonacci work ***Liber Abaci*** (1202).

*Treviso* is particularly significant because it is the earliest-known printed mathematics book in Europe. It also promoted the use of the Hindu Arabic numeral system and computational algorithms. Because commerce of the time began to have a wide international component, prospective businessmen had urgent needs to become facile with mathematics. Today, scholars are fascinated by *Treviso* because it provides a portal into the methods of teaching mathematics in fifteenth-century Europe. Similarly, because the problems involve calculating payment for exchanging goods, fabric cutting, saffron trading, alloy mixtures in coins, currency exchange, and calculating shares of profits derived from partnerships, readers come to understand the contemporary concerns with respect to cheating, usury, and determination of interest charges.

**SEE ALSO** Rhind Papyrus (1650 B.C.), *Ganita Sara Samgraha* (850), Fibonacci's *Liber Abaci* (1202), and *Sumario Compendioso* (1556).

*Merchants weighing their goods in a marketplace, circa 1400, drawn after a fifteenth-century stained-glass window in Chartres Cathedral, France.* Treviso Arithmetic, *the earliest-known printed mathematics book in Europe, includes problems involving merchants, investment, and trade.*

# Discovery of Series Formula for π

**Gottfried Wilhelm Leibniz** (1646–1716), **James Gregory** (1638–1675), **Nilakantha Somayaji** (1444–1544)

An infinite series is the sum of infinitely many numbers, and it plays an important role in mathematics. For a series such as 1 + 2 + 3 +…, the sum is infinite, and the series is said to diverge. An *alternating* series is one in which every other term is negative. One particular alternating series has intrigued mathematicians for centuries.

Pi, symbolized by the Greek letter π, is the ratio of a circle's circumference to its diameter and can be expressed by a remarkably simple formula: $\pi/4 = 1 - 1/3 + 1/5 - 1/7 + \ldots$. Note also that the arctan function in trigonometry can be expressed by $\arctan(x) = x - x^3/3 + x^5/5 - x^7/7 + \ldots$. Using the arctan series, the series for π/4 is obtained by setting $x = 1$.

Ranjan Roy notes that the independent discovery of the infinite series for π "by different persons living in different environments and cultures gives us insight into the character of mathematics as a universal discipline." The series was discovered by German mathematician Gottfried Wilhelm Leibniz, Scottish mathematician and astronomer James Gregory, and an Indian mathematician of the fourteenth or fifteenth century whose identity is not definitively known, although the result is usually ascribed to Nilakantha Somayaji. Leibniz discovered the formula in 1673, and Gregory discovered it in 1671. Roy writes, "Leibniz's discovery of the infinite series for π was Leibniz's first greatest achievement." Dutch mathematician Christiaan Huygens told Leibniz that this remarkable property of the circle would be celebrated among mathematicians forever. Even Newton said that the formula revealed Leibniz's genius.

Gregory's discovery involving the arctan formula came before Leibniz, although Gregory did not note the special case of the arctan formula for π/4. This arctan infinite series was also given in Somayaji's 1500 book *Tantrasangraha*. Somayaji was aware that a *finite* series of rational numbers could never suffice to represent π.

**SEE ALSO** π (c. 250 B.C.), Zeno's Paradoxes (c. 445 B.C.), Harmonic Series Diverges (c. 1350), and Euler-Mascheroni Constant (1735).

*The number π, which can be approximated by the digits shown in this illustration, also can be expressed by a remarkably simple formula: $\pi/4 = 1 - 1/3 + 1/5 - 1/7 + \ldots$.*

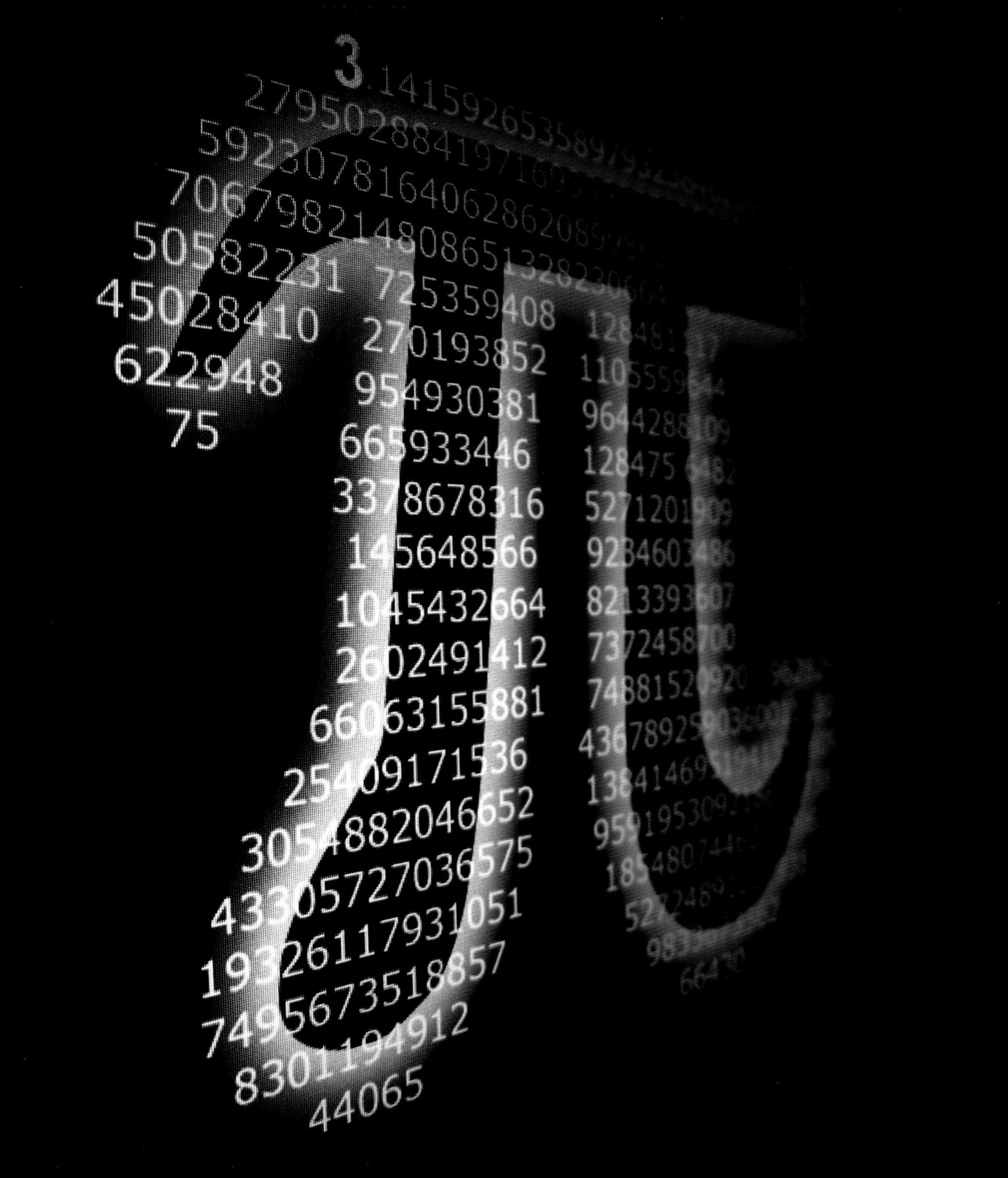

1509

# Golden Ratio

**Fra Luca Bartolomeo de Pacioli** (1445–1517)

In 1509, Italian mathematician Luca Pacioli, a close friend of Leonardo da Vinci, published *Divina Proportione*, a treatise on a number that is now widely known as the "Golden Ratio." This ratio, symbolized by ø, appears with amazing frequency in mathematics and nature. We can understand the proportion most easily by dividing a line into two segments so that the ratio of the whole segment to the longer part is the same as the ratio of the longer part to the shorter part, or $(a + b)/b = b/a = 1.61803\ldots$

If the lengths of the sides of a rectangle are in the golden ratio, then the rectangle is a "golden rectangle." It's possible to divide a golden rectangle into a square and a golden rectangle. Next, we can cut the smaller golden rectangle into a smaller square and golden rectangle. We may continue this process indefinitely, producing smaller and smaller golden rectangles.

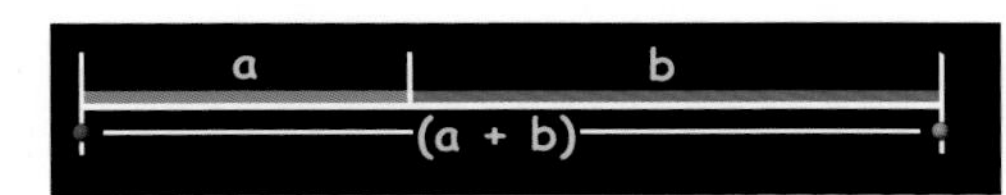


If we draw a diagonal from the top right of the original rectangle to the bottom left, and then from the bottom right of the baby (that is, the next smaller) golden rectangle to the top left, the intersection point shows the point to which all the baby golden rectangles converge. Moreover, the lengths of the diagonals are in golden ratio to each another. The point to which all the golden rectangles converge is sometimes called the "Eye of God."

The golden rectangle is the *only* rectangle from which a square can be cut so that the remaining rectangle will always be similar to the original rectangle. If we connect the vertices in the diagram, we approximate a logarithmic spiral that "envelops" the Eye of God. Logarithmic spirals are everywhere—seashells, animal horns, the cochlea of the ear—anywhere nature needs to fill space economically and regularly. A spiral is strong and uses a minimum of materials. While expanding, it alters its size but never its shape.

**SEE ALSO** Archimedes' Spiral (225 B.C.), Fermat's Spiral (1636), Logarithmic Spiral (1638), and Squaring a Rectangle (1925).

*Artistic depiction of golden ratios. Note that the two diagonal lines intersect at a point to which all the baby golden rectangles will converge.*

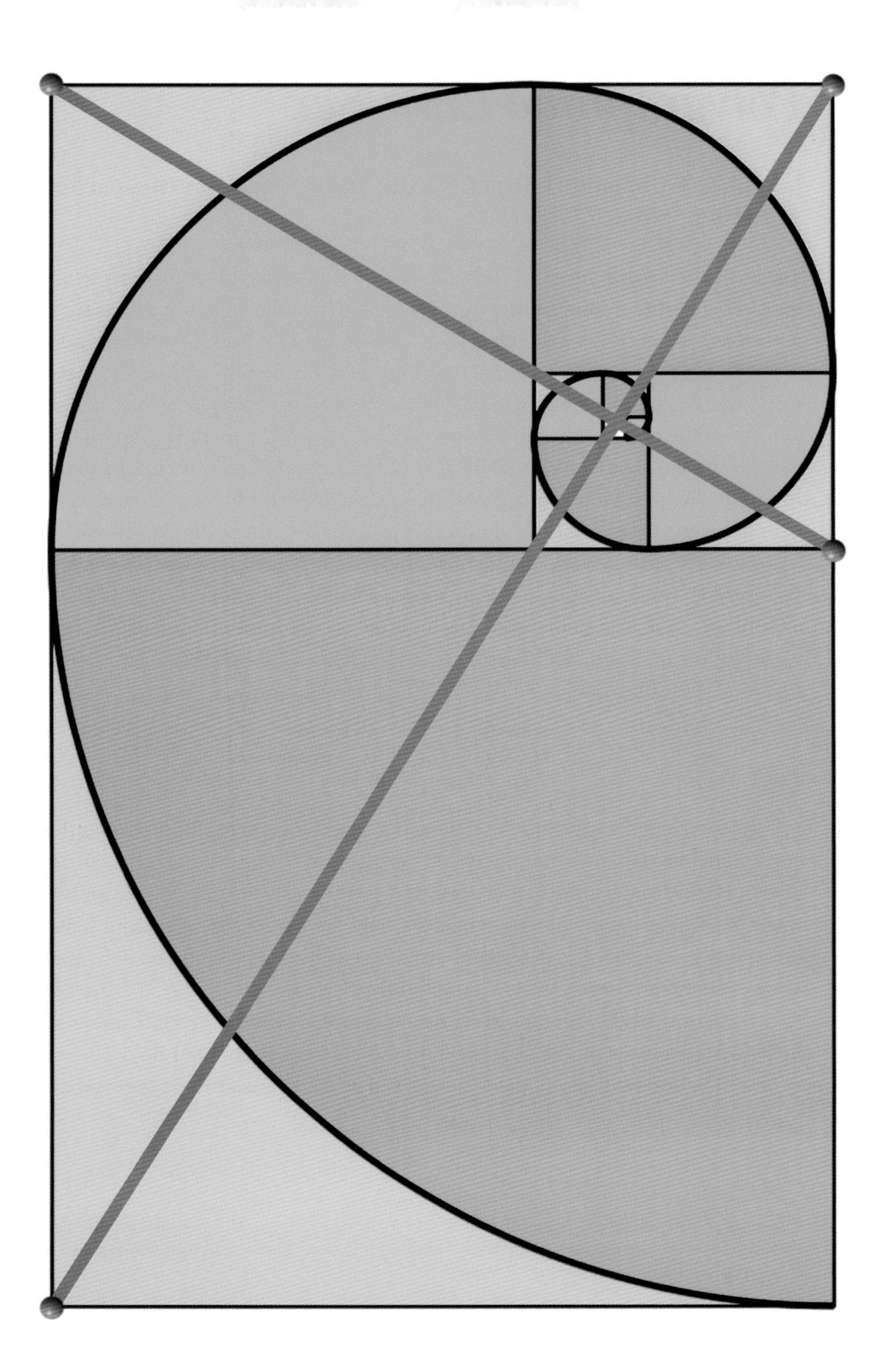

# *Polygraphiae Libri Sex*

**Johannes Trithemius** (1462–1516), **Abu Yusuf Yaqub ibn Ishaq al-Sabbah Al-Kindi** (c. 801–c. 873)

Today, mathematical theory has become central to cryptography. However, in ancient times, simple substitution ciphers were often used in which letters in a message were replaced by other letters. For example, CAT becomes DBU when substituting each letter in CAT with the one following it in the alphabet. Of course, such simple ciphers became easy to break after the discovery of frequency analysis, for example, by the Arab scholar al-Kindi in the ninth century. This method analyzes which letters occur most frequently in a language—such as ETAOIN SHRDLU in English—and uses this information to solve substitution codes. More complex statistics can be used, such as considering counts of pairs of letters. For example, *Q* almost always occurs together with *U* in English.

The first printed book on cryptography, *Polygraphiae Libri Sex* (*Six Books of Polygraphy*), was written by German abbot Johannes Trithemius and published in 1518 after his death. *Polygraphiae* contains hundreds of columns of Latin words, arranged in two columns per page. Each word stands for a letter of the alphabet. For example, the first page starts like this:

| | |
|---|---|
| a: Deus | a: clemens |
| b: Creator | b: clementissimus |
| c: Conditor | c: pius |

To encode a message, one uses a word to stand for a letter. Remarkably, Trithemius constructed the tables so that the coded passage appears to make sense as an actual prayer. For example, if the first two letters in a message were CA, the prayer would start with *Conditor clemens* (Merciful Creator) as the first two words in a Latin sentence. The remaining books of *Polygraphiae* present more sophisticated cryptographic methods, along with tables, for creatively hiding information.

Trithemius's other famous work, *Steganographia* (written in 1499 and published in 1606), was placed on the Catholic Church's "List of Prohibited Books" because it appeared to be a book about black magic, but in reality was just another code book!

**SEE ALSO** Law of Cosines (c. 1427) and Public-Key Cryptography (1977).

*Engraving of German abbot Johannes Trithemius, by André de Thevet (1502–1590). Trithemius's* Polygraphiae, *the first printed book on cryptography, provided various Latin words that may be used to encode secret messages that appear to be ordinary prayers if intercepted.*

1537

# Loxodrome

**Pedro Nunes** (1502–1578)

For the purposes of terrestrial navigation, the loxodromic spiral—also known as a spherical helix, loxodrome, or rhumb line—goes through the north-south meridians of the Earth at a constant angle. The loxodrome coils like a gigantic snake around the Earth and spirals around the poles without reaching them.

One way of sailing the Earth is to attempt to travel the shortest path between points, which is along an arc of a great circle around the Earth. However, even though this is the shortest path, a navigator must continually make adjustments to the course based on compass readings, a nearly impossible task for early navigators.

On the other hand, a loxodromic path allows the navigator to continually direct the vessel to the same point of the compass even though the path to the destination is longer. For example, using this approach to travel from New York to London, a voyager could head at a constant bearing of 73° east of north. A loxodrome is represented as a straight line on a Mercator projection map.

The loxodrome was invented by Portuguese mathematician and geographer Pedro Nunes. Nunes lived during a time when the Inquisition struck fear in the heart of Europe. Many Jews in Spain were forcibly converted to Roman Catholicism, and Nunes was converted as a child. The main targets of the later Spanish Inquisition were descendants of these *conversos*, such as Nunes's grandsons in the early 1600s. Gerardus Mercator (1512–1594), a Flemish cartographer, was imprisoned by the Inquisition because of his Protestant faith and wide travels, and he narrowly escaped execution.

Some Muslim groups in North America make use of a loxodrome line to Mecca (southeastward) as their *qibla* (praying direction) instead of using the traditional shortest path. In 2006, the Malaysian National Space Agency (MYNASA) sponsored a conference to determine the proper qibla for Muslims in orbit.

**SEE ALSO** Archimedes' Spiral (225 B.C.), Mercator Projection (1569), Fermat's Spiral (1636), Logarithmic Spiral (1638), and Voderberg Tilings (1936).

*Computer-graphics artist Paul Nylander created this attractive double spiral by applying a stereographic projection to a loxodrome curve. (A stereographic projection maps a sphere onto a plane.)*

# Cardano's *Ars Magna*

**Gerolamo Cardano** (1501–1576), **Niccolo Tartaglia** (1500-1557), **Lodovico Ferrari** (1522–1565)

Italian Renaissance mathematician, physician, astrologer, and gambler Gerolamo Cardano is most famous for his work on algebra titled *Artis magnae, sive de regulis algebraicis* (*Of the Great Art, or The Rules of Algebra*)—also referred to more succinctly as *Ars magna*. Although the book sold well, Jan Gullberg notes, "No single publication has promoted interest in algebra like Cardano's *Ars magna*, which, however, provides very boring reading to a present-day peruser by consistently devoting pages of verbose rhetoric to a solution.... With the untiring industry of an organ grinder, Cardano monotonously reiterates the same solution for a dozen or more near-identical problems where just one would do."

Nevertheless, Cardano's impressive work revealed solutions to different types of cubic and quartic equations—that is, equations with variables raised to the third and fourth powers, respectively. Italian mathematician Niccolo Tartaglia had earlier communicated to Cardano the solution to the cubic, $x^3 + ax = b$, and he attempted to ensure that Cardano would never publish the solution by making Cardano swear an oath to God. Cardano published the solution anyway, after he seemed to discover that Tartaglia was not the first to solve the cubic equation using radicals. The general quartic equation was solved by Cardano's student Lodovico Ferrari.

In *Ars magna*, Cardano explored the existence of what are now called **imaginary numbers**, which are based on the square root of −1, although he did not fully appreciate their properties. In fact, he presents the first calculation with complex numbers when he writes: "Dismissing mental tortures, and multiplying $5 + \sqrt{-15}$ by $5 - \sqrt{-15}$, we obtain $25 - (-15)$. Therefore, the product is 40."

In 1570, as a result of the Inquisition, Cardano was thrown in jail for several months on the charge of heresy, because he had cast the horoscope of Jesus Christ. According to legend, Cardano correctly predicted the exact date of his own death, a prophecy he is said to have ensured by killing himself on this date.

**SEE ALSO** Omar Khayyam's *Treatise* (1070), Imaginary Numbers (1572), and Group Theory (1832).

*Italian mathematician Gerolamo Cardano, well known for his work on algebra titled* Artis magnae, sive de regulis algebraicis, *also referred to as* Ars Magna (The Great Art).

# HIERONYMI CAR DANI, PRÆSTANTISSIMI MATHE MATICI, PHILOSOPHI, AC MEDICI,

# ARTIS MAGNÆ,

## SIVE DE REGVLIS ALGEBRAICIS,

Lib. unus. Qui & totius operis de Arithmetica, quod

OPVS PERFECTVM

inscripsit, est in ordine Decimus.

HAbes in hoc libro, studiose Lector, Regulas Algebraicas (Itali, de la Cossa uocant) nouis adinuentionibus, ac demonstrationibus ab Authore ita locupletatas, ut pro pauculis antea uulgò tritis, iam septuaginta euaserint. Neq; solum, ubi unus numerus alteri, aut duo uni, uerum etiam, ubi duo duobus, aut tres uni ęquales fuerint, nodum explicant. Hunc aũt librum ideo seorsim edere placuit, ut hoc abstrusissimo, & planè inexhausto totius Arithmeticæ thesauro in lucem eruto, & quasi in theatro quodam omnibus ad spectandum exposito, Lectores incitarẽtur, ut reliquos Operis Perfecti libros, qui per Tomos edentur, tanto auidius amplectantur, ac minore fastidio perdiscant.

1556

# *Sumario Compendioso*

**Juan Diez** (1480–1549)

The *Sumario compendioso*, published in Mexico City in 1556, is the first work on mathematics printed in the Americas. The publication of *Sumario compendioso* in the New World preceded by many decades the emigration of the Puritans to North America and the settlement in Jamestown, Virginia. The author, Brother Juan Diez, was a companion of Hernándo Cortes, the Spanish conquistador, during Cortes's conquests of the Aztec Empire.

Diez wrote the book primarily for people buying gold and silver recovered from the mines of Peru and Mexico. In addition to providing tables to make it easy for merchants to obtain numerical values without much calculation, he also devoted part of the work to algebra related to the quadratic equation—that is, equations of the form $ax^2 + bx + c = 0$ with $a \neq 0$. For example, one of the problems may be translated as: "Find the square from which if 15¾ is subtracted, the result is its own root." This is equivalent to solving $x^2 - 15\frac{3}{4} = x$.

The full title of Diez's work is *Sumario compendioso de las quentas de plata y oro que en los reynos del Piru son necessarias a los mercaderes y todo genero de tratantes. Con algunas reglas tocantes al Arithmetica*, which translates to *Comprehensive Summary of the Counting of Silver and Gold, Which, in the Kingdoms of Peru, Are Necessary for Merchants and All Kinds of Traders*. The printing press and paper were shipped from Spain and then carried to Mexico City. Only four known copies of *Sumario compendioso* exist today.

According to Shirley Gray and C. Edward Sandifer, "The New World's first mathematics book in *English* was not published until 1703. . . . Of all the colonial mathematics books, the ones in Spanish are the most interesting, because they were mostly written in America for use by people living in America."

**SEE ALSO** Diophantus's *Arithmetica* (250), Al-Khwarizmi's *Algebra* (830), and *Treviso Arithmetic* (1478).

*The* Sumario Compendioso *is the first work on mathematics printed in the Americas.*

¶ Sumario cōpēdioso delas quētas de plata y oro q̃ en los reynos del Piru son necessarias a los mercáderes: y todo genero de tratantes. Cō algunas reglas tocantes al Arithmetica.

Fecho por Juan Diez freyle.

# Mercator Projection

**Gerardus Mercator** (1512–1594), **Edward Wright** (c. 1558–1615)

Many of the ancient Greek ideas for representing a spherical Earth on a flat map were lost during the Middle Ages. John Short explains that, in the fifteenth century, "the value of sea charts rivaled gold as a primary target for buccaneer captains. Later, maps became status symbols among wealthy merchants who built vast fortunes thanks to thriving trade routes made possible by reliable sea navigation."

One of the most famous map projections in history is the Mercator map (1569), which became commonly used for nautical voyages and is named after Flemish cartographer Gerardus Mercator. Norman Thrower writes, "Like several other projections, the Mercator is conformal (shapes around a point are correct), but it also has a unique property: straight lines are rhumb lines or **loxodromes** (lines of constant compass bearing)." This latter quality was invaluable to marine navigators who chose routes using compasses and other devices to indicate geographical directions and to steer the ships. The use of the Mercator map increased in the 1700s, after the invention of the accurate marine chronometer, a timekeeping device used to determine longitude by means of celestial navigation.

Although Mercator was the first mapmaker to create his projection in which compass lines intersect meridians at a constant, given angle, he probably used graphical methods and little mathematics. English mathematician Edward Wright provided an analysis of the fascinating properties of the map in his *Certaine Errors in Navigation* (1599). For the mathematically inclined reader, the Mercator map projection with $x$ and $y$ coordinates may be created from latitude $\varphi$ and longitude $\lambda$ values by: $x = \lambda - \lambda_0$ and $y = \sinh^{-1}(\tan(\varphi))$, where $\lambda_0$ is the longitude at the center of the map. The Mercator projection does have its imperfections; for example, it exaggerates the size of areas far from the equator.

**SEE ALSO** Loxodrome (1537), Projective Geometry (1639), and Three-Armed Protractor (1801).

*The Mercator map has been commonly used for nautical voyages. However, the map creates distortions. For example, Greenland appears to be roughly the same as Africa, even though Africa's area is 14 times that of Greenland.*

1572

# Imaginary Numbers

**Rafael Bombelli** (1526–1572)

An imaginary number is one whose square has a negative value. The great mathematician Gottfried Leibniz called imaginary numbers "a wonderful flight of God's Spirit; they are almost an amphibian between being and not being." Because the square of any real number is positive, for centuries many mathematicians declared it impossible for a negative number to have a square root. Although various mathematicians had inklings of imaginary numbers, the history of imaginary numbers started to blossom in sixteenth-century Europe. The Italian engineer Rafael Bombelli, well known during his time for draining swamps, is today famous for his *Algebra*, published in 1572, that introduced a notation for $\sqrt{-1}$, which would be a valid solution to the equation $x^2 + 1 = 0$. He wrote, "It was a wild thought in the judgment of many." Numerous mathematicians were hesitant to "believe" in imaginary numbers, including Descartes, who actually introduced the term *imaginary* as a kind of insult.

Leonhard Euler in the eighteenth century introduced the symbol $i$ for $\sqrt{-1}$ —for the first letter of the Latin word *imaginarius*—and we still use Euler's symbol today. Key advances in modern physics would not have been possible without the use of imaginary numbers, which have aided physicists in a vast range of computations, including efficient calculations involving alternating currents, relativity theory, signal processing, fluid dynamics, and quantum mechanics. Imaginary numbers even play a role in the production of gorgeous **fractal** artworks that show a wealth of detail with increasing magnifications.

From string theory to quantum theory, the deeper one studies physics, the closer one moves to pure mathematics. Some might even say that mathematics "runs" reality in the same way that Microsoft's operating system runs a computer. Schrödinger's wave equation—which describes basic reality and events in terms of wave functions and probabilities—may be thought of as the evanescent substrate on which we all exist, and it relies on imaginary numbers.

**SEE ALSO** Cardano's *Ars Magna* (1545), Euler's Number, *e* (1727), Quaternions (1843), Riemann Hypothesis (1859), Boole's *Philosophy and Fun of Algebra* (1909), and Fractals (1975).

*Imaginary numbers play a role in the production of Jos Leys's gorgeous fractal artworks that show a wealth of detail with increasing magnifications. Early mathematicians were so suspicious of the usefulness of imaginary numbers that they insulted those who suggested their existence.*

# Kepler Conjecture

**Johannes Kepler** (1571–1630), **Thomas Callister Hales** (b. 1958)

Imagine that your goal is to fill a large box with as many golf balls as possible. Close the lid tightly when finished. The density of balls is determined from the proportion of the volume of the box that contains a ball. In order to stuff the most balls into the box, you need to discover an arrangement with the highest possible density. If you simply drop balls into the box, you'll only achieve a density of roughly 65 percent. If you are careful, and create a layer at the bottom in a hexagonal arrangement, and then put the next layer of balls in the indentations created by the bottom layer, and continue, you'll be able to achieve a packing density of $\pi/\sqrt{18}$, which is about 74 percent.

In 1611, German mathematician and astronomer Johannes Kepler wrote that no other arrangement of balls has a higher average density. In particular, he conjectured in his monograph *The Six-Cornered Snowflake* that it is impossible to pack identical spheres in three dimensions greater than the packing found in face-centered (hexagonal) cubic packing. In the nineteenth century, Karl Friedrich Gauss proved that the traditional hexagonal arrangement was the most efficient for a *regular* 3-D grid. Nevertheless, the Kepler conjecture remained, and no one was sure if a denser packing could be achieved with an *irregular* packing.

Finally, in 1998, American mathematician Thomas Hales stunned the world when he presented a proof that Kepler had been right. Hales's equation and its 150 variables expressed every conceivable arrangement of 50 spheres. Computers confirmed that no combination of variables led to a packing efficiency higher than 74 percent.

The *Annals of Mathematics* agreed to publish the proof, provided it was accepted by a panel of 12 referees. In 2003, the panel reported that they were "99 percent certain" of the correctness of the proof. Hales estimates that to produce a complete formal proof will take around 20 years of work.

**SEE ALSO** Sangaku Geometry (c. 1789), Four-Color Theorem (1852), and Hilbert's 23 Problems (1900).

*Fascinated by Kepler's famous conjecture, Princeton University scientists Paul Chaikin, Salvatore Torquato, and colleagues studied the packing of M&M chocolate candies. They discovered that the candies had a packing density of about 68 percent, or 4 percent greater than for randomly packed spheres.*

1614

# Logarithms

**John Napier** (1550–1617)

Scottish mathematician John Napier is famous as the inventor and promoter of logarithms in his 1614 book *A Description of the Marvelous Rule of Logarithms*. This method has since contributed to countless advances in science and engineering by making difficult calculations possible. Before electronic calculators became widely available, logarithms and tables of logarithms were commonly used in surveying and navigation. Napier was also the inventor of Napier's bones, rods carved with multiplication tables that could be arranged in patterns in order to aid in calculations.

A logarithm (to a base $b$) of a number $x$ is expressed as $\log_b(x)$ and equals the exponent $y$ that satisfies $x = b^y$. For example, because $3^5 = 3 \times 3 \times 3 \times 3 \times 3 = 243$, we say that the log of 243 (base 3) is 5, or $\log_3(243) = 5$. As another example, $\log_{10}(100) = 2$. For practical purposes, consider that a multiplication such as $8 \times 16 = 128$ can be rewritten as $2^3 \times 2^4 = 2^7$, thereby converting the calculations into ones involving the simple additions of the powers ($3 + 4 = 7$). Prior to calculators, in order to multiply two numbers, an engineer often looked up the logarithms of both numbers in a table, added them, and then looked up the result in the table to find the product. This could often be faster than multiplying by hand and is the principle on which **slide rules** are based.

Today, various quantities and scales in science are expressed as logarithms of other quantities. For example, the pH scale in chemistry, the bel unit of measurement in acoustics, and the Richter scale used for measuring earthquake intensity all involve a base-10 logarithmic scale. Interestingly, the discovery of logarithms just prior to the era of Isaac Newton had an impact on science comparable to the invention of the computer in the twentieth century.

**SEE ALSO** Slide Rule (1621), Logarithmic Spiral (1638), and Stirling's Formula (1730).

*John Napier, the discoverer of logarithms, created a calculation device known as Napier's rods or bones. Napier's rotatable rods reduced multiplication to a sequence of simple additions.*

| 0 | 1 | 2 | 3 | 4 | 5 | 6 | 7 | 8 | 9 | 10 | 11 |
|---|---|---|---|---|---|---|---|---|---|---|---|
| 1 | 2 | 3 | 4 | 5 | 6 | 7 | 8 | 9 | 10 | 11 | 12 |
| 2 | 3 | 4 | 5 | 6 | 7 | 8 | 9 | 10 | 11 | 12 | 13 |
| 3 | 4 | 5 | 6 | 7 | 8 | 9 | 10 | 11 | 12 | 13 | 14 |
| 4 | 5 | 6 | 7 | 8 | 9 | 10 | 11 | 12 | 13 | 14 | 15 |
| 5 | 6 | 7 | 8 | 9 | 10 | 11 | 12 | 13 | 14 | 15 | 16 |
| 6 | 7 | 8 | 9 | 10 | 11 | 12 | 13 | 14 | 15 | 16 | 17 |
| 7 | 8 | 9 | 10 | 11 | 12 | 13 | 14 | 15 | 16 | 17 | 18 |
| 8 | 9 | 10 | 11 | 12 | 13 | 14 | 15 | 16 | 17 | 18 | 19 |
| 9 | 10 | 11 | 12 | 13 | 14 | 15 | 16 | 17 | 18 | 19 | 20 |
| 10 | 11 | 12 | 13 | 14 | 15 | 16 | 17 | 18 | 19 | 20 | 21 |
| 11 | 12 | 13 | 14 | 15 | 16 | 17 | 18 | 19 | 20 | 21 | 22 |

1621

# Slide Rule

**William Oughtred** (1574–1660)

Those of you who went to high school before the 1970s may recall that the slide rule once seemed to be as common as the typewriter. In just seconds, engineers could multiply, divide, find square roots, and do much more. The earliest version with sliding pieces was invented in 1621 by English mathematician and Anglican minister William Oughtred, based on the **logarithms** of Scottish mathematician John Napier. Oughtred may not have initially recognized the value of his work, because he did not quickly publish his findings. According to some accounts, one of his students stole the idea and published a pamphlet on the slide rule, which emphasized its portability, and raved that the device was "fit for use on horseback as on foot." Oughtred was outraged by his student's duplicity.

In 1850, a 19-year-old French artillery lieutenant modified the original design of the slide rule, and the French army used it to perform projectile calculations when fighting the Prussians. During World War II, American bombers often used specialized slide rules.

Slide-rule guru Cliff Stoll writes, "Consider the engineering achievements that owe their existence to rubbing two sticks together: the Empire State Building; the Hoover Dam; the curves of the Golden Gate Bridge; hydrodynamic automobile transmissions, transistor radios; the Boeing 707 airliner." Wernher Von Braun, the designer of the German V-2 rocket, relied on slide rules made by the German company Nestler, as did Albert Einstein. Pickett slide rules were aboard Apollo space missions in case the computers failed!

In the twentieth century, 40 million slide rules were produced worldwide. Given the crucial role that this device played from the Industrial Revolution until modern times, the device deserves a place in this book. Literature from the Oughtred Society states, "For a span of 3.5 centuries, it was used to perform design calculations for virtually all the major structures built on this earth."

**SEE ALSO** Abacus (c. 1200), Logarithms (1614), Curta Calculator (1948), HP-35: First Scientific Pocket Calculator (1972), and Mathematica (1988).

*The slide rule played a crucial role from the Industrial Revolution until modern times. In the twentieth century, 40 million slide rules were produced and were used by engineers for countless applications.*

# Fermat's Spiral

**Pierre de Fermat** (1601–1665), **René Descartes** (1596–1650)

In the early 1600s, Pierre de Fermat, a French lawyer and mathematician, made brilliant discoveries in number theory and other areas of mathematics. His 1636 manuscript "*Ad locos planos et solidos lisagoge*" ("Introduction to Plane and Solid Loci") went beyond René Descartes' work in analytical geometry and allowed Fermat to define and study many important curves that included the cycloid and the Fermat spiral.

The Fermat spiral, or parabolic spiral, can be created using the polar equation $r^2 = a^2\theta$. Here, $r$ is the distance of the curve from the origin, $a$ is a constant that determines how tightly wound the spiral is, and $\theta$ is the polar angle. For any given positive value of $\theta$, negative and positive values for $r$ exist, which leads to a curve that is symmetrical about the origin. Fermat studied the relationship of the area enclosed by an arm of the spiral and the $x$-axis as the spiral turns.

Today, computer graphics specialists sometimes use this curve to model the arrangement of seed heads in flowers. For example, we may draw spots that have center positions determined by polar coordinate values, $r(i) = ki^{1/2}$, and angles $\theta$ defined by $\theta(i) = 2i\pi/\tau$. Here, $\tau$ is the golden number $(1+\sqrt{5})/2$, and $i$ is simply a counter that steps as 1, 2, 3, 4,...

This graphics approach produces many different spiral arms that twist in one direction or another. It is possible to trace various sets of symmetrical spirals, radiating from the center of the pattern, for example a set of 8, 13, or 21 spiral arms, and these numbers of arms are all Fibonacci numbers (see entry **Fibonacci's *Liber Abaci***).

Michael Mahoney writes, "Fermat had been working with spirals for some time before he encountered one in *Galileo's Dialogue*. In a letter of June 3, 1636, he described to Mersenne the spiral $r^2 = a^2\theta$...."

**SEE ALSO** Archimedes' Spiral (225 B.C.), Fibonacci's *Liber Abaci* (1202), Golden Ratio (1509), Loxodrome (1537), Fermat's Last Theorem (1637), Logarithmic Spiral (1638), Voderberg Tilings (1936), Ulam Spiral (1963), and Spidrons (1979).

*The Fermat spiral, or parabolic spiral, can be created using the polar equation* $r^2 = a^2\theta$. *For any given positive value of* $\theta$, *two values for* r *exist, which leads to a curve that is symmetrical about the origin, which is located at the center of this artistic rendition.*

# Fermat's Last Theorem

**Pierre de Fermat** (1601–1665), **Andrew John Wiles** (b. 1953), **Johann Dirichlet** (1805– 1859), **Gabriel Lamé** (1795–1870)

In the early 1600s, Pierre de Fermat, a French lawyer, made brilliant discoveries in number theory. Although he was an "amateur" mathematician, he created mathematical challenges such as Fermat's Last Theorem (FLT), which was not solved until 1994 by British-American mathematician Andrew Wiles. Wiles spent seven years of his life trying to prove the famous theorem, which may have generated more attempts at proofs than any other theorem.

FLT states that $x^n + y^n = z^n$ has no non-zero integer solutions for $x$, $y$, and $z$ when $n > 2$. Fermat stated his theorem in 1637 when he wrote in his copy of Diophantus's *Arithmetica*, "I have a truly marvelous proof of this proposition which this margin is too narrow to contain." Today, we believe that Fermat had no such proof.

Fermat was no ordinary lawyer, indeed. He is considered, along with Blaise Pascal (1623–1662), the founder of probability theory. As the co-inventor of analytic geometry, along with René Descartes (1596–1650), he is regarded as one of the first modern mathematicians. He once pondered if it was possible to find a right triangle whose hypotenuse and sums of legs were squares. Today, we know that the *smallest* three numbers satisfying these conditions are quite large: 4,565,486,027,761, 1,061,652,293,520, and 4,687,298,610,289.

Since Fermat's time, FLT has spawned significant mathematical research and completely new methods. In 1832, Johann Dirichlet published a proof of Fermat's Last Theorem for $n = 14$. Gabriel Lamé proved it for $n = 7$ in 1839. Amir Aczel writes that FLT "would become the world's most baffling mathematical mystery. Simple, elegant, and [seemingly] utterly impossible to prove, Fermat's Last Theorem captured the imaginations of amateur and professional mathematicians for over three centuries. For some it became a wonderful passion. For others it was an obsession that led to deceit, intrigue, or insanity."

**SEE ALSO** Pythagorean Theorem and Triangles (c. 600 B.C.), Diophantus's *Arithmetica* (250), Fermat's Spiral (1636), Descartes' *La Géométrie* (1637), Pascal's Triangle (1654), and Catalan Conjecture (1844).

*Pierre de Fermat by French painter Robert Lefèvre (1756–1831).*

1637

# Descartes' *La Géométrie*

**René Descartes** (1596–1650)

In 1637, French philosopher and mathematician René Descartes published *La géométrie*, which shows how geometrical shapes and figures can be analyzed using algebra. Descartes' work influenced the evolution of analytical geometry, a field of mathematics that involves the representation of positions in a coordinate system and in which mathematicians algebraically analyze such positions. *La géométrie* also shows how to solve mathematical problems and discusses the representation of points of a plane through the use of real numbers, and the representation and classification of curves through the use of equations.

Interestingly, *La géométrie* does not actually use "Cartesian" coordinate axes or any other coordinate system. The book pays as much attention to representing algebra in geometric forms as vice versa. Descartes believed that algebraic steps in a proof should usually correspond to a geometrical representation.

Jan Gullberg writes, "*La géométrie* is the earliest mathematical text that a modern student of mathematics could read without stumbling over an abundance of obsolete notations. . . . Along with Newton's *Principia*, it is one of the most influential scientific texts of the seventeenth century." According to Carl Boyer, Descartes desired to "free geometry" from the use of diagrams through algebraic procedures and to give meaning to the operations of algebra through geometric interpretation.

More generally, Descartes was groundbreaking in his proposal to unite algebra and geometry into a single subject. Judith Grabiner writes, "Just as the history of Western philosophy has been viewed as a series of footnotes to Plato, so the past 350 years of mathematics can be viewed as a series of footnotes to Descartes' *Geometry*. . . and the triumph of Descartes' methods of problem solving."

Boyer concludes, "In terms of mathematical ability, Descartes probably was the most able thinker of his day, but he was at heart not really a mathematician." His geometry was only one facet of a full life that revolved around science, philosophy, and religion.

**SEE ALSO** Pythagorean Theorem and Triangles (c. 600 B.C.), Quadrature of the Lune (c. 440 B.C.), Euclid's *Elements* (300 B.C.), Pappus's Hexagon Theorem (c. 340), Projective Geometry (1639), and Fractals (1975).

The Ancient of Days *(1794), a watercolor etching by William Blake. European medieval scholars often associated geometry and the laws of nature with the divine. Through the centuries, geometry's focus on compass and straightedge constructions became more abstract and analytical.*

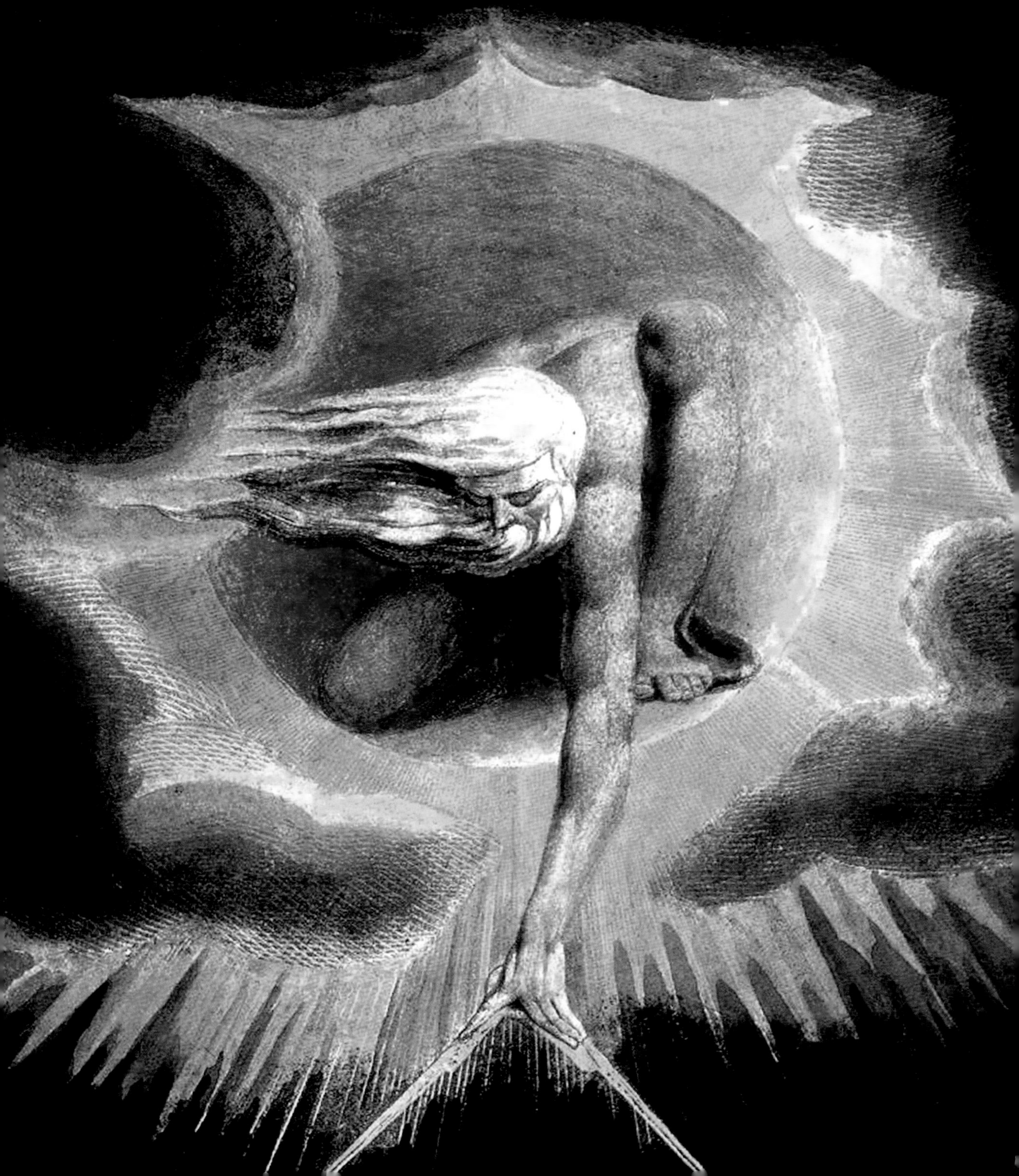

1637

# Cardioid

**Albrecht Dürer** (1471–1528), **Étienne Pascal** (1588–1640), **Ole Rømer** (1644–1710), **Philippe de La Hire** (1640–1718), **Johann Castillon** (1704–1791)

The heart-shaped cardioid has fascinated mathematicians for centuries due to its mathematical properties, graphic beauty, and practical applications. The curve can be produced simply by tracking a point on a circle as it rolls around on another (fixed) circle of the same radius. The name is derived from the Greek word for heart, and its polar equation may be written as $r = a(1 - \cos\theta)$. The area of the cardioid is $(3/2)\pi a^2$, and its perimeter is $8a$.

The cardioid may also be generated by drawing a circle C and fixing a point P on it. Next, draw a set of various circles centered on the circumference of C and passing through P. These circles trace out a cardioid shape. The cardioid appears in a wide range of seemingly disparate mathematical areas, from the caustics in the field of optics to the central shape of the **Mandelbrot Set** in **fractal** geometry.

Many dates may be associated with the cardioid. French lawyer and amateur mathematician Étienne Pascal, father of mathematician Blaise Pascal, formally studied the more general case of the curve, called the Limaçon of Pascal, around 1637. However, even earlier, German painter and mathematician Albrecht Dürer provided a method for drawing the Limaçon in *Underweysung der Messung* (*Instruction in Measurement*), published in 1525. In 1674, Danish astronomer Ole Rømer contemplated the cardioid when considering shapes for gear teeth. French mathematician Philippe de La Hire determined its length in 1708. Interestingly, the cardioid was not given its evocative name until 1741, when Johann Castillon named it in his treatise in the *Philosophical Transactions of the Royal Society*.

Glen Vecchione explains the practical side of cardioids, when he writes that they can show "the interference and congruence patterns of waves that radiate concentrically from a point source. In doing so, they can identify the areas of greatest sensitivity on microphones or antennas.…A cardioid microphone is sensitive to front sound and minimizes rear sound."

**SEE ALSO** Cissoid of Diocles (c. 180 B.C.), The Length of Neile's Semicubical Parabola (1657), Astroid (1674), Fractals (1975), and Mandelbrot Set (1980).

*A cardioid form is traced out by straight lines that connect one point of a circle to another, the front end of the line going twice as fast around the circle as the rear end. (This rendering is by Jos Leys.)*

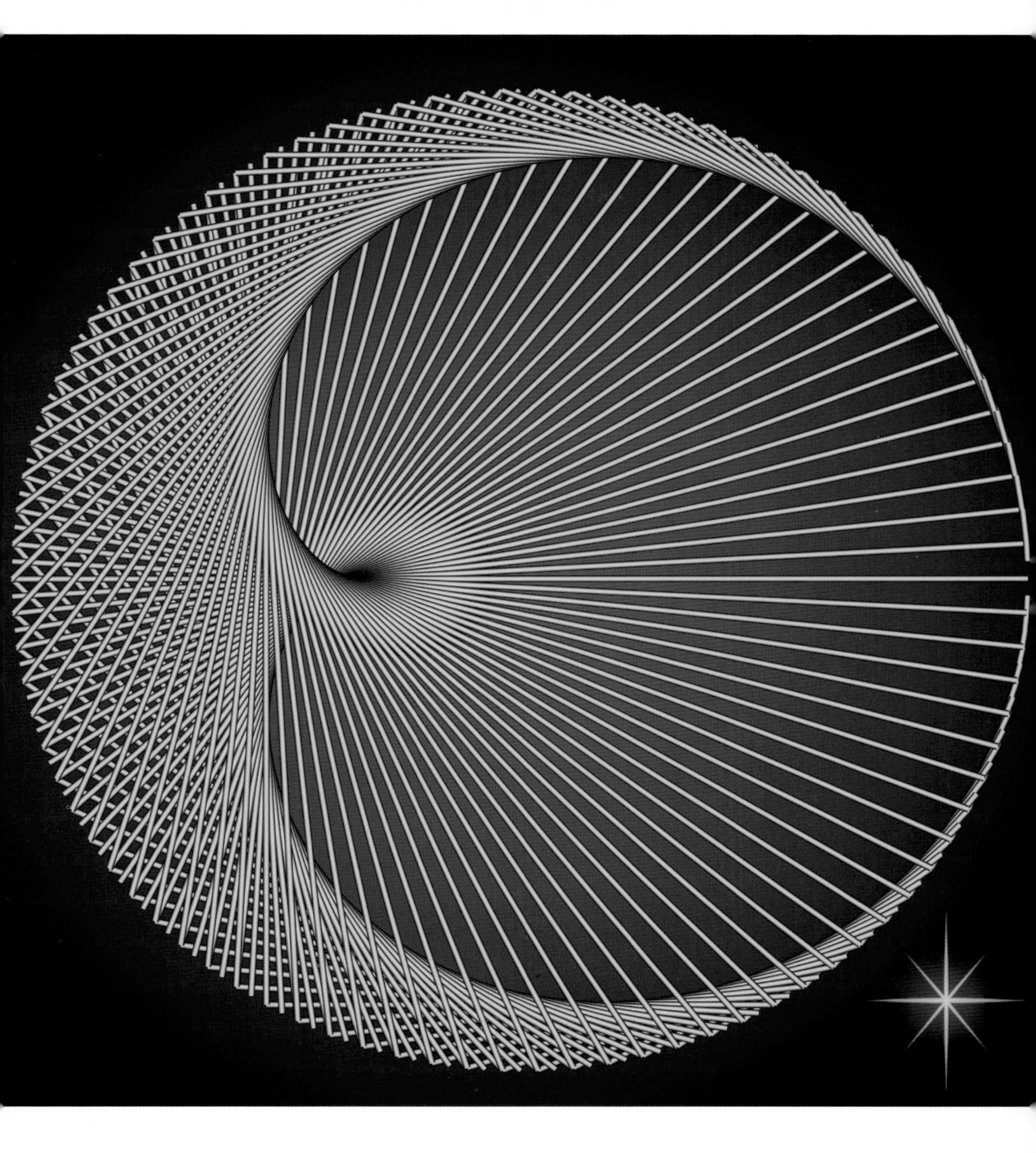

# Logarithmic Spiral

**René Descartes** (1596–1650), **Jacob Bernoulli** (1654–1705)

Logarithmic spirals in nature are ubiquitous and have a range of botanical and zoological manifestations. Probably the most common examples are the logarithmic spirals of nautilus shells and other seashells, the horns of a variety of mammals, the arrangement of seeds of many plants (such as the sunflower and daisy), and the scales of a pinecone. Martin Gardner has noted that *Eperia*, a common variety of spider, spins a web in which a strand coils around the center in a logarithmic spiral.

The logarithmic spiral (also known as the equiangular spiral or Bernoulli spiral) can be expressed as $r = ke^{a\theta}$, where $r$ is the distance from the origin. The angle $\theta$ between a tangent line to the curve and a radial line drawn to $(r, \theta)$ is constant. The spiral was first discussed by French mathematician and philosopher René Descartes in 1638 in letters written to French theologian and mathematician Marin Mersenne. Later, the spiral was studied more extensively by Swiss mathematician Jacob Bernoulli.

The most impressive appearance of the logarithmic spiral is in the huge arms of many galaxies, and the traditional view is that it is necessary to have a long-range interaction like gravity to create such vast order. In spiral galaxies, the spiral arms are sites of active star formation.

Spiral patterns often occur spontaneously in matter that is organized through symmetry transformations: change of size (growth) and rotation. Form follows function, and the spiral form can allow for the compaction of a relatively long length. Long yet compact tubes are useful in mollusks and cochleas for obvious reasons, including physical strength and increased surface area. As a member of a species grows to maturity, it generally transforms in such a way that its parts maintain approximately the same proportion with respect to each other, and this is probably a reason why nature often exhibits self-similar spiral growth.

**SEE ALSO** Archimedes' Spiral (225 B.C.), Golden Ratio (1509), Loxodrome (1537), Logarithms (1614), Fermat Spiral (1636), The Length of Neile's Semicubical Parabola (1657), Voderberg Tilings (1936), Ulam Spiral (1963), and Spidrons (1979).

*The nautilus seashell exhibits a logarithmic spiral form. The shell is internally divided into chambers, the number of which can grow to 30 or more in adult creatures.*

# Projective Geometry

**Leon Battista Alberti** (1404–1472), **Gérard Desargues** (1591–1661), **Jean-Victor Poncelet** (1788–1867)

Projective geometry generally concerns the relationships between shapes and their mappings, or "images," that result from projecting the shapes onto a surface. Projections may often be visualized as the shadows cast by objects.

The Italian architect Leon Battista Alberti was one of the first individuals to experiment with projective geometry through his interest in perspective in art. More generally, Renaissance artists and architects were concerned with methods for representing three-dimensional objects in two-dimensional drawings. Alberti sometimes placed a glass screen between himself and the landscape, closed one eye, and marked on the glass certain points that appeared to be in the image. The resulting 2-D drawing gave a faithful impression of the 3-D scene.

French mathematician Gérard Desargues was the first professional mathematician to formalize projective geometry while searching for ways to extend Euclidean geometry. In 1636, Desargues published *Exemple de l'une des manières universelles du S.G.D.L. touchant la pratique de la perspective* (*Example of a Universal Method by Sieur Girard Desargues Lyonnais Concerning the Practice of Perspective*), in which he presented a geometric method for constructing perspective images of objects. Desargues also examined the properties of shapes that were preserved under perspective mappings. Painters and engravers made use of his approach.

Desargues' most important work, *Brouillon project d'une atteinte aux événements des rencontres d'un cône avec un plan* (*Rough Draft of Attaining the Outcome of Intersecting a Cone with a Plane*), published in 1639, treats the theory of conic sections using projective geometry. In 1882, French mathematician and engineer Jean-Victor Poncelet (1788–1867) published a treatise that revitalized interest in projective geometry.

In projective geometry, elements such as points, lines, and planes generally remain points, lines, and planes when projected. However, lengths, ratios of lengths, and angles may change under projection. In projective geometry, parallel lines in Euclidean geometry intersect at infinity in the projection.

**SEE ALSO** Pappus's Hexagon Theorem (c. 340), Mercator Projection (1569), and Descartes' *La Géométrie* (1637).

*Drawing by Jan Vredeman de Vries (1527–c. 1607), a Dutch Renaissance architect and engineer who experimented with the principles of perspective in his artwork. Projective geometry grew from the principles of perspective art established during the European Renaissance.*

# Torricelli's Trumpet

**Evangelista Torricelli** (1608–1647)

Your friend hands you a gallon of red paint and asks how you would completely paint an infinite surface with this gallon of paint. What surface would you choose? Many possible answers to this question exist, but Torricelli's trumpet is one famous shape to consider—a hornlike object created by revolving $f(x) = 1/x$ for $x \in$ of $[1, \infty)$ about the $x$-axis. Standard calculus methods can be used to demonstrate that the Torricelli's trumpet has *finite* volume but *infinite* surface area!

John dePillis explains that, mathematically speaking, pouring red paint into the Torricelli's trumpet could fill the funnel and, in so doing, you could paint the entire interior infinite surface, even though you have a finite number of paint molecules. This seeming paradox can be partly resolved by remembering that the Torricelli's trumpet is actually a mathematical construct, and our finite number of paint molecules that "fills" the horn is an approximation to the actual finite volume of the horn.

For what values of $a$ does $f(x) = 1/x^a$ produce a horn with finite volume and infinite area? This is something for you to ponder with your mathematical friends.

Torricelli's trumpet is sometimes called Gabriel's horn and is named after Italian physicist and mathematician Evangelista Torricelli, who discovered it in 1641. He was astounded by this trumpet that seemed to be an infinitely long solid with an infinite-area surface and a finite volume. Torricelli and his colleagues thought that it was a deep paradox and unfortunately did not have the tools of calculus to fully appreciate and understand the object. Today, Torricelli is remembered for the telescopic astronomy he did with Galileo and for his invention of the barometer. The name "Gabriel's horn" conjures visions of Archangel Gabriel blowing his horn to announce Judgment Day, thereby associating the infinite with the powers of God.

**SEE ALSO** Discovery of Calculus (c. 1665), Minimal Surface (1774), Beltrami's Pseudosphere (1868), and Cantor's Transfinite Numbers (1874).

*Torricelli's trumpet encloses a* finite *volume but has an* infinite *surface area. This form is also sometimes called* Gabriel's horn, *which conjures visions of Archangel Gabriel blowing his horn to announce Judgment Day. (This rendering is by Jos Leys and rotated by 180°.)*

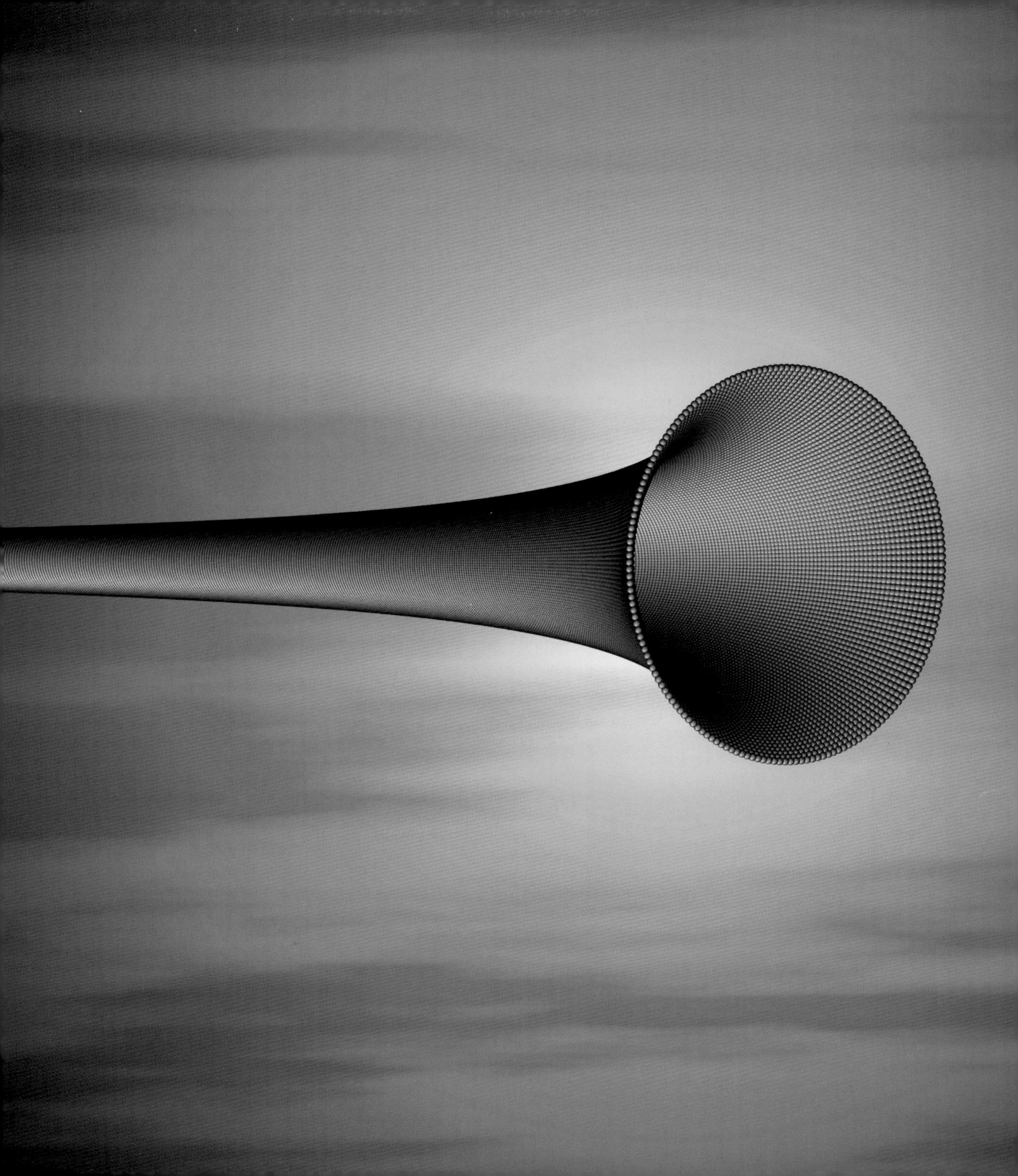

1654

# Pascal's Triangle

**Blaise Pascal** (1623–1662), **Omar Khayyam** (1048–1131)

One of the most famous integer patterns in the history of mathematics is Pascal's triangle. Blaise Pascal was the first to write a treatise about this progression in 1654, although the pattern had been known by the Persian poet and mathematician Omar Khayyam as far back as A.D. 1100, and even earlier to the mathematicians of India and ancient China. The first seven rows of Pascal's triangle are depicted at upper right.

Each number in the triangle is the sum of the two above it. Mathematicians have discussed the role that Pascal's triangle plays in probability theory, in the expansion of binomials of the form $(x + y)^n$, and in various number theory applications for years. Mathematician Donald Knuth (b. 1938) once indicated that there are so many relations and patterns in Pascal's triangle that when someone finds a new identity, there aren't many people who get excited about it anymore, except the discoverer. Nonetheless, fascinating studies have revealed countless wonders, including special geometric patterns in the diagonals, the existence of perfect square patterns with various hexagonal properties, and an extension of the triangle and its patterns to negative integers and to higher dimensions.

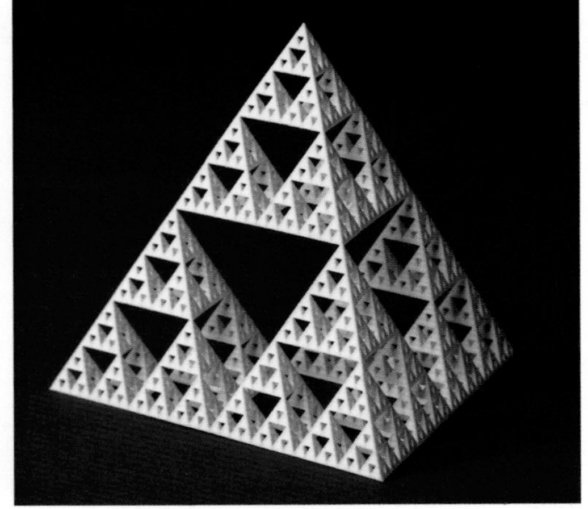

When even numbers in the triangle are replaced by dots and odd numbers by gaps, the resulting pattern is a **fractal**, with intricate repeating patterns on different size scales. These fractal figures may have a practical importance in that they can provide models for materials scientists to help produce new structures with novel properties. For example, in 1986, researchers created wire gaskets on the micron size scale almost identical to Pascal's triangle, with holes for the odd numbers. The area of their smallest triangle was about 1.38 microns squared, and the scientists investigated many unusual properties of their superconducting gasket in a magnetic field.

**SEE ALSO** Omar Khayyam's *Treatise* (1070), Normal Distribution Curve (1733), and Fractals (1975).

LEFT: *George W. Hart created this nylon model of Pascal's pyramid using a physical process known as* selective laser sintering. RIGHT: *The fractal Pascal triangle discussed in the text. The number of cells in the central red triangles is always even (6, 28, 120, 496, 2,016....) and includes all perfect numbers (numbers that are the sum of their proper positive divisors).*

1
1 1
1 2 1
1 3 3 1
1 4 6 4 1
1 5 10 10 5 1
1 6 15 20 15 6 1

1657

# The Length of Neile's Semicubical Parabola

**William Neile** (1637–1670), **John Wallis** (1616–1703)

In 1657, British mathematician William Neile became the first person to "rectify," or find the arc length of, a nontrivial algebraic curve. This special curve is called a *semicubical parabola*, defined by $x^3 = ay^2$. When written as $y = \pm ax^{3/2}$, it is easier to see how it might have been considered "half a cubic" and hence the genesis of the term *semicubic*. A report of Neile's work appeared in British mathematician John Wallis's *De Cycloide* in 1659. Interestingly, only the arc lengths of transcendental curves, such as the logarithmic spiral and cycloid, had been calculated before 1659.

Because attempts to rectify the ellipse and hyperbola were unsuccessful, some mathematicians, such as French philosopher and mathematician René Descartes (1596–1650), had conjectured that few curves could be rectified. However, Italian physicist and mathematician Evangelista Torricelli (1608–1647) rectified the logarithmic spiral, which was the first curved line (other than the circle) whose length was determined. The cycloid was the next curve rectified, by English geometer and architect Sir Christopher Wren (1632–1723) in 1658.

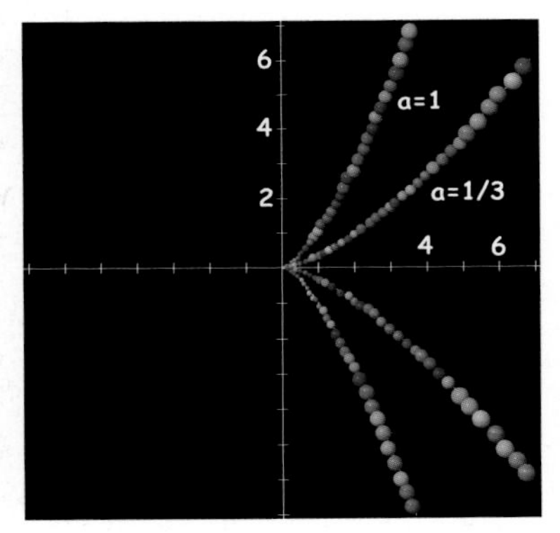


Around 1687, Dutch mathematician and physicist Christiaan Huygens (1629–1695) showed that the semicubical parabola is a curve along which a particle may descend under the force of gravity so that it moves equal vertical distances in equal times. The semicubical parabola can also be expressed as a pair of equations: $x = t^2$ and $y = at^3$. Given this form, the length of the curve as a function of $t$ is $(1/27) \times (4 + 9t^2)^{3/2} - 8/27$. In other words, the curve has this length on the interval from 0 to $t$. In the literature, we sometimes see Neile's parabola referred to as the curve for $y^3 = ax^2$, which places the cusp of the curve pointing downward along the $y$-axis instead of to the left on the $x$-axis.

**SEE ALSO** Cissoid of Diocles (c. 180 B.C.), Descartes' *La Géométrie* (1637), Logarithmic Spiral (1638), Torricelli's Trumpet (1641), Tautochrone Problem (1673), and Transcendental Numbers (1844).

LEFT: *Semicubical parabolas defined by* $x^3 = ay^2$ *for two different values of* a. RIGHT: *Some curves are non-rectifiable, i.e., they have infinite length. Shown here is a graph of the function defined by* $f(x) = x \cdot \sin(1/x)$, *for* $0 < x < 1$. *The curve has infinite length for any open set with* $x = 0$ *as one of its delimiters and* $f(0) = 0$.

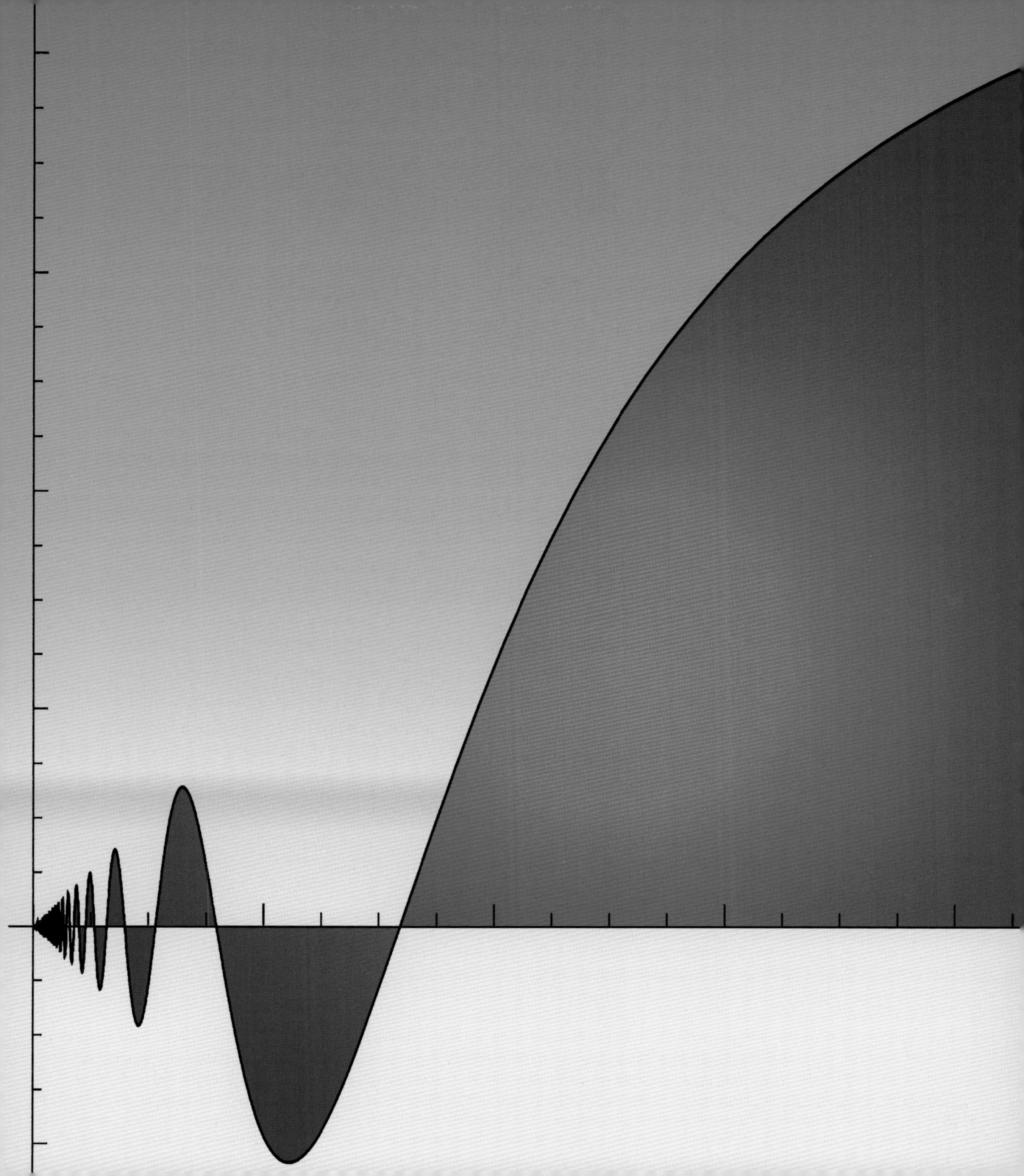

1659

# Viviani's Theorem

**Vincenzo Viviani** (1622–1703)

Place a point inside an equilateral triangle. From this point, draw a line to each of the sides so that these three lines are perpendicular to each side. No matter where you placed your point, the sum of the perpendicular distances from the point to the sides is equal to the height of the triangle. The theorem is named after Italian mathematician and scientist Vincenzo Viviani. Galileo was so impressed by Viviani's talent that he took him into his house in Arcetri, Italy, as a collaborator.

Researchers have found ways to extend Viviani's theorem to problems in which the point is placed outside the triangle and have also explored the application of the theorem to any regular $n$-sided polygon. In this case, the sum of the perpendicular distances from an interior point to the $n$ sides is $n$ times the apothem of the polygon. (An *apothem* is the distance from the center to a side.) The theorem can also be studied in higher dimensions.

When Galileo died, Viviani wrote Galileo's biography and hoped to publish a complete edition of Galileo's works. Alas, the Church prohibited this effort, which hurt Viviani's reputation and was a blow to science in general. Viviani published the Italian version of Euclid's *Elements* in 1690.

Not only is the theorem interesting mathematically due to its numerous different proofs, the theorem is used for teaching children various aspects of geometry. Some teachers have placed the problem in a real-world setting by casting it in the context of a surfer stranded on an island in the shape of an equilateral triangle. The surfer wants to build a hut where the sum of the distances to the sides is a minimum, because she surfs on each of the three beaches an equal amount of time. Students are intrigued to learn that the placement of the hut does not matter.

**SEE ALSO** Pythagorean Theorem and Triangles (c. 600 B.C.), Euclid's *Elements* (300 B.C.), Law of Cosines (c. 1427), Morley's Trisector Theorem (1899), and Ball Triangle Picking (1982).

*Place a point* anywhere *inside an equilateral triangle. Draw lines, as shown, to the sides of the triangle. The sum of the perpendicular distances from the point to the sides is always equal to the height of the triangle.*

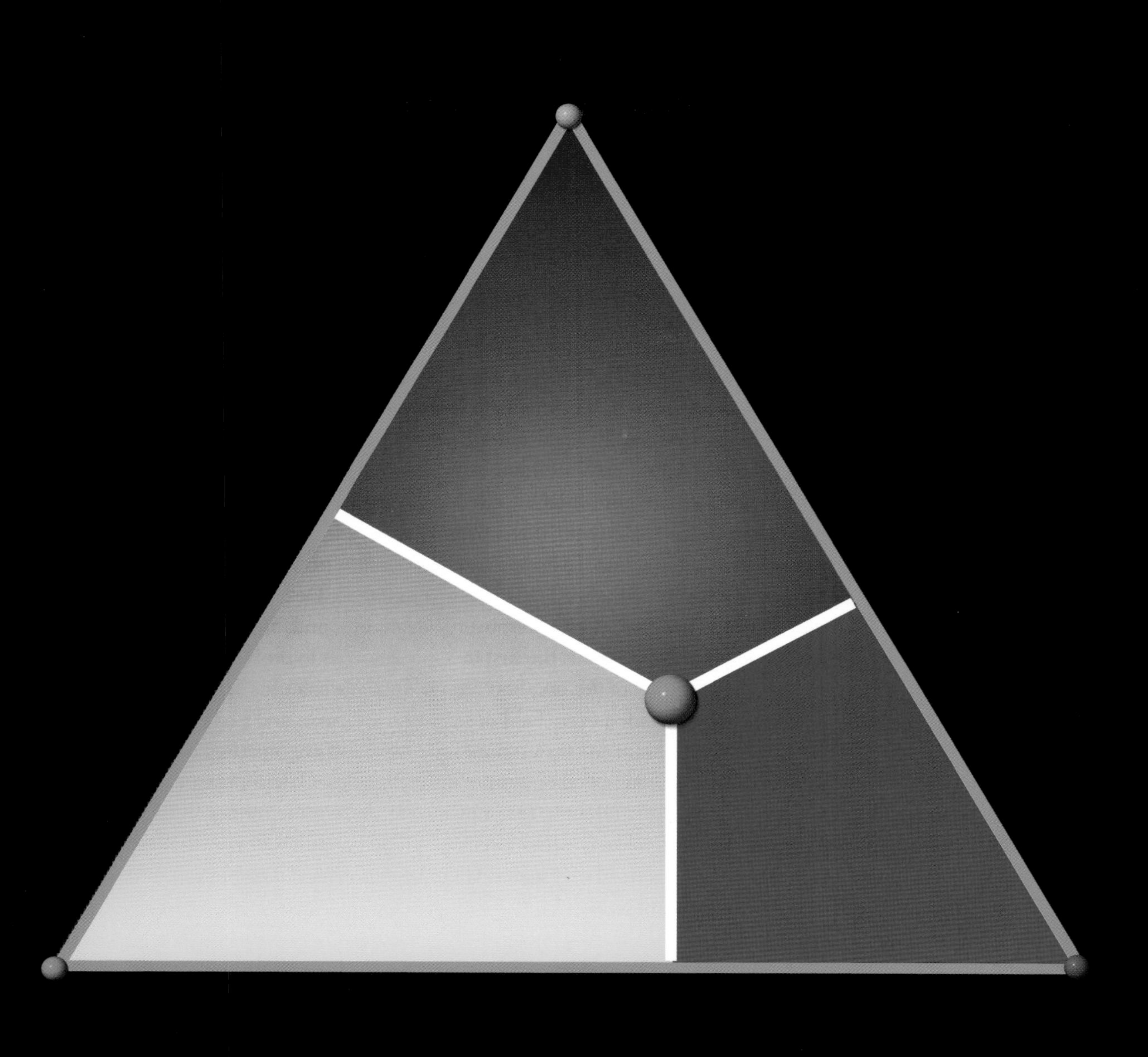

c.1665

# Discovery of Calculus

**Isaac Newton** (1642–1727), **Gottfried Wilhelm Leibniz** (1646–1716)

English mathematician Isaac Newton and German mathematician Gottfried Wilhelm Leibniz are usually credited with the invention of calculus, but various earlier mathematicians explored the concept of rates and limits, starting with the ancient Egyptians who developed rules for calculating the volume of pyramids and approximating the areas of circles.

In the 1600s, both Newton and Leibniz puzzled over problems of tangents, rates of change, minima, maxima, and infinitesimals (unimaginably tiny quantities that are almost but not quite zero). Both men understood that differentiation (finding the tangent to a curve at a point—that is, a straight line that "just touches" the curve at that point) and integration (finding the area under a curve) are inverse processes. Newton's discovery (1665–1666) started with his interest in infinite sums; however, he was slow to publish his findings. Leibniz published his discovery of differential calculus in 1684 and of integral calculus in 1686. He said, "It is unworthy of excellent men, to lose hours like slaves in the labor of calculation.... My new calculus... offers truth by a kind of analysis and without any effort of imagination." Newton was outraged. Debates raged for many years on how to divide the credit for the discovery of calculus, and, as a result, progress in calculus was delayed. Newton was the first to apply calculus to problems in physics, and Leibniz developed much of the notation seen in modern calculus books.

Today, calculus has invaded every field of scientific endeavor and plays invaluable roles in biology, physics, chemistry, economics, sociology, and engineering, and in any field where some quantity, like speed or temperature, changes. Calculus can be used to help explain the structure of a rainbow, teach us how to make more money in the stock market, guide a spacecraft, make weather forecasts, predict population growth, design buildings, and analyze the spread of diseases. Calculus has caused a revolution. It has changed the way we look at the world.

**SEE ALSO** Zeno's Paradoxes (c. 445 B.C.), Torricelli's Trumpet (1641), L'Hôpital's *Analysis of the Infinitely Small* (1696), Agnesi's *Instituzioni Analitiche* (1748), Laplace's *Théorie Analytique des Probabilités* (1812), and Cauchy's *Le Calcul Infinitésimal* (1823).

*William Blake's* Newton *(1795). Blake, a poet and artist, portrays Isaac Newton as a kind of divine geometer, gazing at technical diagrams drawn on the ground as he ponders mathematics and the cosmos.*

1669

# Newton's Method

**Isaac Newton** (1642–1727)

The use of computational techniques based on recurrence relationships, in which each term of a sequence is defined as a function of the preceding term, can be traced back to the dawn of mathematics. The Babylonians used such techniques to compute the square root of a positive number, and the Greeks to approximate pi. Today, many important special functions of mathematical physics may be computed by recurrence formulas.

Numerical analysis is often concerned with obtaining approximate solutions to difficult problems. Newton's method is one of the most famous numerical methods for solving equations of the form $f(x) = 0$, some solutions of which may be difficult to find using simple algebraic methods. The problem of finding the zeros, or roots, of a function by these kinds of methods occurs frequently in science and engineering.

To apply Newton's method, one starts with a numerical guess for the solution of the root, and then the function is approximated by its tangent line, which is a straight line that "just touches" the graph of the function at a point. After determining the $x$-intercept of this line, which is often a better approximation to the nearby root than the initial guess, the method can be iterated (repeated) to produce successively accurate approximations. The precise formula for Newton's method is $x_{n+1} = x_n - f(x_n)/f'(x_n)$, where the prime symbol (′) indicates the first derivative of the function $f$.

When the method is applied to functions with complex values, computer graphics renditions are sometimes used to give an indication of where the method can be relied upon and where it behaves strangely. The resulting graphics often reveal chaotic behavior and beautiful **fractal** patterns.

The mathematical seeds for Newton's method were described by Isaac Newton in *De analysi per aequationes numero terminorum infinitas* (*On Analysis by Equations with an Infinite Number of Terms*), written in 1669 and published by William Jones in 1711. In 1740, British mathematician Thomas Simpson refined the approach and described Newton's method as an iterative method for solving general nonlinear equations using calculus.

**SEE ALSO** Discovery of Calculus (c. 1665), Chaos and the Butterfly Effect (1963), and Fractals (1975).

*Computer graphics can be useful for revealing the intricate behavior of Newton's method when applied to finding the complex-number roots of an equation. Paul Nylander generated this image by employing the method in order to find the solutions of* $z^5 - 1 = 0$.

1673

# Tautochrone Problem

**Christiaan Huygens** (1629–1695)

In the 1600s, mathematicians and physicists sought a curve that specified the shape of a special kind of ramp. In particular, objects are placed on the ramp, one at a time, and they must slide down to the very bottom, always in the same amount of time and no matter where they start on the ramp. The objects are accelerated by gravity, and the ramp is considered to have no friction.

Dutch mathematician, astronomer, and physicist Christiaan Huygens discovered a solution in 1673 and published it in his *Horologium oscillatoriumi* (*The Pendulum Clock*). Technically speaking, the tautochrone is a cycloid—that is, a curve defined by the path of a point on the edge of a circle as the circle rolls along a straight line. The tautochrone is also called the brachistochrone when referring to the curve that gives a frictionless object the fastest rate of descent when the object slides down from one point to another.

Huygens attempted to use his discovery to design a more accurate pendulum clock. The clock made use of portions of tautochrone surfaces near where the string pivoted to ensure that the string followed the optimum curve no matter where the pendulum started swinging. (Alas, the friction due to the surfaces introduced significant errors.)

The special property of the tautochrone is mentioned in *Moby Dick* in a discussion on a try-pot, a bowl used for rendering blubber to produce oil: "[The try-pot] is also a place for profound mathematical meditation. It was in the left-hand try-pot of the *Pequod*, with the soapstone diligently circling round me, that I was first indirectly struck by the remarkable fact, that in geometry all bodies gliding along a cycloid, my soapstone, for example, will descend from any point in precisely the same time."

**SEE ALSO** The Length of Neile's Semicubical Parabola (1657).

*Under the influence of gravity, three balls run along the tautochrone curve starting from different positions, yet the balls will arrive at the bottom at the same time. (The balls are placed on the ramp, one at a time.)*

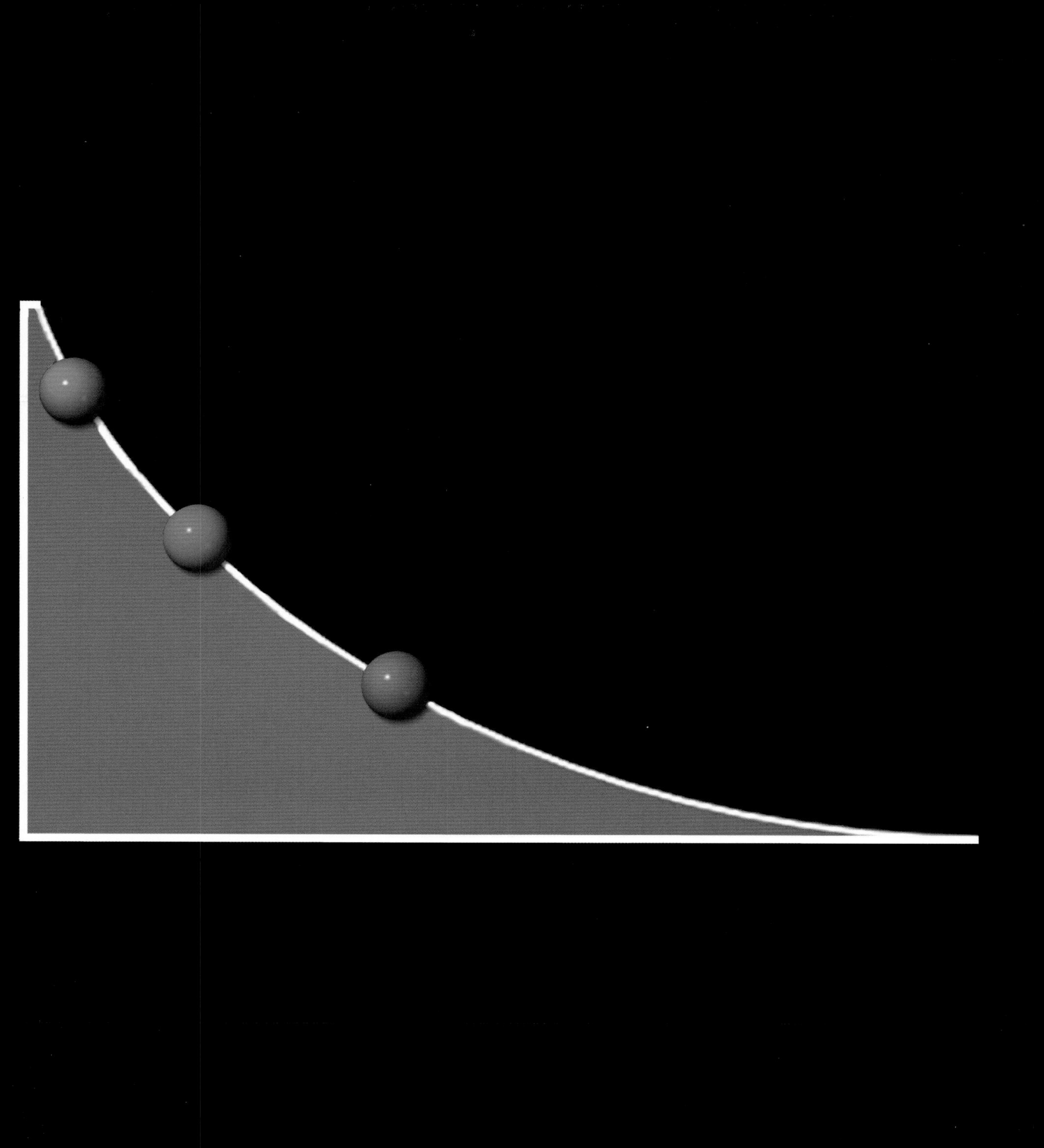

1674

# Astroid

**Ole Christensen Rømer** (1644–1710)

The astroid is a curve with four cusps that is traced by a point on a circle that rolls like a gear along the inside of a larger circle. This larger circle is four times the diameter of the small circle. The astroid is notable for the diversity of famous mathematicians who researched its intriguing properties. The curve was first studied by the Danish astronomer Ole Rømer in 1674, as a result of his quest for gear teeth with more useful shapes. Swiss mathematician Johann Bernoulli (1691), German mathematician Gottfried Leibniz (1715), and French mathematician Jean d'Alembert (1748) all became fascinated by the curve.

The astroid has the equation $x^{2/3} + y^{2/3} = R^{2/3}$, where $R$ is the radius of the stationary outer circle, and $R/4$ is the radius of the inner rolling circle. The length of the astroid is $6R$, and the area is $3\pi R^2/8$. Interestingly, its $6R$ circumference has no dependence on $\pi$, despite the involvement of circles that are used for generating the astroid.

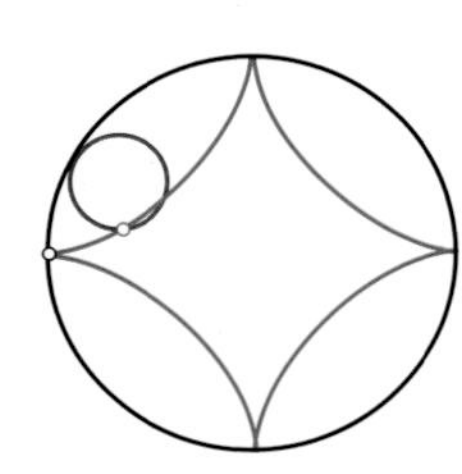

In 1725, mathematician Daniel Bernoulli discovered that an astroid is also traced by an inner circle that has ¾ the diameter of the fixed circle. In other words, this traces out the same curve as the inner circle with only ¼ the diameter of the larger one.

In physics, the Stoner-Wohlfarth astroid curve is used to characterize various properties of energy and magnetism. U.S. Patent 4,987,984 describes the use of an astroid for mechanical roller clutches: "The astroid curve provides the same good dispersal of stresses that the equivalent circular arc would, but removes less cam race material, giving a stronger structure."

Interestingly, tangent lines along the astroid curve, when extended until they touch the $x$- and $y$-axes, all have the same length. You can visualize this by imagining a ladder leaning at all possible angles against a wall, which traces out a portion of the astroid curve.

**SEE ALSO** Cissoid of Diocles (c. 180 B.C.), Cardioid (1637), The Length of Neile's Semicubical Parabola (1657), Reuleaux Triangle (1875), and Superegg (c. 1965).

*Artistic depiction of an astroid as the "envelope" of a family of ellipses. (In geometry, an envelope of a family of curves is a curve that is tangent to each member of the family at some point.)*

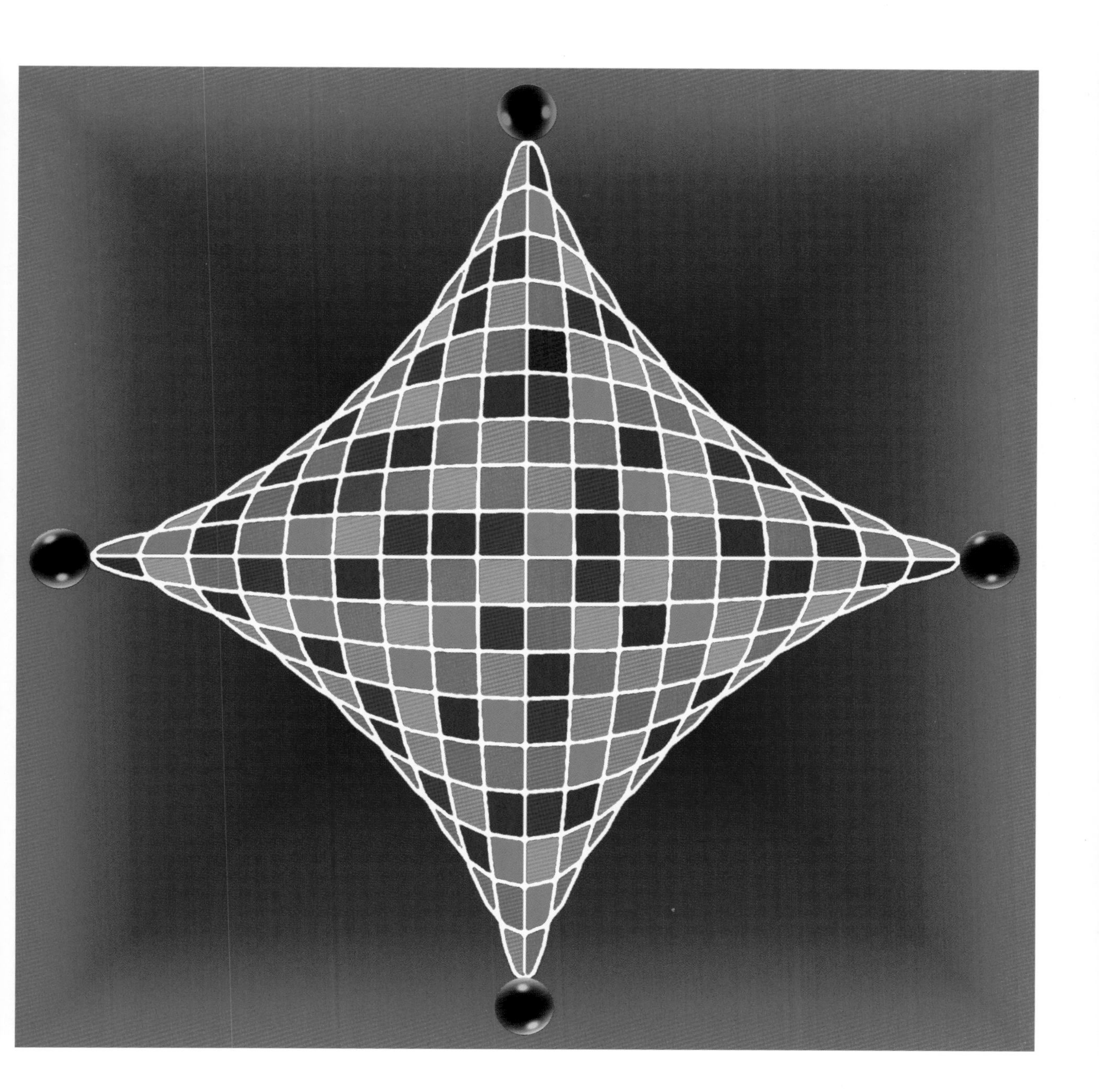

# L'Hôpital's *Analysis of the Infinitely Small*

**Guillaume François Antoine, Marquis de l'Hôpital** (1661–1704)

In 1696, French mathematician Marquis de l'Hôpital published Europe's first calculus textbook, *Analyse des infiniment petits, pour l'intelligence des lignes courbes* (*Analysis of the Infinitely Small, for the Understanding of Curves*). He intended the book to be a vehicle to promote understanding of the techniques of the differential calculus. Calculus had been invented a few years earlier by Isaac Newton and Gottfried Leibniz and refined by the Bernoulli brothers, mathematicians Jacob and Johann. Keith Devlin writes, "In fact, until the appearance of l'Hôpital's book, Newton, Leibniz, and the two Bernoullis were pretty well the only people on the face of the earth who knew much about calculus."

In the early 1690s, l'Hôpital hired Johann Bernoulli to teach him calculus. L'Hôpital was so intrigued by calculus that he learned quickly, and soon consolidated his knowledge into his comprehensive textbook. Rouse Ball writes of l'Hôpital's book, "the credit of putting together the first treatise which explained the principles and use of the method is due to l'Hôpital.... This work had a wide circulation; it brought the differential notation into general use in France, and helped make it known in Europe."

Aside from his textbook, l'Hôpital is known for the rule of calculus, included in his book, for calculating the limiting value of a fraction whose numerator and denominator either both approach zero or both approach infinity. He initially had planned a military career, but poor eyesight caused him to switch to mathematics.

Today, we know that l'Hôpital, in 1694, paid Bernoulli 300 francs a year to tell him of his discoveries, which l'Hôpital described in his book. In 1704, after l'Hôpital's death, Bernoulli began to speak of the deal and claimed that many of the results in *Analysis of the Infinitely Small* were due to him.

**SEE ALSO** Discovery of Calculus (c. 1665), Agnesi's *Instituzioni Analitiche* (1748), and Cauchy's *Le Calcul Infinitésimal* (1823).

*Frontispiece for Europe's first calculus textbook:* Analyse des infiniment petits, pour l'intelligence des lignes courbes (Analysis of the Infinitely Small, for the Understanding of Curves).

# ANALYSE

DES

# INFINIMENT PETITS,

*POUR*

## L'INTELLIGENCE DES LIGNES COURBES.

*Par Mr le Marquis* DE L'HOSPITAL.

SECONDE EDITION.

A PARIS,

Chez FRANÇOIS MONTALANT à l'entrée du Quay des Augustins du côté du Pont S. Michel.

---

MDCCXVI.

*AVEC APPROBATION ET PRIVILEGE DU ROY.*

# Rope Around the Earth Puzzle

**William Whiston** (1667–1752)

Although this puzzle is not a mathematical milestone on par with most of the others in this book, this little gem from 1702 is worthy of mention simply because it has intrigued schoolchildren and adults for more than two centuries and is a metaphor for how simple mathematics may help analysts reason beyond the limits of their own intuition.

Imagine that you are given a rope that tightly encircles the equator of a basketball. How much longer would you have to make the rope so that it is one foot from the surface of the basketball at all points? What is your guess?

Next, imagine that we have the rope around the equator of a sphere the size of the Earth, which would make the rope around 25,000 miles long! How much longer would you *now* have to make the rope so that it is one foot off the ground all the way around the equator?

The answer, which is a surprise to most people, is $2\pi$ or approximately 6.28 feet for both the basketball and the Earth—only about the length of an adult man. If $R$ is the radius of the Earth, and $1 + R$ is the radius, in feet, of the enlarged circle, we can compare the rope circumference before ($2\pi R$) and after $2\pi(1 + R)$, which shows that the difference is $2\pi$ feet, independent of the radius of either the Earth or the basketball.

A puzzle very similar to this appeared in William Whiston's *The Elements of Euclid*, a book written in 1702 for students. Whiston—an English theologian, historian, and mathematician—was probably most famous for his A *New Theory of the Earth from its Original to the Consummation of All Things* (1696), in which he suggests that Noah's Flood was caused by a comet.

**SEE ALSO** Euclid's *Elements* (300 B.C.), π (c 250 B.C.), and Wheat on a Chessboard (1256).

*A rope or metal band is tightly wrapped around an Earth-size sphere at the equator (or along another great circle). How much longer would the band be if it were enlarged and now one foot off the ground all the way around?*

1713

# Law of Large Numbers

**Jacob Bernoulli** (1654–1705)

In 1713, Swiss mathematician Jacob Bernoulli's proof of his Law of Large Numbers (LLN) was presented in a posthumous publication, *Ars Conjectandi* (*The Art of Conjecturing*). The LLN is a theorem in probability that describes the long-term stability of a random variable. For example, when the number of observations of an experiment (such as the tossing of a coin) is sufficiently large, then the proportion of an outcome (such as the occurrence of heads) will be close to the probability of the outcome, for example 0.5. Stated more formally, given a sequence of independent and identically distributed random variables with a finite population mean and variance, the average of these observations will approach the theoretical population mean.

Imagine you are tossing a standard six-sided die. We expect the mean of the values obtained by tossing to be the average, or 3.5. Imagine that your first three tosses happen to be 1, 2, and 6, giving a mean of 3. With more tosses, the value of the average eventually settles to the expected value of 3.5. Casino operators love the LLN because they can count on stable results in the long run and can plan accordingly. Insurers rely on the LLN to cope with and plan for variations in losses.

In *Ars Conjectandi*, Bernoulli estimates the proportion of white balls in an urn filled with an unknown number of black and white balls. By drawing balls from the urn and "randomly" replacing a ball after each draw, he estimates the proportion of white balls by the proportion of balls drawn that are white. By doing this enough times, he obtains any desired accuracy for the estimate. Bernoulli writes, "If observations of all events were to be continued throughout all eternity (and, hence, the ultimate probability would tend toward perfect certainty), everything in the world would be perceived to happen in fixed ratios.... Even in the most accidental... occurrences, we would be bound to recognize... a certain fate."

**SEE ALSO** Dice (c. 3000 B.C.), Normal Distribution Curve (1733), St. Petersburg Paradox (1738), Bayes' Theorem (1761), Buffon's Needle (1777), Laplace's *Théorie Analytique des Probabilités* (1812), Benford's Law (1881), and Chi-Square (1900).

*Swiss commemorative stamp of mathematician Jacob Bernoulli, issued in 1994. The stamp features both a graph and a formula related to his Law of Large Numbers.*

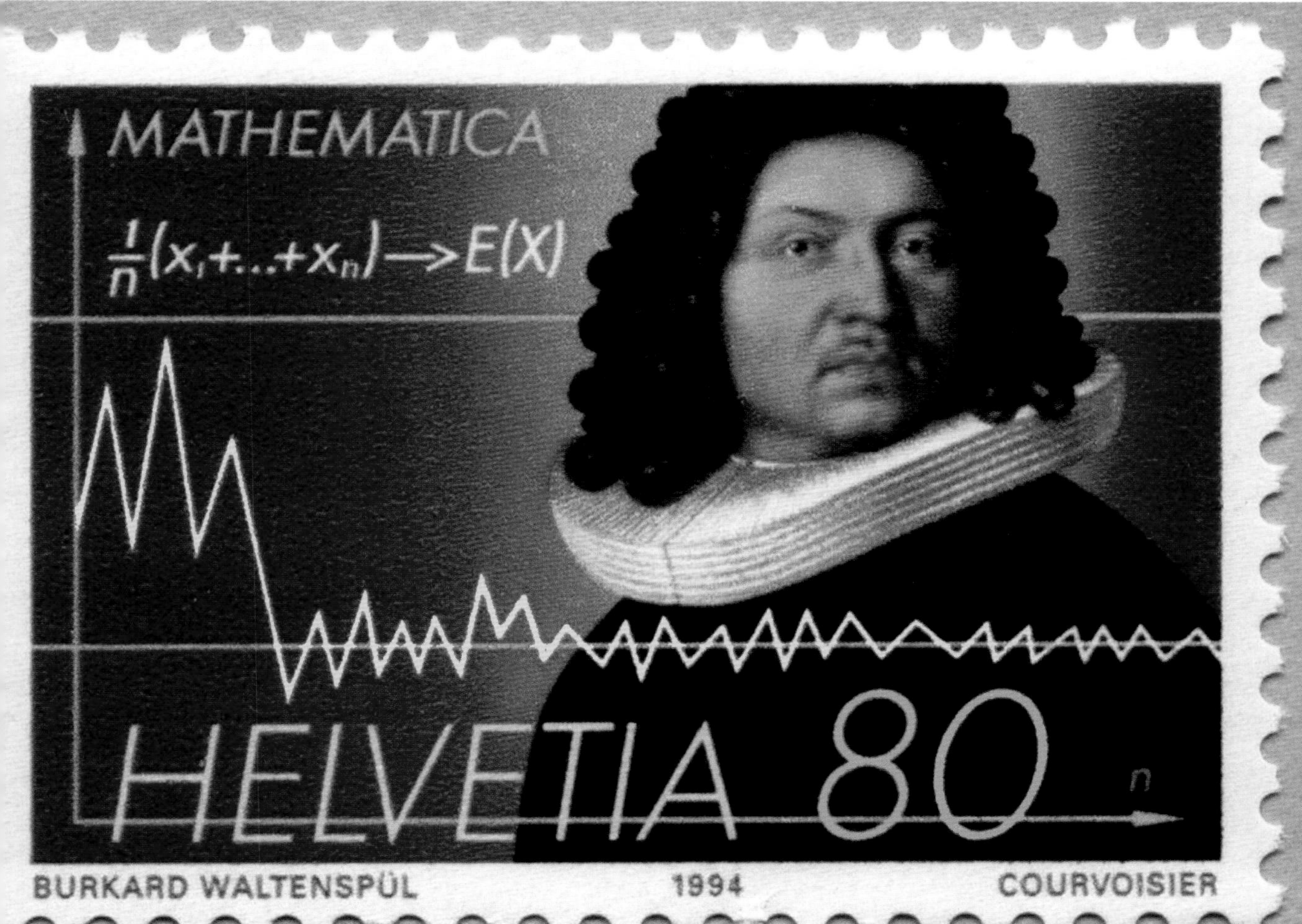
MATHEMATICA
$\frac{1}{n}(x_1+\ldots+x_n)\longrightarrow E(X)$
HELVETIA 80
n
BURKARD WALTENSPÜL
1994
COURVOISIER

# Euler's Number, $e$

## **Leonhard Paul Euler** (1707–1783)

British science writer David Darling writes that the number $e$ is "possibly the most important number in mathematics. Although pi is more familiar to the layperson, $e$ is far more significant and ubiquitous in the higher reaches of the subject."

The number $e$, which is approximately equal to 2.71828, can be calculated in many ways. For example, it is the limit value of the expression $(1 + 1/n)$ raised to the $n$th power, when $n$ increases indefinitely. Although mathematicians like Jacob Bernoulli and Gottfried Leibniz were aware of the constant, Swiss mathematician Leonhard Euler was among the first to extensively study the number, and he was the first to use the symbol $e$ in letters written in 1727. In 1737, he showed that $e$ is irrational—that is, it cannot be expressed as a ratio of two integers. In 1748, he calculated 18 of its digits, and today more than 100,000,000,000 digits of $e$ are known.

$e$ is used in diverse areas, such as in the formula for the catenary shape of a hanging rope supported at its two ends, in the calculation of compound interest, and in numerous applications in probability and statistics. It also appears in one of the most amazing mathematical relationships ever discovered, $e^{i\pi} + 1 = 0$, which unites the five most important symbols of mathematics: 1, 0, $\pi$, $e$, and $i$ (the square root of minus one). Harvard mathematician Benjamin Pierce said that "we cannot understand [the formula], and we don't know what it means, but we have proved it, and therefore we know it must be the truth." Several surveys among mathematicians have placed this formula at the top of the list for the most beautiful formula in mathematics. Kasner and Newman note, "We can only reproduce the equation and not stop to inquire into its implications. It appeals equally to the mystic, the scientist, and the mathematician."

**SEE ALSO** $\pi$ (c. 250 B.C.), Imaginary Numbers (1572), Euler-Mascheroni Constant (1735), Transcendental Numbers (1844), and Normal Number (1909).

*The St. Louis Gateway Arch is in the shape of an upside-down catenary. A catenary can be described by the formula* $y = (a/2)\cdot(e^{x/a} + e^{-x/a})$. *The Gateway Arch is the world's tallest monument, with a height of 630 feet (192 meters).*

1730

# Stirling's Formula

**James Stirling** (1692–1770)

These days, factorials are everywhere in mathematics. For non-negative integers $n$, "$n$ factorial" (written as $n!$), is the product of all positive integers less than or equal to $n$. For example, $4! = 1 \times 2 \times 3 \times 4 = 24$. The notation $n!$ was introduced by French mathematician Christian Kramp in 1808. Factorials are important in combinatorics, for example, when determining the number of different ways of arranging objects in a sequence. They also occur in number theory, probability, and calculus.

Because factorial values grow so large (for example, 70! is greater than $10^{100}$, and 25,206! is greater than $10^{100,000}$), convenient methods for approximating large factorials are extremely useful. Stirling's formula, $n! \approx \sqrt{2\pi}\, e^{-n} n^{n+1/2}$, provides an accurate estimate for $n$ factorial. Here, the $\approx$ symbol means "approximately equal to," and $e$ and $\pi$ are the mathematical constants $e \approx 2.71828$ and $\pi \approx 3.14159$. For large values of $n$, this expression results in an even simpler-looking approximation, $\ln(n!) \approx n\ln(n) - n$, which can also be written as $n! \approx n^n e^{-n}$.

In 1730, Scottish mathematician James Stirling presented his approximation for the value of $n!$ in his most important work, *Methodus Differentialis*. Stirling began his career in mathematics amidst political and religious conflicts. He was friends with Newton, but devoted most of his life after 1735 to industrial management.

Keith Ball writes, "To my mind, this is one of the quintessential discoveries of eighteenth-century mathematics. A formula like this gives us some idea of the astonishing transformation of mathematics that took place in the seventeenth and eighteenth centuries. Logarithms were not invented until about 1600. Newton's *Principia*, setting out the principles of calculus, appeared 90 years later. Within another 90 years, mathematicians were producing formulae like Stirling's formula of a subtlety that would have been unimaginable without a formalization of the calculus. Mathematics was no longer a game for amateurs—it had become a job for professionals."

**SEE ALSO** Logarithms (1614), Pigeonhole Principle (1834), Transcendental Numbers (1844), and Ramsey Theory (1928).

*Stirling's formula, surrounded by precisely 4!, or 24, beetles.*

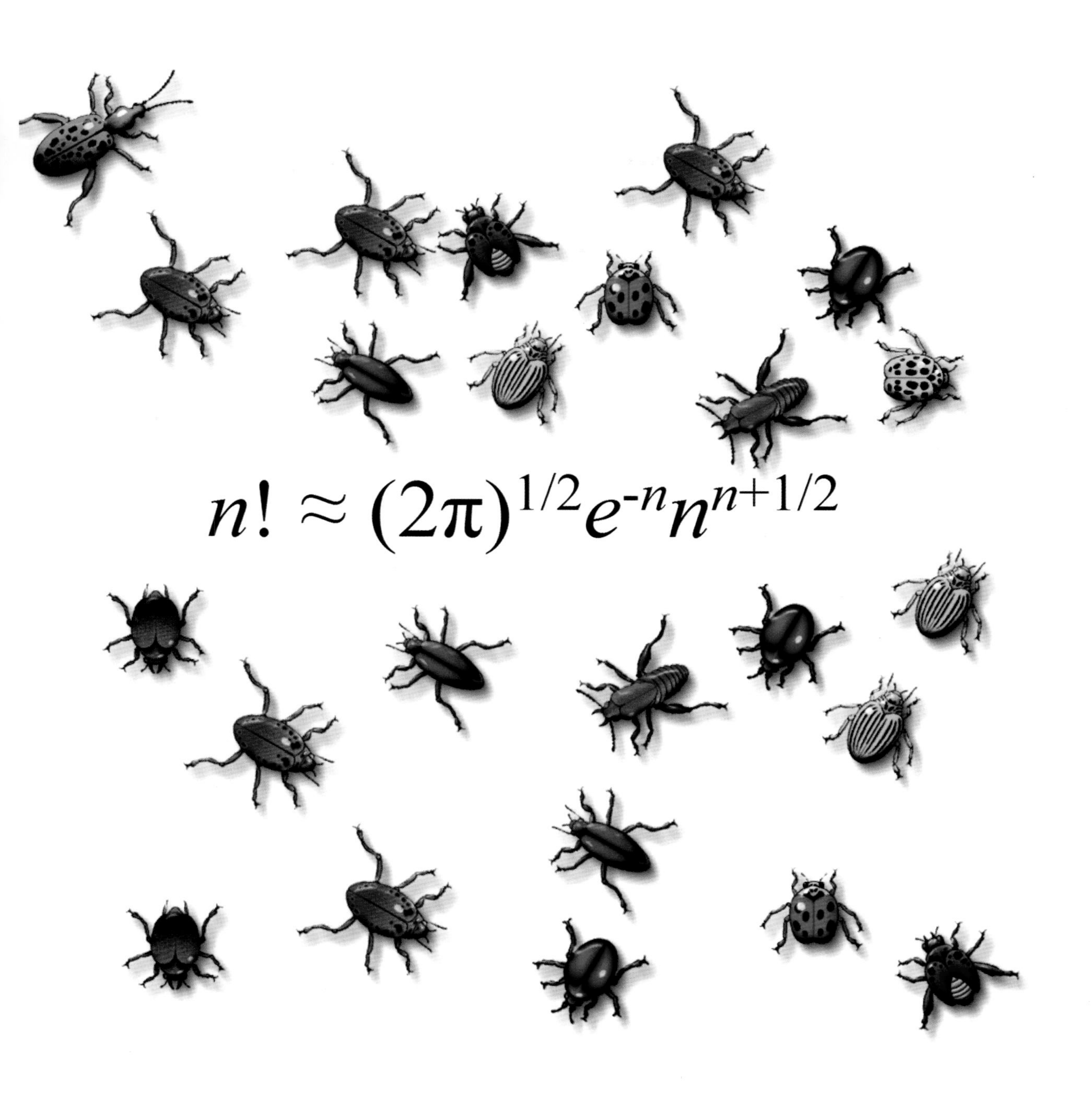
$n! \approx (2\pi)^{1/2} e^{-n} n^{n+1/2}$

# Normal Distribution Curve

**Abraham de Moivre** (1667–1754), **Johann Carl Friedrich Gauss** (1777–1855), **Pierre-Simon Laplace** (1749–1827)

In 1733, French mathematician Abraham de Moivre was the first to describe the normal distribution curve, or law of errors, in *Approximatio ad summam terminorum binomii* $(a+b)^n$ *in seriem expansi* ("Approximation to the Sum of the Terms of a Binomial $(a+b)^n$ Expanded as a Series"). Throughout his life, de Moivre remained poor and earned money on the side by playing chess in coffeehouses.

The normal distribution—also called the Gaussian distribution, in honor of Carl Friedrich Gauss, who studied the curve years later—represents an important family of continuous probability distributions that are applied in countless fields in which observations are made. These fields include studies of population demographics, health statistics, astronomical measurements, heredity, intelligence, insurance statistics, and any fields in which variation exists in experimental data and observed characteristics. In fact, early in the eighteenth century, mathematicians began to realize that a vast number of different measurements tended to show a similar form of scattering or distribution.

The normal distribution is defined by two key parameters, the mean (or average) and the standard deviation, which quantifies the spread or variability of the data. The normal distribution, when graphed, is often called the *bell curve* because of its symmetric bell-like shape with values more concentrated in the middle than in the tails at the sides of the curve.

De Moivre researched the normal distribution during his studies of approximations to the binomial distribution, which arises, for example, in coin toss experiments. Pierre-Simon Laplace used the distribution in 1783 to study measurement errors. Gauss applied it in 1809 to study astronomical data.

The anthropologist Sir Francis Galton wrote of the normal distribution, "I know of scarcely anything so apt to impress the imagination as the wonderful form of cosmic order expressed by the 'Law of Frequency of Error.' The law would have been personified by the Greeks and deified, if they had known of it. It reigns with serenity and in complete self-effacement amidst the wildest confusion."

**SEE ALSO** Omar Khayyam's *Treatise* (1070), Pascal's Triangle (1654), Law of Large Numbers (1713), Buffon's Needle (1777), Laplace's *Théorie Analytique des Probabilités* (1812), and Chi-Square (1900).

*A deutsche mark banknote featuring Carl Friedrich Gauss and a graph and formula of the normal probability function.*

DEUTSCHE BUNDESBANK
Banknote
10
Carl Friedr. Gauß
1777–1855
AY7831976K1
10

1735

# Euler-Mascheroni Constant

**Leonhard Paul Euler** (1707–1783), **Lorenzo Mascheroni** (1750–1800)

The Euler-Mascheroni constant, denoted by the Greek letter $\gamma$, has a numerical value of 0.5772157.…This number links the exponentials and logarithms to number theory, and it is defined by the limit of $(1 + 1/2 + 1/3 + \ldots + 1/n - \log n)$ as $n$ approaches infinity. The reach of $\gamma$ is far and wide, as it plays roles in such diverse areas as infinite series, products, probability, and definite integral representations. For example, the average number of divisors of all numbers from 1 to $n$ is very close to $\ln n + 2\gamma - 1$.

Calculating $\gamma$ has not attracted the same public interest as calculating $\pi$, but $\gamma$ has still inspired many ardent devotees. While we presently know $\pi$ to 1,241,100,000,000 decimal places, in 2008, only about 10,000,000,000 places of $\gamma$ were known. The evaluation of $\gamma$ is considerably more difficult than $\pi$. Here are the first few digits: 0.57721566490153286060651209008240243104215933593992.…

This mathematical constant has a long and fascinating history, just as do other famous constants like $\pi$ and $e$. Swiss mathematician Leonhard Euler discussed $\gamma$ in a paper, "*De Progressionibus harmonicis observationes*" ("Observations about Harmonic Progressions"), published in 1735, but he was only able to calculate it to six decimal places at the time. In 1790, Italian mathematician and priest Lorenzo Mascheroni computed additional digits. Today, we don't know if the number can be expressed as a fraction (in the way that a number like 0.142857142857…can be expressed as 1/7). Julian Havil, who devoted an entire book to $\gamma$, tells of stories in which the English mathematician G. H. Hardy offered to give up his Savilian Chair at Oxford to anyone who proved $\gamma$ could not be expressed as a fraction.

**SEE ALSO** $\pi$ (c. 250 B.C.), Discovery of Series Formula for $\pi$ (c. 1500), and Euler's Number, $e$ (1727).

*A 1737 portrait of Leonhard Euler by Johann Georg Brucker.*

1736

# Königsberg Bridges

**Leonhard Paul Euler** (1707–1783)

Graph theory is an area of mathematics that concerns how objects are connected and often simplifies problems by representing them as dots connected by lines. One of the oldest problems in graph theory involves the seven Königsberg bridges of Germany (now part of Russia). People in old Königsberg loved to take walks along the river, bridges, and islands. In the early 1700s, people still wondered if it was possible to take a journey across all seven bridges without having to cross any bridge more than once, and return to the starting location. Finally, in 1736, Swiss mathematician Leonhard Euler proved that such a tour was impossible.

Euler represented the bridges by a graph in which land areas are represented by dots and bridges by lines. He showed that one could traverse such a graph by going through every segment just once only if the graph had fewer than three vertices of odd *valence*. (The valence of a vertex is the number of lines that start or stop at the vertex.) The Königsberg bridges did not have the proper graph characteristics; thus, it is not possible to traverse the graph without going through a line more than once. Euler generalized his findings to journeys on any network of bridges.

The Königsberg bridge problem is important in the history of mathematics because Euler's solution corresponds to the first theorem in graph theory. Today, graph theory is used in countless fields, from the study of chemical pathways and car traffic flow to the social networks of Internet users. Graph theory can even explain how sexually transmitted diseases spread. Euler's very simple representations of the bridges' connectivity, without regard to the specifics of bridge lengths, was a forerunner of topology, the mathematical field concerned with shapes and their relationships to one another.

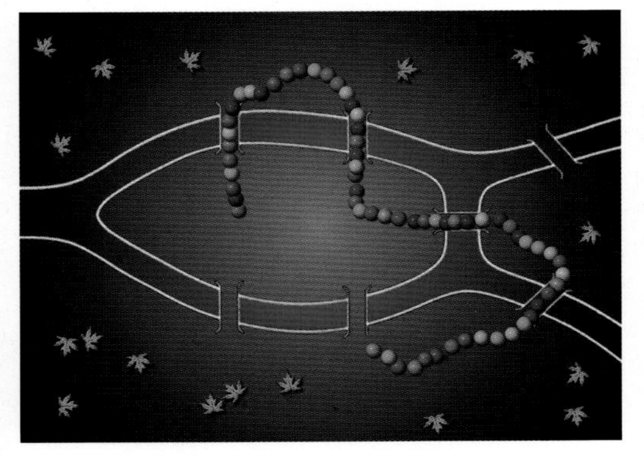

**SEE ALSO** Euler's Formula for Polyhedra (1751), Icosian Game (1857), The Möbius Strip (1858), Poincaré Conjecture (1904), Jordan Curve Theorem (1905), and Sprouts (1967).

LEFT: *One possible route through four of the seven Königsberg bridges.* RIGHT: *Matt Britt's partial map of the Internet. The lengths of the lines are indicative of the delay between two nodes. Colors indicate node type—for example, commercial, government, military, or educational.*

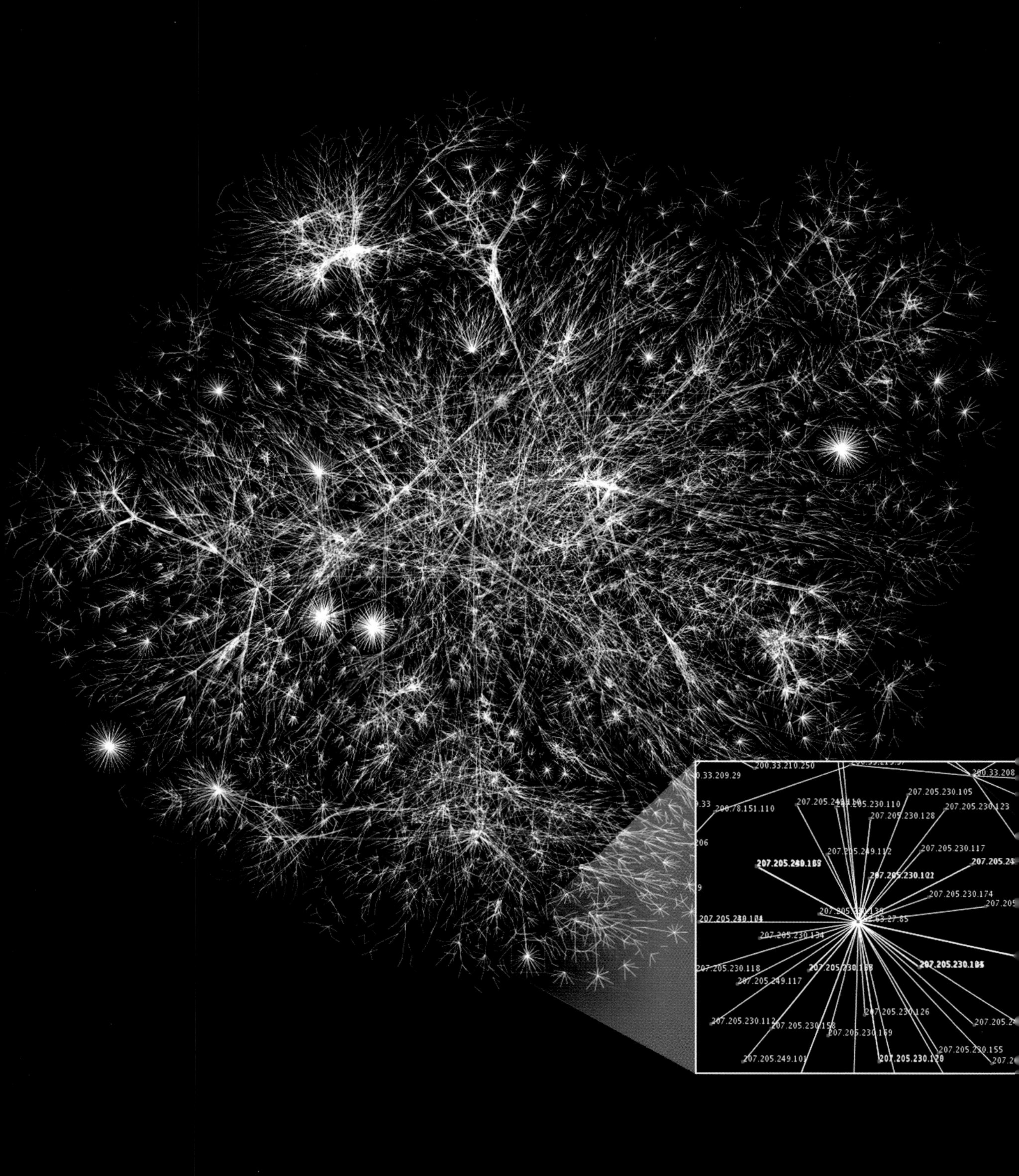
200.33.210.250
200.78.151.110
207.205.230.110
207.205.230.105
207.205.230.123
207.205.230.128
207.205.249.112
207.205.230.117
207.205.230.174
207.205.230.134
207.205.230.118
207.205.249.117
207.205.230.126
207.205.230.112
207.205.230.159
207.205.230.155
207.205.249.101

1738

# St. Petersburg Paradox

**Daniel Bernoulli** (1700–1782)

Daniel Bernoulli, the Dutch-born Swiss mathematician, physicist, and medical doctor, wrote a fascinating paper on probability, which was finally published in 1738 in *Commentaries of the Imperial Academy of Science of Saint Petersburg*. The paper described a paradox now known as the St. Petersburg paradox, and it can be expressed in terms of coin flips and money that a gambler is to receive, depending on the outcome of the flips. Philosophers and mathematicians have long discussed what the fair price should be for joining the game. How much would you be ready to pay for joining this activity?

Here's one way to view the St. Petersburg scenario. Flip a penny until it lands tails. The total number of flips, $n$, determines the prize, which equals $\$2^n$. Thus, if the penny lands on tails the first time, the prize is $\$2^1 = \$2$, and the game ends. If the penny comes up heads the first time, it is flipped again. If it comes up tails the second time, the prize is $\$2^2$, = \$4, and the game ends. And so on. A detailed discussion of the paradox of this game is beyond the scope of this book, but, according to game theory, a "rational gambler" would enter a game if and only if the price of entry was less than the expected value of the financial payoff. In some analyses of the St. Petersburg game, any finite price of entry is smaller than the expected value of the game, and a rational gambler might desire to play the game no matter how large we set a finite entry price!

Peter Bernstein comments on the profundity of Bernoulli's paradox: "His paper is one of the most profound documents ever written, not just on the subject of risk but on human behavior as well. Bernoulli's emphasis on the complex relationships between measurement and gut touches on almost every aspect of life."

**SEE ALSO** Zeno's Paradoxes (c. 445 B.C.), Aristotle's Wheel Paradox (c. 320 B.C.), Law of Large Numbers (1713), Barber Paradox (1901), Banach-Tarski Paradox (1924), Hilbert's Grand Hotel (1925), Birthday Paradox (1939), Coastline Paradox (c. 1950), Newcomb's Paradox (1960), and Parrondo's Paradox (1999).

*Since the 1730s, philosophers and mathematicians have pondered the St. Petersburg paradox. According to some analyses, a player may be expected to win an unlimited amount of money, but how much would you really pay for joining this game?*

THE UNITED STATES OF AMERICA
ONE DOLLAR
ONE

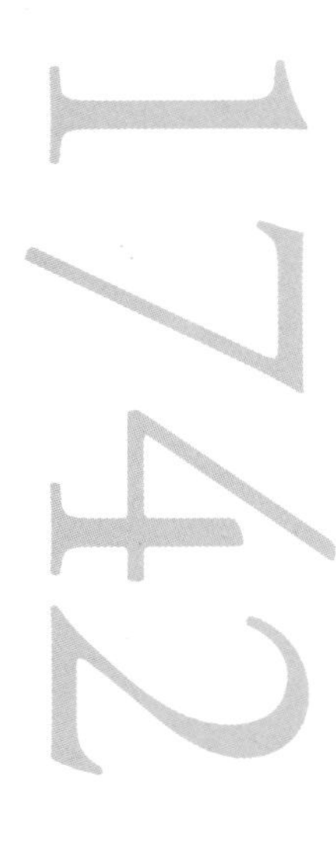

# Goldbach Conjecture

**Christian Goldbach** (1690–1764), **Leonhard Paul Euler** (1707–1783)

Sometimes the most challenging problems in mathematics are among the easiest and simplest to state. In 1742, Prussian historian and mathematician Christian Goldbach conjectured that every integer greater than 5 can be written as the sum of three prime numbers, such as $21 = 11 + 7 + 3$. (A prime number is a number larger than 1, such as 5 or 13, that is divisible only by itself or 1.) As re-expressed by Swiss mathematician Leonhard Euler, an equivalent conjecture (called the "strong" Goldbach conjecture) asserts that all positive even integers greater than 2 can be expressed as the sum of two primes. In order to promote the novel *Uncle Petros and Goldbach's Conjecture*, publishing giant Faber and Faber offered a $1,000,000 prize to anyone who proved Goldbach's conjecture between March 20, 2000, and March 20, 2002, but the prize went unclaimed, and the conjecture remains open. In 2008, Tomás Oliveira e Silva, a researcher at the University of Aveiro, Portugal, ran a distributed computer search that has verified the conjecture up to $12 \cdot 10^{17}$.

Of course, no amount of computing power can confirm the conjecture for every number; thus, mathematicians hope for an actual proof that Goldbach's intuition was right. In 1966, Chen Jing-Run, a Chinese mathematician, made some progress when he proved that every sufficiently large even number is the sum of one prime, plus a number that is the product of at most two primes. So, for example, 18 is equal to $3 + (3 \times 5)$. In 1995, French mathematician Olivier Ramaré showed that every even number greater than or equal to 4 is the sum of at most six primes.

**SEE ALSO** Cicada-Generated Prime Numbers (c. 1 Million B.C.), Sieve of Eratosthenes (c. 240 B.C.), Constructing a Regular Heptadecagon (1796), Gauss's *Disquisitiones Arithmeticae* (1801), Riemann Hypothesis (1859), Proof of the Prime Number Theorem (1896), Brun's Constant (1919), Gilbreath's Conjecture (1958), Ulam Spiral (1963), Erdös and Extreme Collaboration (1971), Public-Key Cryptography (1977), and Andrica's Conjecture (1985).

*Goldbach's comet illustrates the number of ways (x-axis) to write an even number* n *(y-axis) as the sum of two primes* ($4 \leq n \leq 1,000,000$)*. The star at bottom left is at 0,0. The* x*-axis goes from 0 to approximately 15,000.*

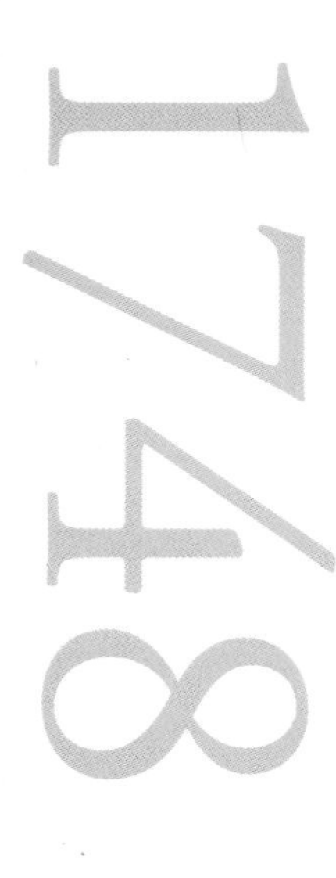

# Agnesi's *Instituzioni Analitiche*

**Maria Gaetana Agnesi** (1718–1799)

Italian mathematician Maria Agnesi is the author of *Instituzioni analitiche* (*Analytical Institutions*), the first comprehensive textbook that covered both differential and integral calculus, and the first surviving mathematical work written by a woman. Dutch mathematician Dirk Jan Struik referred to Agnesi as "the first important woman mathematician since Hypatia (A.D. fifth century)."

Agnesi was a child prodigy, speaking at least seven languages by age 13. For much of her life, she avoided social interactions and devoted herself entirely to the study of mathematics and religion. Clifford Truesdell writes, "She did ask her father's permission to become a nun. Horrified that his dearest child should desire to leave him, he begged her to change her mind." She agreed to continue living with her father so long as she could live in relative seclusion.

The publication of *Instituzioni analitiche* caused a sensation in the academic world. The committee of the Académie des Sciences in Paris wrote, "It took much skill and sagacity to reduce…to almost uniform methods these discoveries scattered among the works of modern mathematicians and often presented by methods very different from each other. Order, clarity, and precision reign in all parts of this work.…We regard it as the most complete and best made treatise." The book also includes a discussion of the cubic curve now known as the Witch of Agnesi and expressed as $y = 8a^3/(x^2 + 4a^2)$.

The president of the Academy of Bologna invited Agnesi to accept the Chair of Mathematics at the University of Bologna. According to some accounts, she never actually went to Bologna because, by this time, she was devoting herself entirely to religion and charitable work. Nonetheless, this made her the second woman to be appointed professor at a university; the first was Laura Bassi (1711–1778). Agnesi spent all her money on helping the poor, and she died in total poverty in a poorhouse.

**SEE ALSO** The Death of Hypatia (415), Discovery of Calculus (c. 1665), L'Hôpital's *Analysis of the Infinitely Small* (1696), and The Doctorate of Kovalevskaya (1874).

*Frontispiece from* Instituzioni analitiche (Analytical Institutions), *the first comprehensive textbook that covers both differential and integral calculus, and the first surviving mathematical work written by a woman.*

# INSTITUZIONI ANALITICHE

*AD USO*

DELLA GIOVENTU' ITALIANA

DI D.[NA] MARIA GAETANA AGNESI

MILANESE

*Dell' Accademia delle Scienze di Bologna.*

TOMO I.

IN MILANO, MDCCXLVIII.

---

NELLA REGIA-DUCAL CORTE.

*CON LICENZA DE' SUPERIORI.*

1751

# Euler's Formula for Polyhedra

**Leonhard Paul Euler** (1707–1783), **René Descartes** (1596–1650), **Paul Erdös** (1913–1996)

Euler's formula for polyhedra is considered to be one of the most beautiful formulas in all of mathematics and one of the first great formulas of topology—the study of shapes and their interrelationships. A survey conducted of *Mathematical Intelligencer* readers ranked the formula as the second most beautiful formula in history, second to Euler's $e^{i\pi} + 1 = 0$, discussed in the entry **Euler's Number, *e*** (1727).

In 1751, Swiss mathematician and physicist Leonhard Euler discovered that any convex polyhedron (an object with flat faces and straight edges), with $V$ vertices, $E$ edges, and $F$ faces, satisfies the equation $V - E + F = 2$. A polyhedron is *convex* if it has no indentations or holes, or more formally, if every line segment connecting interior points is entirely contained within the interior of the figure.

For example, the surface of a cube has six faces, twelve edges, and eight vertices. Plugging these values into Euler's formula, we obtain $8 - 12 + 6 = 2$. For a dodecahedron with its 12 faces, we have $20 - 30 + 12 = 2$. Interestingly, around 1639, René Descartes discovered a related polyhedral formula that may be converted to Euler's formula through several mathematical steps.

The polyhedron formula was later generalized to the study of networks and graphs, and to help mathematicians understand a wide range of shapes with holes and in higher dimensions. The formula also facilitates many practical applications such as helping computer specialists find ways to arrange wire paths in electrical circuits and cosmologists ponder models for the shape of our universe.

Euler is second only to Hungarian Paul Erdös as the most prolific mathematician in history in terms of number of publications. Sadly, Euler went blind toward the end of his life. However, British science writer David Darling notes, "the quantity of his output seemed to be inversely proportional to the quality of his sight, because his rate of publication increased after he became almost totally blind in 1766."

**SEE ALSO** Platonic Solids (c. 350 B.C.), Archimedean Semi-Regular Polyhedra (c. 240 B.C.), Euler's Number, *e* (1727), Königsberg Bridges (1736), Icosian Game (1857), Pick's Theorem (1899), Geodesic Dome (1922), Császár Polyhedron (1949), Erdös and Extreme Collaboration (1971), Szilassi Polyhedron (1977), Spidrons (1979), and Solving of the Holyhedron (1999).

*Non-convex polyhedra, such as this small stellated dodecahedron by Teja Krašek, can have Euler characteristics other than 2, where the characteristic is equal to* $V - E + F$. *Here,* $F = 12$, $E = 30$, *and* $V = 12$, *so that the characteristic is* $-6$.

1751

# Euler's Polygon Division Problem

**Leonhard Paul Euler** (1707–1783)

In 1751, Swiss mathematician Leonhard Paul Euler posed the following problem to Prussian mathematician Christian Goldbach (1690–1764): In how many ways $E_n$ can a planar convex polygon of $n$ sides be divided into triangles by diagonals? Or, stated more informally: How many ways can you divide a polygonal pie into triangles, starting your straight, downward knife cuts at one corner and ending at another? Your cuts can't intersect one another. The formula Euler found was this:

$$E_n = \frac{2 \cdot 6 \cdot 10 \ldots (4n-10)}{(n-1)!}$$

A polygon is *convex* if, for every pair of points that belong to the shape, the shape contains the entire straight line segment connecting the two points. Author and mathematician Heinrich Dörrie writes, "This problem is of the greatest interest because it involves many difficulties in spite of its innocuous appearance, as many a surprised reader will discover.... Euler himself said, 'The process of induction I employed was quite laborious.'"

For example, for a square, we have $E_4 = 2$, which corresponds to the two diagonals. For a pentagon, we have $E_5 = 5$. In fact, earlier experimenters were inclined to use graphic representations to get a feel for the solution, but this visual approach soon becomes intractable as the number of polygon sides grows. By the time we get to a 9-sided polygon, we have 429 ways to divide the polygon into triangles by diagonals.

The polygon division problem has attracted much attention. In 1758, Carpatho-German mathematician Johann Andreas Segner (1704–1777) developed a recurrence formula for determining the values: $E_n = E_2E_{n-1} + E_3E_{n-2} + \ldots + E_{n-1}E_2$. A *recurrence* formula is one in which each term of the sequence is defined as a function of the preceding terms.

Interestingly, the values of $E_n$ are intimately tied to another class of numbers called Catalan numbers ($E_n = C_{n-1}$). Catalan numbers arise in combinatorics, the field of mathematics concerned with problems of selection, arrangement, and operation within a finite or discrete system.

**SEE ALSO** Archimedes: Sand, Cattle & Stomachion (c. 250 B.C.), Goldbach Conjecture (1742), Morley's Trisector Theorem (1899), and Ramsey Theory (1928).

*A regular pentagon can be divided by diagonals into triangles in five different ways.*

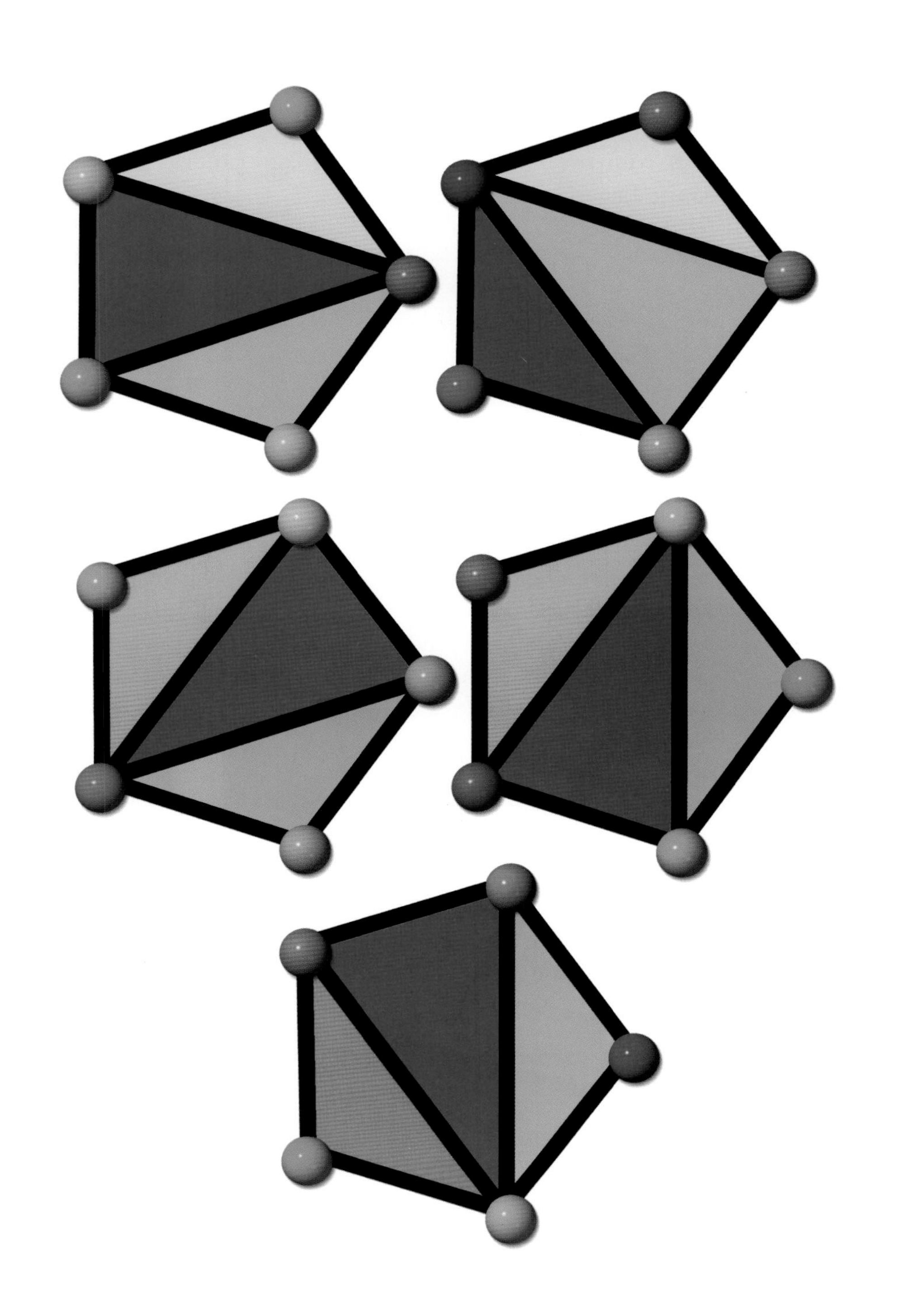

# Knight's Tours

**Abraham de Moivre** (1667–1754), **Leonhard Paul Euler** (1707–1783), **Adrien-Marie Legendre** (1752–1833)

To create a Knight's Tour, a chess knight must jump exactly once to every square on the 8 × 8 chessboard in a complete tour. Various kinds of Knight's Tours have fascinated mathematicians for centuries. The earliest recorded solution was provided by Abraham de Moivre, the French mathematician better known for the **normal distribution curve** and his theorems about complex numbers. In his solution, the knight ends its tour on a square that is far away from its starting position. The French mathematician Adrien-Marie Legendre "improved" on this and found a solution in which the first and last squares are a single move apart so that the tour closes up on itself into a single loop of 64 knight moves. Such a tour is said to be *reentrant*. The Swiss mathematician Leonhard Euler found a reentrant tour that visits two halves of the board in turn.

Euler was the first to write a mathematical paper analyzing Knight's Tours. He presented the paper to the Academy of Sciences at Berlin in 1759, but this influential paper was not published until 1766. Interestingly, in 1759, the Academy proposed a prize of 4,000 francs for the best memoir on the Knight's Tour, but the prize was never awarded, perhaps because Euler was the Director of Mathematics at the Berlin Academy and ineligible for the prize.

My favorite Knight's Tour is one over the six surfaces of a cube, each surface being a chessboard. Henry E. Dudeney presented the cube tour in his book *Amusement in Mathematics*, and I believe he based the solution (in which each face is toured in turn) on the earlier work of French mathematician Alexandre-Théophile Vandermonde (1735–1796). The properties of Knight's Tours have since been carefully studied on checkerboards on the surfaces of a cylinder, **Möbius Strip**, torus, and **Klein Bottle**, and even in higher dimensions.

**SEE ALSO** The Möbius Strip (1858), Klein Bottle (1882), and Peano Curve (1890).

*Knight's Tour path on a 30 × 30 chessboard, discovered by computer scientist Dmitry Brant using a neural network consisting of an interconnected group of artificial neurons that work together to produce the solution.*

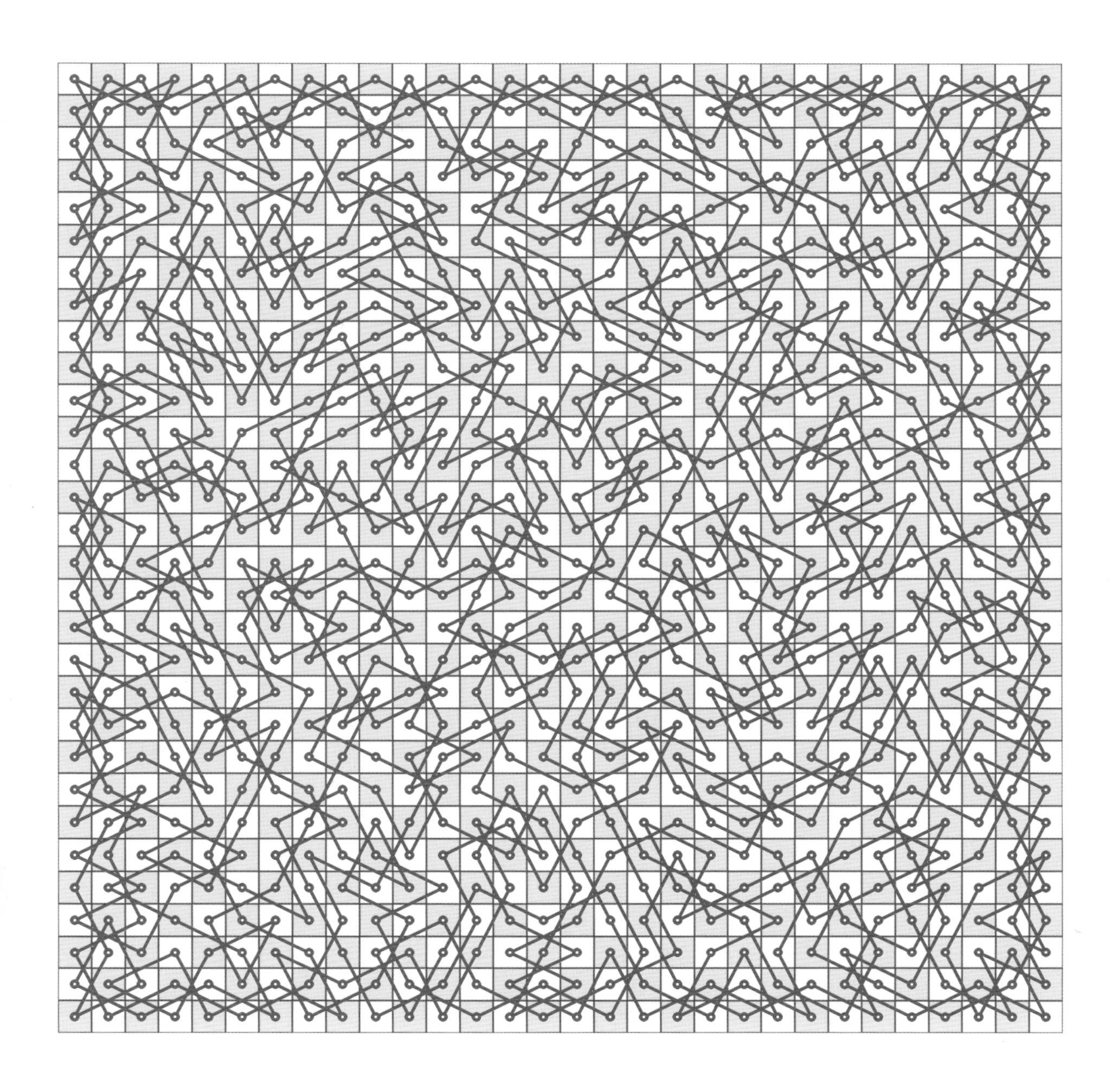

# Bayes' Theorem

**Thomas Bayes** (c. 1702–1761)

Bayes' theorem, formulated by British mathematician and Presbyterian minister Thomas Bayes, plays a fundamental role in science and can be stated by a simple mathematical formula used for calculating conditional probabilities. *Conditional probability* refers to the probability of some event A, given the occurrence of some other event B, written as $P(A|B)$. Bayes' theorem states: $P(A|B) = [P(B|A) \times P(A)]/P(B)$. Here, $P(A)$ is called the prior probability of A because it is the probability of event A without taking into account anything we know about B. $P(B|A)$ is the conditional probability of B given A. $P(B)$ is the prior probability of B.

Imagine we have two boxes. Box 1 has 10 golf balls and 30 billiard balls. Box 2 has 20 of each. You select a box at random and pull out a ball. We assume that the balls are equally likely to be selected. Your ball turns out to be a billiard ball. How probable is it that you chose Box 1? In other words, what is the probability that you chose Box 1, given that you have a billiard ball in your hand?

Event A corresponds to your picking Box 1. Event B is your picking a billiard ball. We want to compute $P(A|B)$. $P(A)$ is 0.5, or 50 percent. $P(B)$ is the probability of picking a billiard ball regardless of any information on the boxes. It is computed as the sum of the probability of getting a billiard ball from a box multiplied by the probability of selecting a box. The probability of picking a billiard ball from Box 1 is 0.75. The probability of picking one from Box 2 is 0.5. The probability of getting a billiard ball overall is $0.75 \times 0.5 + 0.5 \times 0.5 = 0.625$. $P(B|A)$, or the probability of getting a billiard ball given that you selected Box 1, is 0.75. We can use Bayes' formula to find that the probability of your having chosen Box 1, which is $P(A|B) = 0.6$.

**SEE ALSO** Law of Large Numbers (1713) and Laplace's *Théorie Analytique des Probabilités* (1812).

*Box 1 (upper box) and Box 2 (lower box) are shown here. You select a box at random and withdraw a billiard ball. How probable it is that you choose the upper box?*

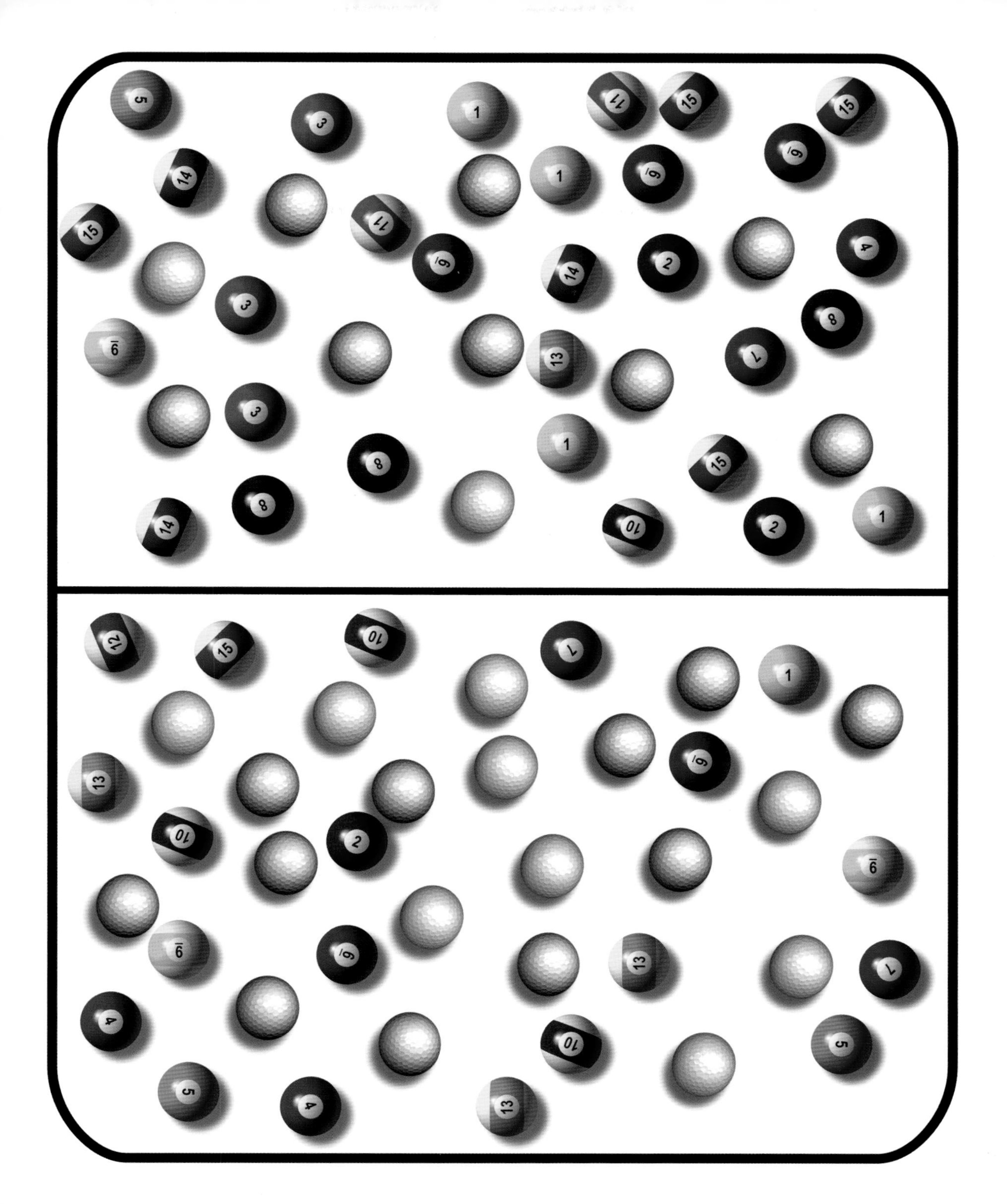

# Franklin Magic Square

**Benjamin Franklin** (1706–1790)

Benjamin Franklin was a scientist, inventor, statesman, printer, philosopher, musician, and economist. In 1769, in a letter to a colleague, he describes a **Magic Square** he had created earlier in his life.

His 8 × 8 magic square is filled with wondrous symmetries, some of which Ben Franklin was probably not aware. Each row and column of the square has a sum of 260. Half of each row or column sums to half of 260. In addition, each of the *bent rows* has the sum 260. See the gray highlighted squares for two examples of bent rows. See the squares with thick black borders for an example of a *broken bent row* (14 + 61 + 64 + 15 + 18 + 33 + 36 + 19), which also sums to 260. Numerous other symmetries can be found—for example, the four corner numbers and the four middle numbers sum to 260. The sum of the numbers in any 2 × 2 subsquare is 130, and the sum of any four numbers that are arranged equidistant from the center of the square also equals 130. When converted to binary numbers, even more startling symmetries are found. Alas, despite all the marvelous symmetries, the main diagonals don't each sum to 260, so this cannot strictly qualify as a magic square according to the common definition that includes the diagonal sums.

| | | | | | | | |
|---|---|---|---|---|---|---|---|
| 52 | 61 | 4 | 13 | 20 | 29 | 36 | 45 |
| 14 | 3 | 62 | 51 | 46 | 35 | 30 | 19 |
| 53 | 60 | 5 | 12 | 21 | 28 | 37 | 44 |
| 11 | 6 | 59 | 54 | 43 | 38 | 27 | 22 |
| 55 | 58 | 7 | 10 | 23 | 26 | 39 | 42 |
| 9 | 8 | 57 | 56 | 41 | 40 | 25 | 24 |
| 50 | 63 | 2 | 15 | 18 | 31 | 34 | 47 |
| 16 | 1 | 64 | 49 | 48 | 33 | 32 | 17 |

We do not know what method Franklin used to construct his squares. Many people have tried to crack the secret, but until the 1990s no *quick* recipe could be found, although Franklin claimed he could generate the squares "as fast as he could write." In 1991, author Lalbhai Patel invented a method to construct the Franklin squares. Although the method seems quite long, Patel has trained himself to quickly carry out the procedure. So many wonderful patterns have been found in Franklin's magic square that this square has become a metaphor for mathematical objects that contain symmetries and other properties that continue to be discovered long after the inventor's death.

**SEE ALSO** Magic Squares (c. 2200 B.C.) and Perfect Magic Tesseract (1999).

*Portrait of Benjamin Franklin (1767) by artist David Martin (1737–1797).*

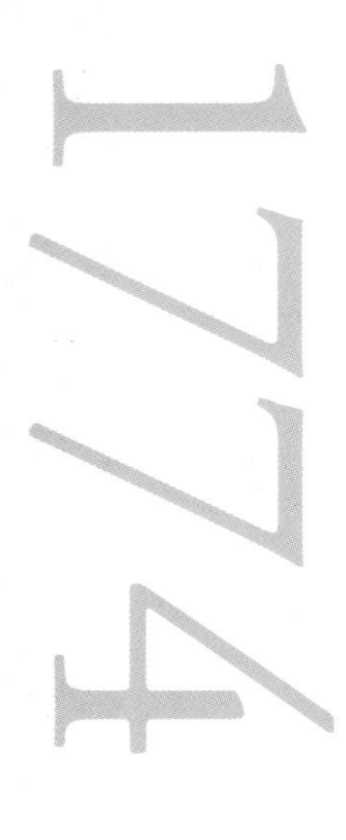

# Minimal Surface

**Leonhard Paul Euler** (1707–1783), **Jean Meusnier** (1754–1793), **Heinrich Ferdinand Scherk** (1798–1885)

Imagine withdrawing a flat wire ring from soapy water. Because the ring contains a disk-shaped soap film that has less area than other shapes that hypothetically may have formed, mathematicians call the surface a minimal surface. More formally, a finite minimal surface is often characterized as having the smallest possible area bounded by a given closed curve or curves. The mean curvature of the surface is zero. The mathematician's quest for minimal surfaces and proofs of their minimality has lasted for more than two centuries. Minimal surfaces with bounding curves that twist into the third dimensions can be both beautiful and complicated.

In 1744, Swiss mathematician Leonhard Euler discovered the catenoid, the first example of a minimal surface beyond mere trivial examples like circular areas. In 1776, French geometer Jean Meusnier discovered the helicoid minimal surface. (Meusnier was also a military general and designer of the first propeller-driven, elliptically shaped balloon for carrying people.)

Another minimal surface wasn't found until 1873 by German mathematician Heinrich Scherk. The same year, the Belgian physicist Joseph Plateau performed experiments that led him to conjecture that soap films always form minimal surfaces. "Plateau's problem" deals with the mathematics required to prove this to be true. (Plateau went blind as a result of staring into the sun for 25 seconds in an experiment dealing with vision physiology.) More recent examples include Costa's minimal surface, which was first described mathematically in 1982 by Brazilian mathematician Celso Costa.

Computers and computer graphics now play a significant role in helping mathematicians construct and visualize minimal surfaces, some of which can be quite complicated. Minimal surfaces may one day have numerous applications in materials science and nanotechnology. For example, certain polymers, when mixed, form interfaces that are minimal surfaces. Knowledge of the interface shapes may help scientists predict the chemical properties of such mixtures.

**SEE ALSO** Torricelli's Trumpet (1641), Beltrami's Pseudosphere (1868), and Boy's Surface (1901).

*A version of Enneper's surface, an example of a minimal surface rendered by Paul Nylander. The surface was discovered around 1863 by German mathematician Alfred Enneper (1830–1885).*

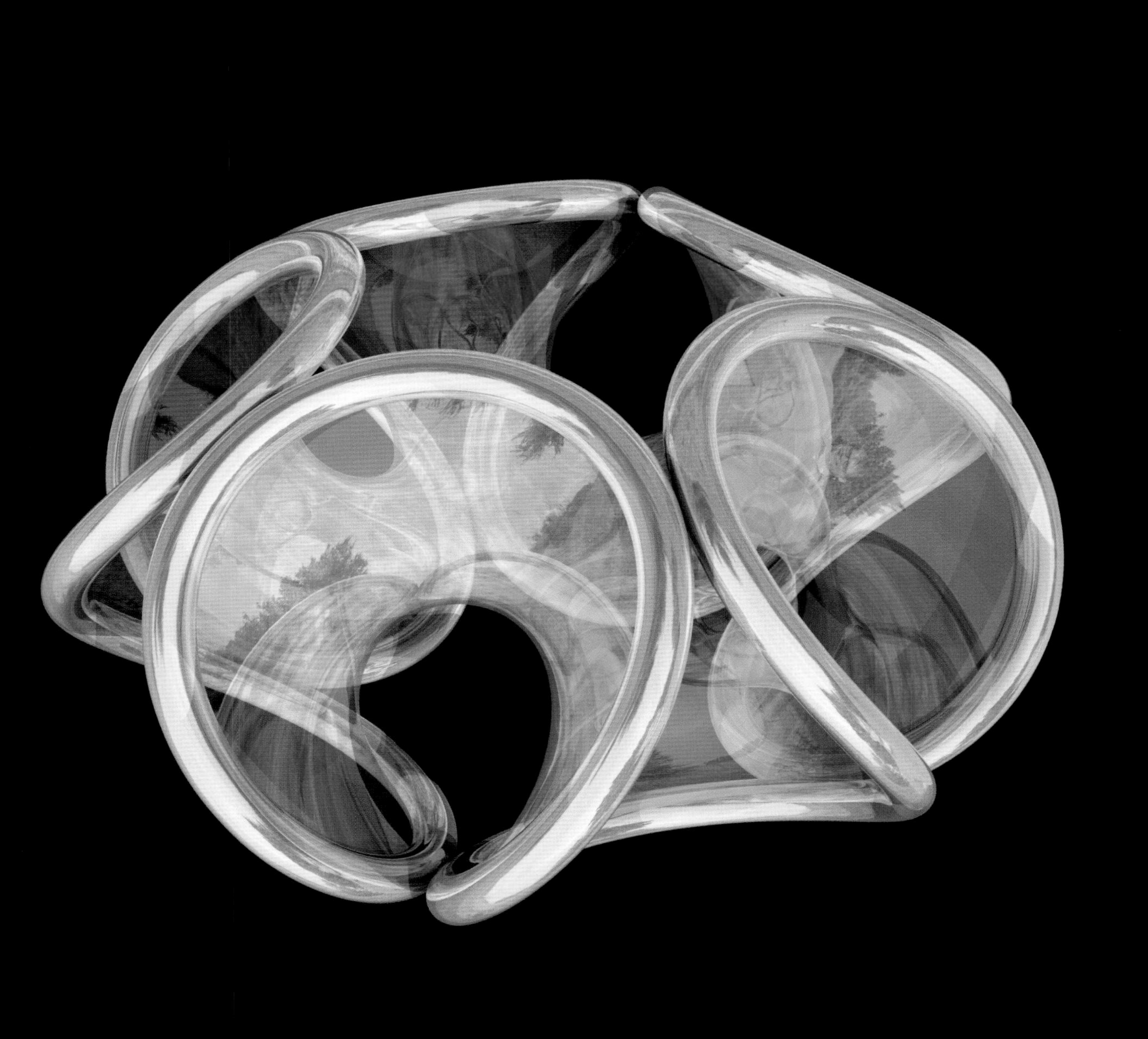

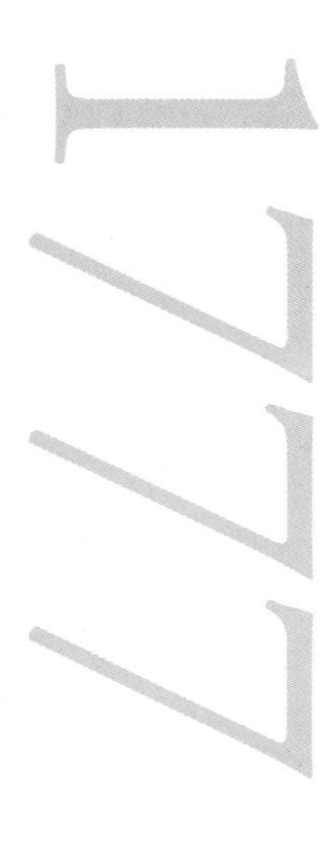


# Buffon's Needle

## Georges-Louis Leclerc, Comte de Buffon (1707–1788)

Named after a district in Monaco famous for its many casinos, Monte Carlo methods play critical roles in mathematics and science, and make use of randomness in order to solve problems ranging from the statistics of nuclear chain reactions to the regulation of traffic flows.

One of the earliest and most famous uses of the method occurred in the eighteenth century when French naturalist and mathematician Comte de Buffon showed that by dropping a needle repeatedly onto a lined sheet of paper, and counting the number of times the needle touched a line, he could obtain an estimate for the value of the mathematical constant pi ($\pi = 3.1415\ldots$). In the simplest case, imagine dropping a toothpick onto a hardwood floor with spacing between the floor lines equal to the toothpick length. To approximate pi from the toothpick drops, simply consider the number of drops and multiply it by 2, and then divide by the number of times the toothpick touched a line.

Buffon was a man of many talents. His 36-volume *Histoire naturelle, générale et particulière* (*Natural History: General and Particular*) includes everything known about the natural world and influenced Charles Darwin and the theory of evolution.

Today, powerful computers can generate huge quantities of pseudorandom numbers per second and allow scientists to take full advantage of Monte Carlo methods in order to understand problems in economics, physics, chemistry, protein structure prediction, galactic formation, artificial intelligence, cancer therapy, stock forecasting, oil-well exploration, the design of aerodynamic shapes, and problems in pure mathematics for which no other methods are available.

In modern times, the Monte Carlo approach was brought to world attention by mathematicians and physicists such as Stanislaw Ulam, John von Neumann, Nicholas Metropolis, and Enrico Fermi. Fermi used the approach to study properties of the neutron. Monte Carlo methods were crucial to the simulations required for the Manhattan Project, America's program to develop an atomic bomb during World War II.

**SEE ALSO** Dice (c. 3000 B.C.), $\pi$ (c. 250 B.C.), Law of Large Numbers (1713), Normal Distribution Curve (1733), Laplace's *Théorie Analytique des Probabilités* (1812), The Rise of Randomizing Machines (1938), Von Neumann's Middle-Square Randomizer (1946), and Ball Triangle Picking (1982).

*Portrait of Georges-Louis Leclerc, Comte de Buffon, by François-Hubert Drouais (1727–1775).*

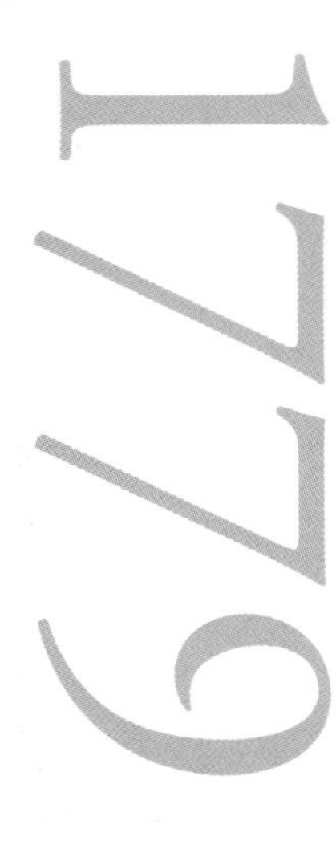

# Thirty-Six Officers Problem

**Leonhard Paul Euler** (1707–1783), **Gaston Tarry** (1843–1913)

Consider six army regiments, each consisting of six officers of different ranks. In 1779, Leonhard Euler asked if it was possible to arrange these 36 officers in a $6 \times 6$ square array so that each of six ranks and each of six regiments are represented once in each row and column. In the language of mathematics, this problem is equivalent to finding two mutually orthogonal Latin squares of order six. Euler correctly conjectured that there was no solution, and French mathematician Gaston Tarry proved this in 1901. Through the centuries, the problem had led to significant work in combinatorics, the area of mathematics concerned with the selection and arrangement of objects. Latin squares also play a role in error-correcting codes and communications.

A Latin square consists of $n$ sets of numbers, 1 to $n$, arranged in such a way that no row or column contains the same two numbers. The numbers of Latin squares starting with order $n = 1$ are 1, 2, 12, 576, 161,280, 812,851,200, 61,479,419,904,000, 108,776,032,459,082,956,800, and so forth.

A pair of Latin squares is said to be orthogonal if the $n^2$ pairs formed by juxtaposing the two arrays are all distinct. (*Juxtaposing* refers to combining the two numbers to form an ordered pair.) For example, two orthogonal Latin squares of order 3 are:

| | | |
|---|---|---|
| 3 | 2 | 1 |
| 2 | 1 | 3 |
| 1 | 3 | 2 |

| | | |
|---|---|---|
| 2 | 3 | 1 |
| 1 | 2 | 3 |
| 3 | 1 | 2 |

Euler conjectured that if $n = 4k + 2$, where $k$ is a positive integer, then no pair of orthogonal $n \times n$ Latin squares exists. This conjecture was not settled for more than a century, until 1959, when mathematicians Bose, Shikhande, and Parker constructed a pair of $22 \times 22$ orthogonal Latin squares. Today, we know that a pair of orthogonal $n \times n$ Latin squares exists for every positive integer $n$ except $n = 2$ and $n = 6$.

**SEE ALSO** Magic Squares (c. 2200 B.C.), Archimedes: Sand, Cattle & Stomachion (c. 250 B.C.), Euler's Polygon Division Problem (1751), and Ramsey Theory (1928).

*An example of a $6 \times 6$ Latin square consisting of six colors, arranged in such a way that no row or column contains the same two colors. Today, we know that 812,851,200 order-six Latin squares exist.*

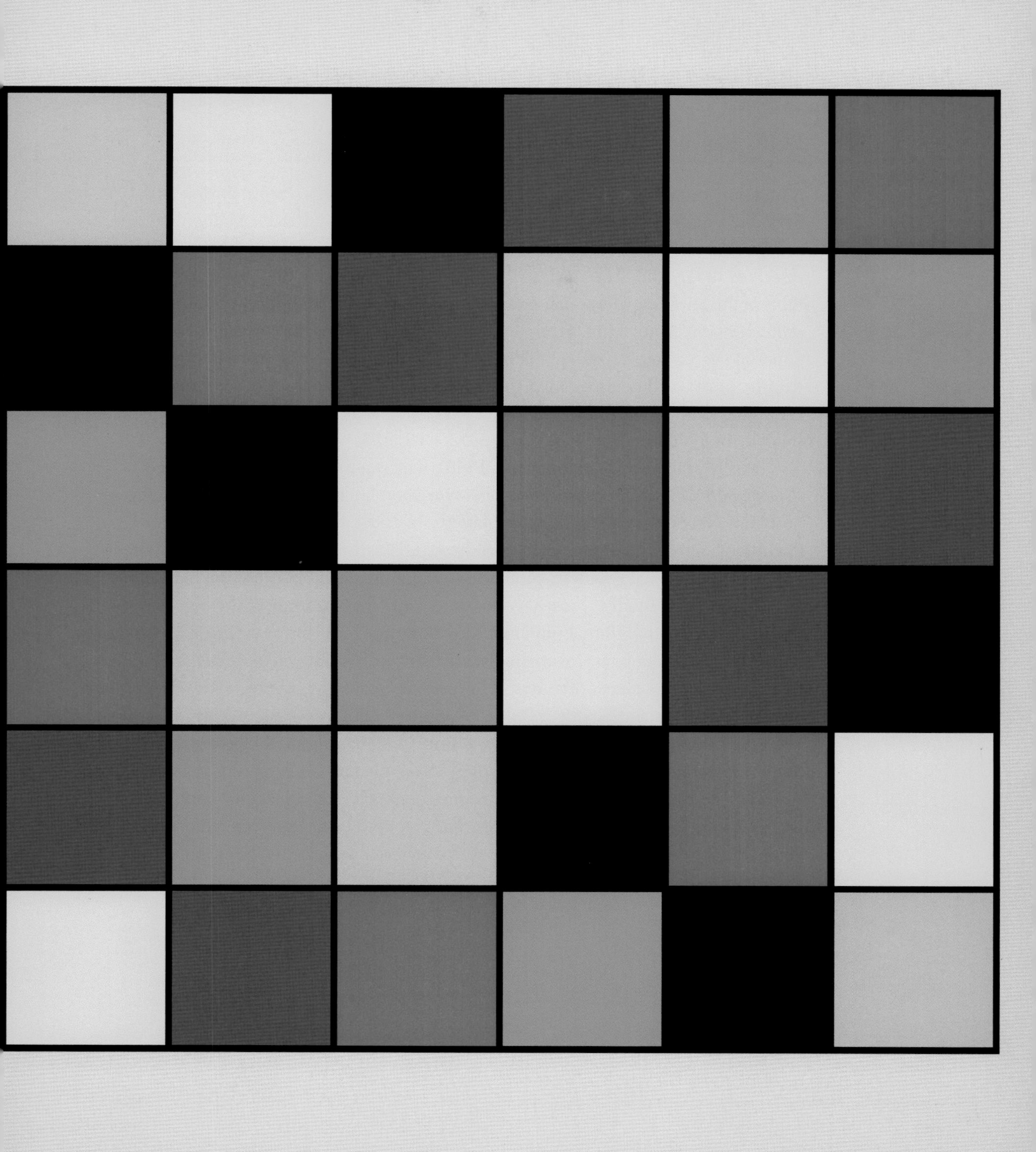

c. 1789

# Sangaku Geometry

**Fujita Kagen** (1765–1821)

A tradition known as Sangaku, or "Japanese temple geometry," arose during Japan's period of isolation from the West, roughly between 1639 and 1854. Mathematicians, farmers, samurai, women, and children solved difficult geometry problems and inscribed the solutions on tablets. These colorful tablets were then hung under the roofs of the temples. More than 800 tablets have survived, and many of them feature problems concerning tangent circles. As one example, consider the figure on the opposite page, a late Sangaku tablet from 1873 created by an 11-year-old boy named Kinjiro Takasaka. The illustration shows a fan, which is one-third of a complete circle. Given the diameter $d_1$ of the yellow-shaded circle, what is the diameter $d_2$ of the green-shaded circle? The answer is $d_2 \approx d_1(\sqrt{3072} + 62)/193$.

In 1789, Japanese mathematician Fujita Kagen published *Shimpeki Sampo* (*Mathematical Problems Suspended before the Temple*), the first collection of Sangaku problems. The oldest surviving tablet dates from 1683, although other historical documents refer to examples from as early as 1668. Most of the Sangaku are strangely different from typical geometry problems found in textbooks because the Sangaku aficionados were usually obsessed with circles and ellipses. Some of the Sangaku problems are so difficult that physicist Tony Rothman and educator Hidetoshi Fukagawa write, "Modern geometers invariably tackle them with advanced methods, including calculus and affine transformations." However, by avoiding calculus, Sangaku problems were, in principle, sufficiently simple that children could solve them with some effort.

Chad Boutin writes, "Perhaps it's not surprising that Sudoku—the number puzzles that everyone seems to be working on these days—first became popular in Japan before spreading across the ocean. The fad is reminiscent of a math craze that swept the islands centuries ago, when ardent enthusiasts went so far as to turn the most beautiful geometrical solutions into finely illustrated wooden tablets, called Sangaku...."

**SEE ALSO** Euclid's *Elements* (300 B.C.), Kepler Conjecture (1611), and Johnson's Theorem (1916).

*A late Sangaku pattern from 1873, created by an 11-year-old boy.*

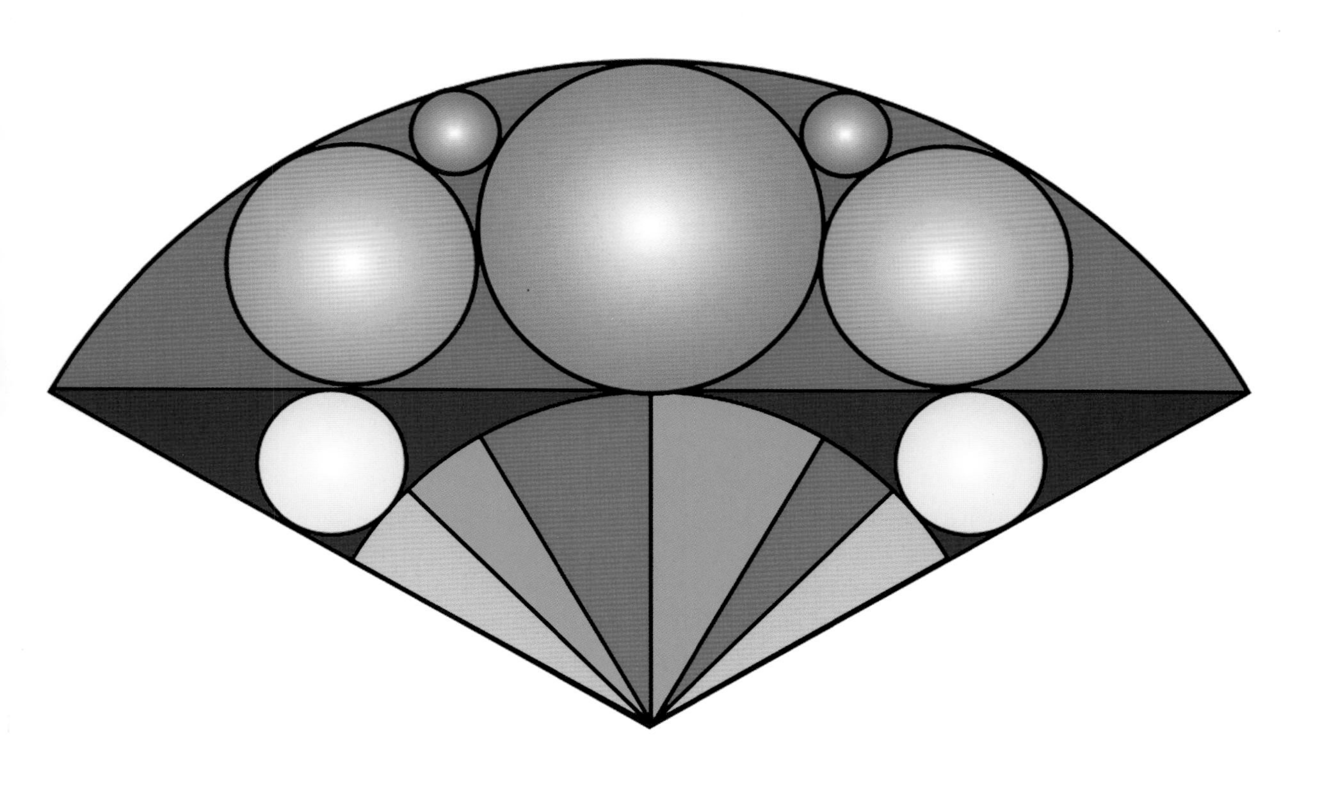

# Least Squares

## Johann Carl Friedrich Gauss (1777–1855)

You enter a cave with marvelous stalactites protruding from the ceiling. You might expect that a correlation exists between the length of a stalactite and its age, although the relationship between these two variables may not be exact. Unpredictable temperature and humidity fluctuations probably affect the growth. However, assuming that chemical or physical methods exist for estimating the age of a stalactite, some kind of trend surely exists between age and length that lets us make rough predictions.

The method of least squares has played a crucial role in science for elucidating and visualizing trends such as these, and today the method is available in most statistical computer packages that draw lines or smooth curves through noisy experimental data. Least squares is a mathematical procedure for finding the "best-fitting" curve for a given set of data points by minimizing the sum of the squares of the offsets of the points from the curve.

In 1795, German mathematician and scientist Carl Friedrich Gauss, at the age of 18, began to develop least-squares analysis. He demonstrated the value of his approach in 1801, when he predicted the future location of the asteroid Ceres. As background, the Italian astronomer Giuseppe Piazzi (1746–1826) had originally discovered Ceres in 1800, but the asteroid later disappeared behind the sun and could not be relocated. Austrian astronomer Franz Xaver von Zach (1754–1832) noted that "without the intelligent work and calculations of Doctor Gauss, we might not have found Ceres again." Interestingly, Gauss kept his methods a secret to maintain an advantage over his contemporaries and to enhance his reputation. Later in his life, he sometimes published scientific results as a cipher, so that he could always prove that he had made various discoveries before others had. Gauss finally published his secret least-squares method in 1809 in his *Theory of the Motion of the Heavenly Bodies*.

**SEE ALSO** Laplace's *Théorie Analytique des Probabilités* (1812) and Chi-Square (1900).

*A least-squares plane. Here, the least-squares procedure is used to find the "best-fitting" plane for a given set of data points by minimizing the sum of the squares of the lengths of the blue segments parallel to the y-axis.*

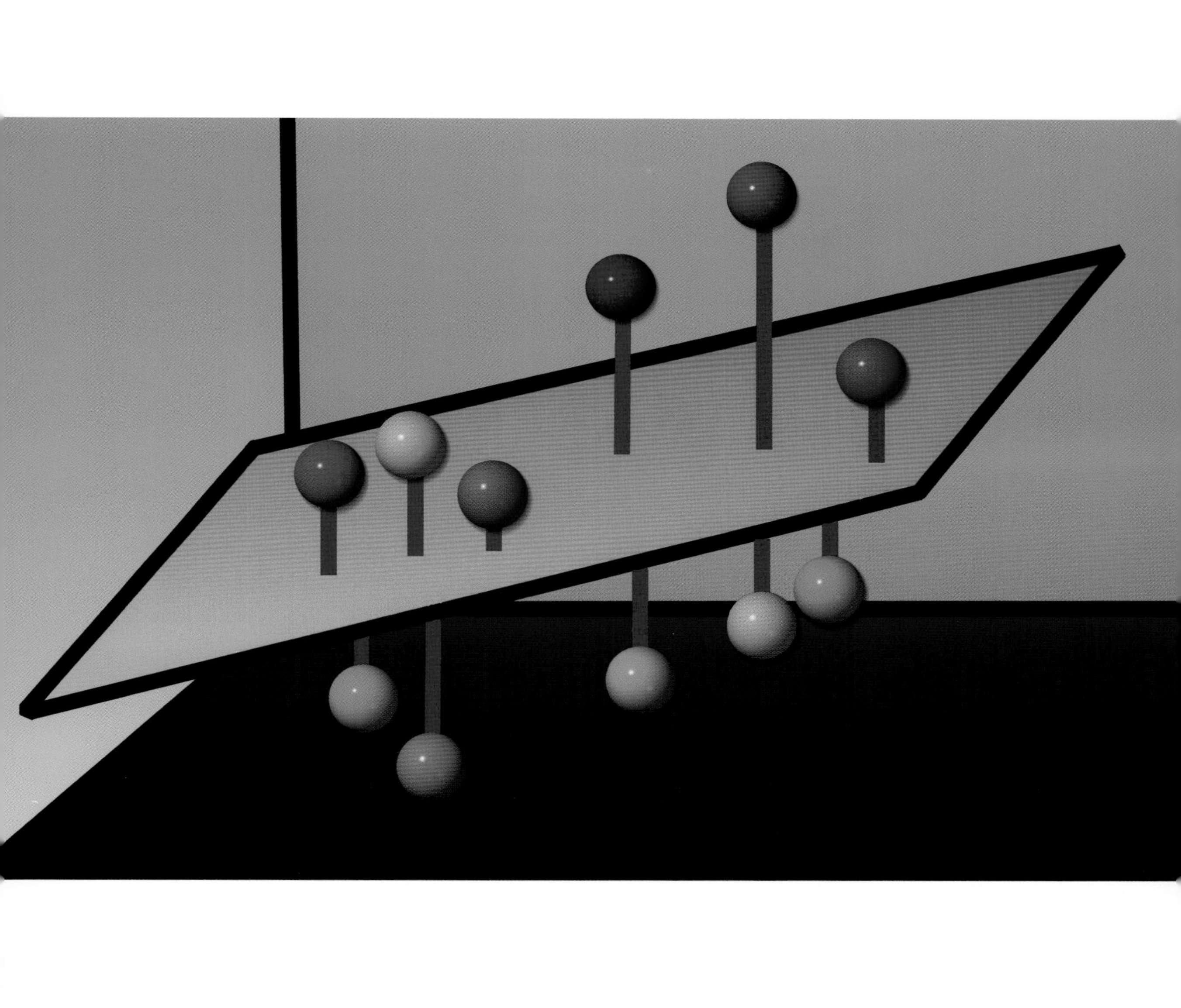

# Constructing a Regular Heptadecagon

**Johann Carl Friedrich Gauss** (1777–1855)

In 1796, when Gauss was still a teenager, he discovered a way to construct a regular 17-sided polygon, also known as a heptadecagon, using just a straightedge and compass. He published the result in his monumental 1801 work, *Disquisitiones Arithmeticae* (*Arithmetic Disquisitions*). Gauss's construction was very significant because only failed attempts had been made since the time of Euclid.

For more than 1,000 years, mathematicians had known how to construct, with a compass and straightedge, regular $n$-gons in which $n$ was a multiple of 3, 5, and powers of 2. Gauss was able to add more polygons to this list, namely those with a prime number of sides of the form $2^{(2^n)} + 1$, where $n$ is an integer. We can make a list of the first few such numbers: $F_0 = 3$, $F_1 = 5$, $F_2 = 17$, $F_3 = 257$, and $F_4 = 65{,}537$. (Numbers of this form are also known as Fermat numbers, and they are not necessarily prime.) A 257-gon was constructed in 1832.

When he was older, Gauss still regarded his 17-gon finding as one of his greatest achievements, and he asked that a regular 17-gon be placed on his tombstone. According to legend, the stonemason declined, stating that the difficult construction would essentially make the 17-gon look like a circle.

The year 1796 was an auspicious year for Gauss, when his ideas gushed like a fountain from a fire hose. Aside from solving the heptadecagon construction (March 30), Gauss invented modular arithmetic and presented his quadratic reciprocity law (April 8) and the prime number theorem (May 31). He proved that every positive integer is represented as a sum of at most three triangular numbers (July 10). He also discovered solutions of polynomials with coefficients in finite fields (October 1). Regarding the heptadecagon, Gauss said he was "amazed" that so little had been discovered with respect to polygon construction since Euclid's time.

**SEE ALSO** Cicada-Generated Prime Numbers (c. 1 Million B.C.), Sieve of Eratosthenes (c. 240 B.C.), Goldbach Conjecture (1742), Gauss's *Disquisitiones Arithmeticae* (1801), Riemann Hypothesis (1859), Proof of the Prime Number Theorem (1896), Brun's Constant (1919), Gilbreath's Conjecture (1958), Ulam Spiral (1963), and Andrica's Conjecture (1985).

*A fish swims and explores in a heptadecagonal pool.*

# Fundamental Theorem of Algebra

**Johann Carl Friedrich Gauss** (1777–1855)

The Fundamental Theorem of Algebra (FTA) is stated in several forms, one of which is that every polynomial of degree $n \geq 1$, with real or complex coefficients, has $n$ real or complex roots. In other words, a polynomial $P(x)$ of degree $n$ has $n$ values $x_i$ (some of which are possibly repeated) for which $P(x_i) = 0$. As background, polynomial equations of degree $n$ are of the form $P(x) = a_n x^n + a_{n-1}x^{n-1} + \ldots + a_1 x + a_0 = 0$ where $a_n \neq 0$.

As an example, consider the quadratic polynomial $f(x) = x^2 - 4$. When plotted, the curve is a parabola with its minimum at $f(x) = -4$. The polynomial has two distinct real roots ($x = 2$ and $x = -2$), which are graphically seen as points where the parabola intersects the $x$-axis.

This theorem is notable, in part, because of the sheer number of attempts at proving it through history. German mathematician Carl Friedrich Gauss is usually credited with the first proof of the FTA, discovered in 1797. In his doctoral thesis, published in 1799, he presented his first proof, which focused on polynomials with real coefficients, and also on his objections to the other previous attempts at proofs. By today's standards, Gauss's proof was not rigorously complete, because he relied on the continuity of certain curves, but it was a significant improvement over all previous attempts at a proof.

Gauss considered the FTA to have great importance, as evidenced by his returning to the problem repeatedly. His fourth proof was in the last paper he ever wrote, which appeared in 1849, exactly 50 years after his dissertation. Note that Jean-Robert Argand (1768–1822) published a rigorous proof of the Fundamental Theorem of Algebra in 1806 for polynomials with complex coefficients. The FTA arises in many areas of mathematics, and the various proofs span fields that range from abstract algebra and complex analysis to topology.

**SEE ALSO** Al-Samawal's *The Dazzling* (c. 1150), Constructing a Regular Heptadecagon (1796), Gauss's *Disquisitiones Arithmeticae* (1801), and Jones Polynomial (1984).

*Greg Fowler's depiction of the three solutions to* $z^3 - 1 = 0$. *These roots (or zeros) are 1,* $-0.5 + 0.86603i$, *and* $-0.5 - 0.86603i$, *and are located at the center of the three large bull's-eyes in this Newton's method rendition of the solutions.*

1801

# Gauss's *Disquisitiones Arithmeticae*

## **Johann Carl Friedrich Gauss** (1777–1855)

Stephen Hawking writes, "When Gauss began work on his epochal *Disquisitiones Arithmeticae* (*Arithmetic Disquisitions*), number theory was merely a collection of isolated results. . . . In the *Disquisitiones*, he introduced the notion of congruence and in so doing unified number theory." Gauss published this monumental work at the age of 24.

The *Disquisitiones* involves modular arithmetic, which relies on congruency relationships. Two integers $p$ and $q$ are "congruent modulo the integer $s$" if and only if $(p - q)$ is evenly divisible by $s$. Such a congruence is written as $p \equiv q \pmod{s}$. Using this compact notation, Gauss restated and proved the famous quadratic reciprocity theorem, which was incompletely proven several years earlier by French mathematician Adrien-Marie Legendre (1752–1833). Consider two distinct odd prime numbers, $p$ and $q$. Consider the statements: (1) $p$ is a square mod $q$, and (2) $q$ is a square mod $p$. According to the theorem: If both $p$ and $q$ are congruent to 3 (mod 4), then exactly one of (1) and (2) is true; otherwise, either both (1) and (2) are true, or neither of them is true. (A *square* is an integer that can be written as the square of some other integer, such as 25, which is $5^2$.)

Thus, this theorem connects the solvability of two related quadratic equations in modular arithmetic. Gauss devoted an entire section of his book to his proof of this theorem. He considered this beloved theorem of quadratic reciprocity to be the "golden theorem" or the "gem of arithmetic," which so enthralled Gauss that he went on to provide eight separate proofs over his lifetime.

Mathematician Leopold Kronecker said, "It is really astonishing to think a single man of such young years was [able] to present such a profound and well-organized treatment of an entirely new discipline." In *Disquisitiones*, Gauss's approach to providing theorems, followed by proofs, corollaries, and examples was used by subsequent authors. *Disquisitiones* was a seed from which flowered the work of many leading nineteenth-century number theorists.

**SEE ALSO** Cicada-Generated Prime Numbers (c. 1 Million B.C.), Sieve of Eratosthenes (c. 240 B.C.), Goldbach Conjecture (1742), Constructing a Regular Heptadecagon (1796), Riemann Hypothesis (1859), Proof of the Prime Number Theorem (1896), Brun's Constant (1919), Gilbreath's Conjecture (1958), Ulam Spiral (1963), Erdös and Extreme Collaboration (1971), Public-Key Cryptography (1977), and Andrica's Conjecture (1985).

*Johann Carl Friedrich Gauss, painted by Danish artist Christian Albrecht Jensen (1792–1870).*

1801

# Three-Armed Protractor

**Joseph Huddart** (1741–1816)

The common protractor of today is an instrument used to construct and measure angles on a plane and to draw lines at various angles. This protractor resembles a semicircular disk marked with degrees, from 0° to 180°. In the seventeenth century, protractors began to be used as stand-alone instruments rather than as parts of other devices, when sailors used them on ocean maps.

In 1801, Joseph Huddart, an English naval captain, invented the three-armed protractor for plotting the position of a boat on navigation maps. This kind of protractor makes use of two outer arms that may rotate with respect to a fixed central arm. The two rotating arms may be clamped so that they can be set at fixed angles.

In 1773, Huddart worked for the East India Company, sailing for St. Helena Island in the South Atlantic Ocean and Bencooleen in Sumatra. During his journey, he made detailed surveys of the west coast of Sumatra. His 1778 chart of St. George's Channel, connecting the Irish Sea to the north and the Atlantic Ocean to the southwest, was a masterpiece of clarity and accuracy. Aside from his later fame as inventor of the three-armed protractor, he also suggested the use of high water marks at the London docks, still in use until the 1960s. He invented steam-driven devices for manufacturing rope that set the quality standard for rope making.

In 1916, the United States Hydrographic Office explained the use of his protractor: "To plot a position, the two angles observed between the three selected [known] objects are set on the instrument, which is then moved over the chart until the three beveled edges pass respectively and simultaneously through the three objects. The center of the instrument will then mark the ship's position, which may be pricked on the chart or marked with a pencil point through the center hole."

**SEE ALSO** Loxodrome (1537) and Mercator Projection (1569).

*Joseph Huddart, English naval captain and inventor of the three-armed protractor, useful for navigation.*

Captn. Joseph Huddart. F.R.S.

Engraved for the European Magazine from an Original Picture in the Possession of Chas. Turner Esqr. by T. Blood.

1807

# Fourier Series

## **Jean Baptiste Joseph Fourier** (1768–1830)

Fourier series are useful in countless applications today, ranging from vibration analysis to image processing—virtually any field in which a frequency analysis is important. For example, Fourier series help scientists characterize and better understand the chemical composition of stars or how the vocal tract produces speech.

Before French mathematician Joseph Fourier discovered his famous series, he accompanied Napoleon on his 1798 expedition of Egypt, where Fourier spent several years studying Egyptian artifacts. Fourier's research on the mathematical theory of heat began around 1804 when he was back in France, and in 1807 he had completed his important memoir *On the Propagation of Heat in Solid Bodies*. One of his interests was heat diffusion in different shapes. For these problems, researchers are usually given the temperatures at points on the surface, as well as at its edges, at time $t = 0$. Fourier introduced a series with sine and cosine terms in order to find solutions to these kinds of problems. More generally, he found that any differentiable function can be represented to arbitrary accuracy by a sum of sine and cosine functions, no matter how bizarre the function may look when graphed.

Biographers Jerome Ravetz and I. Grattan-Guiness note, "Fourier's achievement can be understood by [considering] the powerful mathematical tools he invented for the solutions of the equations, which yielded a long series of descendents and raised problems in mathematical analysis that motivated much of the leading work in that field for the rest of the century and beyond." British physicist Sir James Jeans (1877–1946) remarked, "Fourier's theorem tells us that every curve, no matter what its nature may be, or in what way it was originally obtained, can be exactly reproduced by superposing a sufficient number of simple harmonic curves—in brief, every curve can be built up by piling up waves."

**SEE ALSO** Bessel Functions (1817), Harmonic Analyzer (1876), and Differential Analyzer (1927).

*Molecular model of human growth hormone. Fourier series and corresponding Fourier synthesis methods are used to determine molecular structures from X-ray diffraction data.*

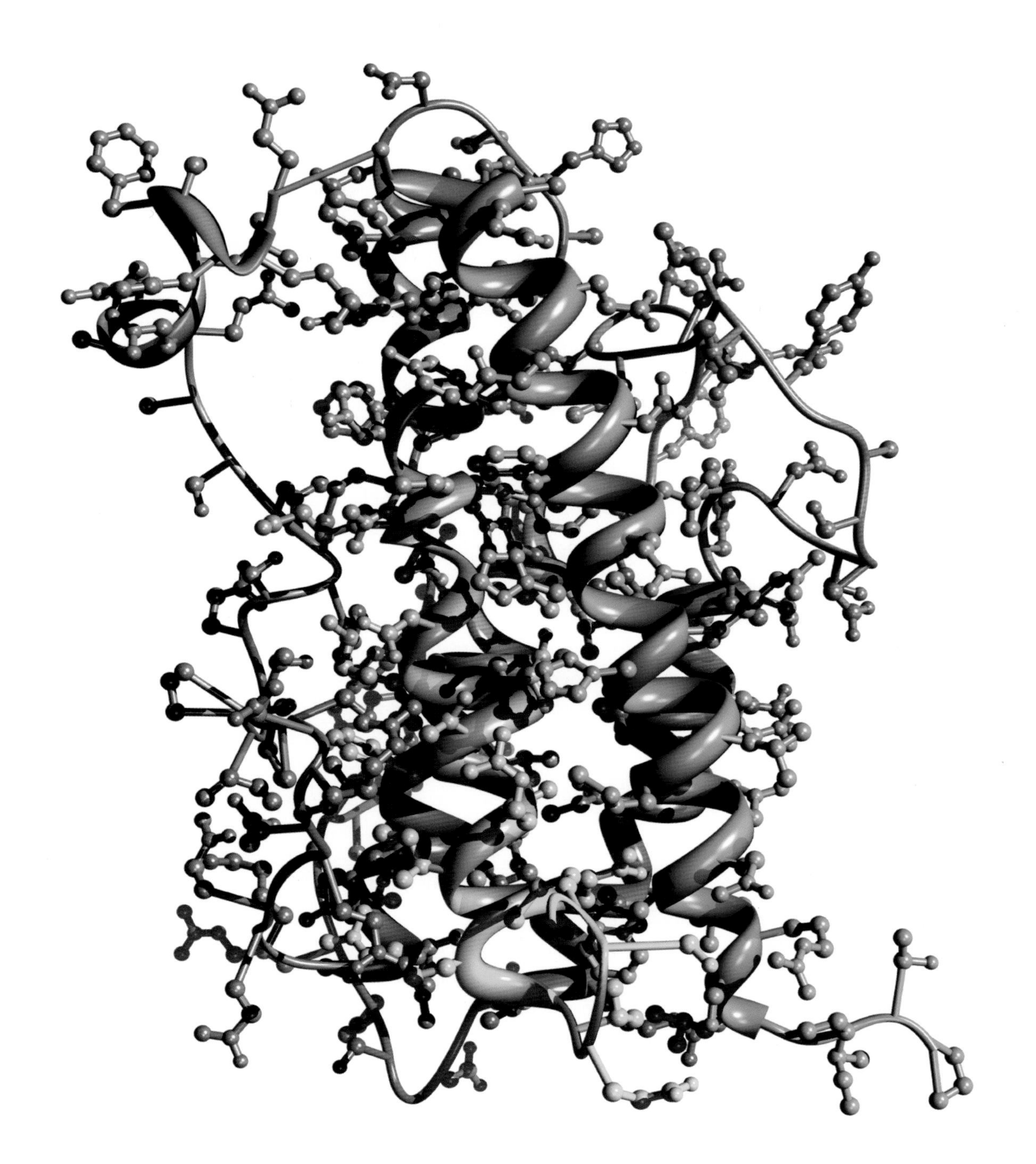

# Laplace's *Théorie Analytique des Probabilités*

**Pierre-Simon, Marquis de Laplace** (1749–1827)

The first major treatise on probability that combines probability theory and calculus was French mathematician and astronomer Pierre-Simon Laplace's *Théorie Analytique des Probabilités* (*Analytical Theory of Probabilities*). Probability theorists focus on random phenomena. Although a single roll of the dice may be considered a random event, after numerous repetitions, certain statistical patterns become apparent, and these patterns can be studied and used to make predictions.

The first edition of Laplace's *Théorie Analytique* was dedicated to Napoleon Bonaparte and discusses methods of finding probabilities of compound events from component probabilities. The book also discusses the method of **least squares** and **Buffon's Needle** and considers many practical applications.

Stephen Hawking calls *Théorie Analytique* a "masterpiece" and writes, "Laplace held that because the world is determined, there can be no probabilities in things. Probability results in our lack of knowledge." According to Laplace, nothing would be "uncertain" for a sufficiently advanced being—a conceptual model that remained strong until the rise of quantum mechanics and chaos theory in the twentieth century.

To explain how probabilistic processes can yield predictable results, Laplace asks readers to imagine several urns arranged in a circle. One urn contains only black balls, while another contains only white balls. The other urns have various ball mixtures. If we withdraw a ball, place it in the adjacent urn, and continue around the circle, eventually the ratio of black to white balls will be approximately the same in all of the urns. Here, Laplace shows how random "natural forces" can create results that have a predictability and order. Laplace writes, "It is remarkable that this science, which originated in the consideration of games of chance, should become the most important object of human knowledge. . . .The most important questions in life are, for the most part, really only problems of probability." Other famous probabilists include Gerolamo Cardano (1501–1576), Pierre de Fermat (1601–1665), Blaise Pascal (1623–1662), and Andrey Nikolaevich Kolmogorov (1903–1987).

**SEE ALSO** Discovery of Calculus (c. 1665), Law of Large Numbers (1713), Normal Distribution Curve (1733), Buffon's Needle (1777), Least Squares (1795), Infinite Monkey Theorem (1913), and Ball Triangle Picking (1982).

*Laplace felt it was remarkable that probability, which originated in analysis of games of chance, should become "the most important object of human knowledge . . ."*

32
0
26
35
12
28

# Prince Rupert's Problem

**Prince Rupert of the Rhine** (1619–1682), **Pieter Nieuwland** (1764–1794)

Prince Rupert's Problem has had a long and fascinating history. Prince Rupert was an inventor, artist, and soldier. He was fluent in virtually all of the major European languages and excelled in mathematics. Soldiers were frightened of the large poodle he took with him during battles, believing that it had supernatural powers.

In the 1600s, Prince Rupert asked a famous geometrical question: What is the largest wooden cube that can pass through a given cube of side length one inch? More precisely, what is the size $R$ of the edge of the largest tunnel (with a square cross section) that can be made through a cube without breaking the cube?

Today, we know the answer to be $R = 3\sqrt{2}/4 = 1.060660\ldots$. In other words, a cube with side length $R$ inches (or smaller) can pass through a cube with a side length of 1 inch. Prince Rupert won a wager that a hole could be made in one of two equally sized cubes large enough for the other cube to slide through. Many thought this could not be accomplished.

Although the first publication of Prince Rupert's Problem was by John Wallis (1616–1703) in his 1685 *De Algebra Tractatus*, the 1.060660 solution was not readily known until Dutch mathematician Pieter Nieuwland solved it over a century *after* Prince Rupert asked the question. His solution was published posthumously in 1816 by his teacher Jan Hendrik van Swinden, who had found Nieuwland's solution among his papers.

If you hold a cube so that one corner points toward you, you will see a regular hexagon. The largest square that will squeeze through a cube has a face that can be inscribed in this hexagon. As reported by mathematicians Richard Guy and Richard Nowakowski, the largest cube that can fit through a hypercube has an edge of 1.007434775..., which is the square root of 1.014924..., the smallest root of $4x^4 - 28x^3 - 7x^2 + 16x + 16$.

**SEE ALSO** Platonic Solids (c. 350 B.C.), Euler's Formula for Polyhedra (1751), Tesseract (1888), and Menger Sponge (1926).

*Prince Rupert of the Rhine won a wager that a hole could be made in one of two equal cubes large enough for the other cube to slide through. Many thought this could not be accomplished.*

1817

# Bessel Functions

**Friedrich Wilhelm Bessel** (1784–1846)

German mathematician Friedrich Bessel, who had no formal education after the age of 14, developed Bessel functions in 1817 for use in his studies of the motion of planets moving under mutual gravitation. Bessel had generalized the earlier findings of mathematician Daniel Bernoulli (1700–1782).

Since the time of Bessel's discoveries, his functions have become indispensable tools in a vast range of mathematics and engineering. Author Boris Korenev writes, "A large number of diverse problems concerning practically all the most important areas of mathematical physics and various technical problems is connected with Bessel functions." Indeed, different aspects of Bessel function theory are used when solving problems involving heat conduction, hydrodynamics, diffusion, signal processing, acoustics, radio and antenna physics, plate vibrations, oscillations in chains, stresses that evolve near cracks in materials, wave propagation in general, and atomic and nuclear physics. In elasticity theory, Bessel functions are useful for solving numerous spatial problems that employ spherical or cylindrical coordinates.

Bessel functions are solutions to specific differential equations, and when graphed, the functions resemble rippling, decaying sinusoidal waves. For example, in the case of a wave equation involving a circular membrane such as a drumhead, one class of solutions involves Bessel functions, and the standing wave solution can be expressed as a Bessel function that is a function of the distance $r$ from the center to the rim of the membrane.

In 2006, researchers at Japan's Akishima Laboratories and Osaka University relied on Bessel function theory to create a device that uses waves to draw actual text and pictures on the surface of water. The device, called AMOEBA (Advanced Multiple Organized Experimental Basin), consists of 50 water wave generators encircling a cylindrical tank 1.6 meters in diameter and 30 cm deep. AMOEBA is capable of spelling out the entire Roman alphabet. Each picture or letter remains on the water surface only for a moment, but they can be produced in succession every few seconds.

**SEE ALSO** Fourier Series (1807), Differential Analyzer (1927), and Ikeda Attractor (1979).

*Bessel functions are useful in studying problems of wave propagation, as well as modes of vibration of a thin circular membrane. (This rendering is by Paul Nylander, who uses Bessel functions to study wave phenomena.)*

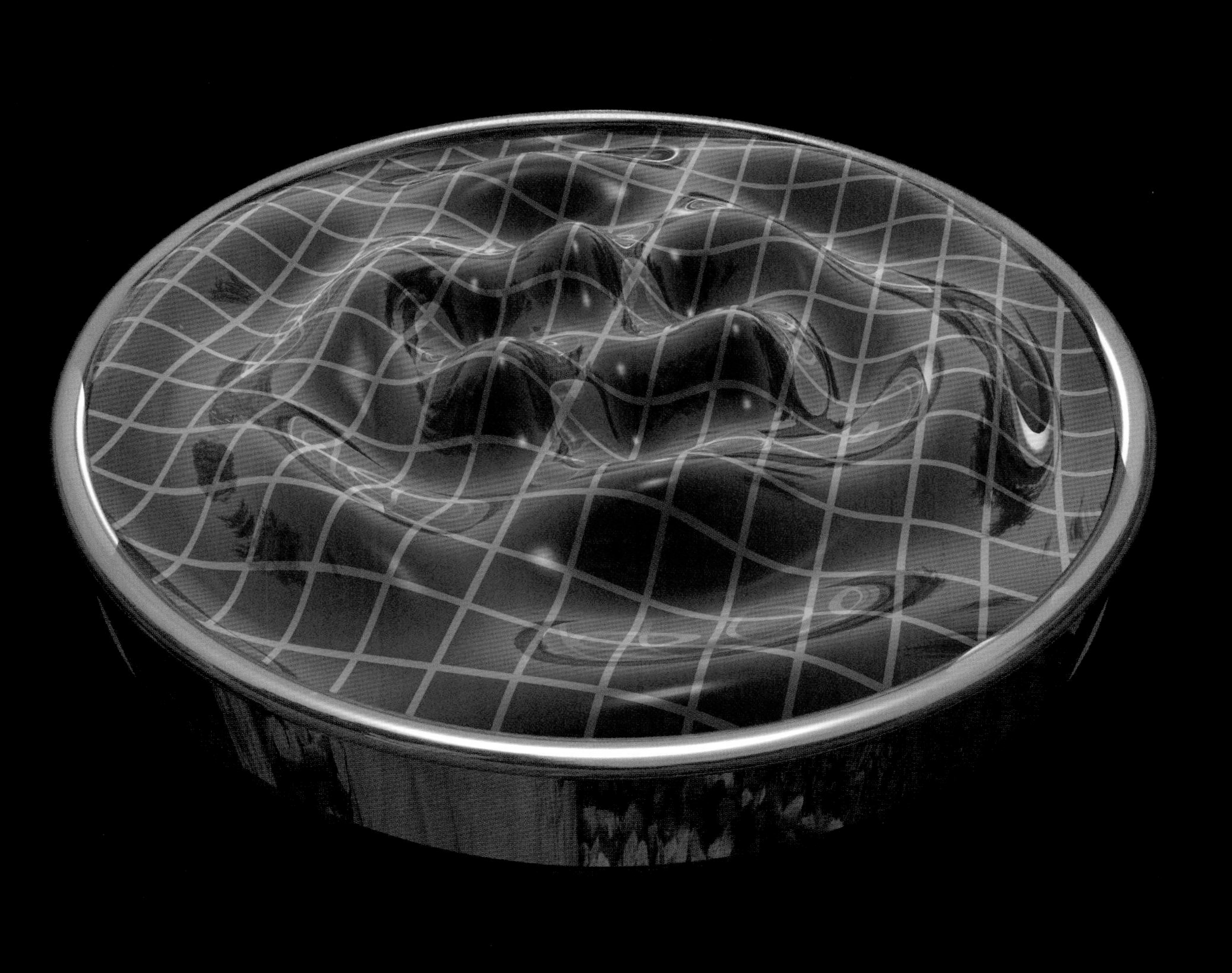

1822

# Babbage Mechanical Computer

**Charles Babbage** (1792–1871), **Augusta Ada King, Countess of Lovelace** (1815–1852)

Charles Babbage was an English analyst, statistician, and inventor who was also interested in the topic of religious miracles. He once wrote, "Miracles are not a breach of established laws, but...indicate the existence of far higher laws." Babbage argued that miracles could occur in a mechanistic world. Just as Babbage could imagine programming strange behaviors on his calculating machines, God could program similar irregularities in nature. While investigating biblical miracles, he suggested that the chance of a man rising from the dead is one in $10^{12}$.

Babbage is often considered the most important mathematician-engineer involved in the prehistory of computers. In particular, he is famous for conceiving an enormous hand-cranked mechanical calculator, an early progenitor of our modern computers. Babbage thought the device would be most useful in producing mathematical tables, but he worried about mistakes that would be made by humans who transcribed the results from its 31 metal output wheels. Today, we realize that Babbage was around a century ahead of his time and that the politics and technology of his era were inadequate for his lofty dreams.

Babbage's Difference Engine, begun in 1822 but never completed, was designed to compute values of polynomial functions, using about 25,000 mechanical parts. He also had plans to create a more general-purpose computer, the Analytical Engine, which could be programmed using punch cards and had separate areas for number storage and computation. Estimates suggest that an Analytical Engine capable of storing 1,000 50-digit numbers would be more than 100 feet (about 30 meters) in length. Ada Lovelace, the daughter of the English poet Lord Byron, gave specifications for a program for the Analytical Engine. Although Babbage provided assistance to Ada, many consider Ada to be the first computer programmer.

In 1990, novelists William Gibson and Bruce Sterling wrote *The Difference Engine*, which asked readers to imagine the consequences of Babbage's mechanical computers becoming available to Victorian society.

**SEE ALSO** Abacus (c. 1200), Slide Rule (1621), Differential Analyzer (1927), ENIAC (1946), Curta Calculator (1948), and HP-35: First Scientific Pocket Calculator (1972).

*Working model of a portion of Charles Babbage's Difference Engine, currently located at the London Science Museum.*

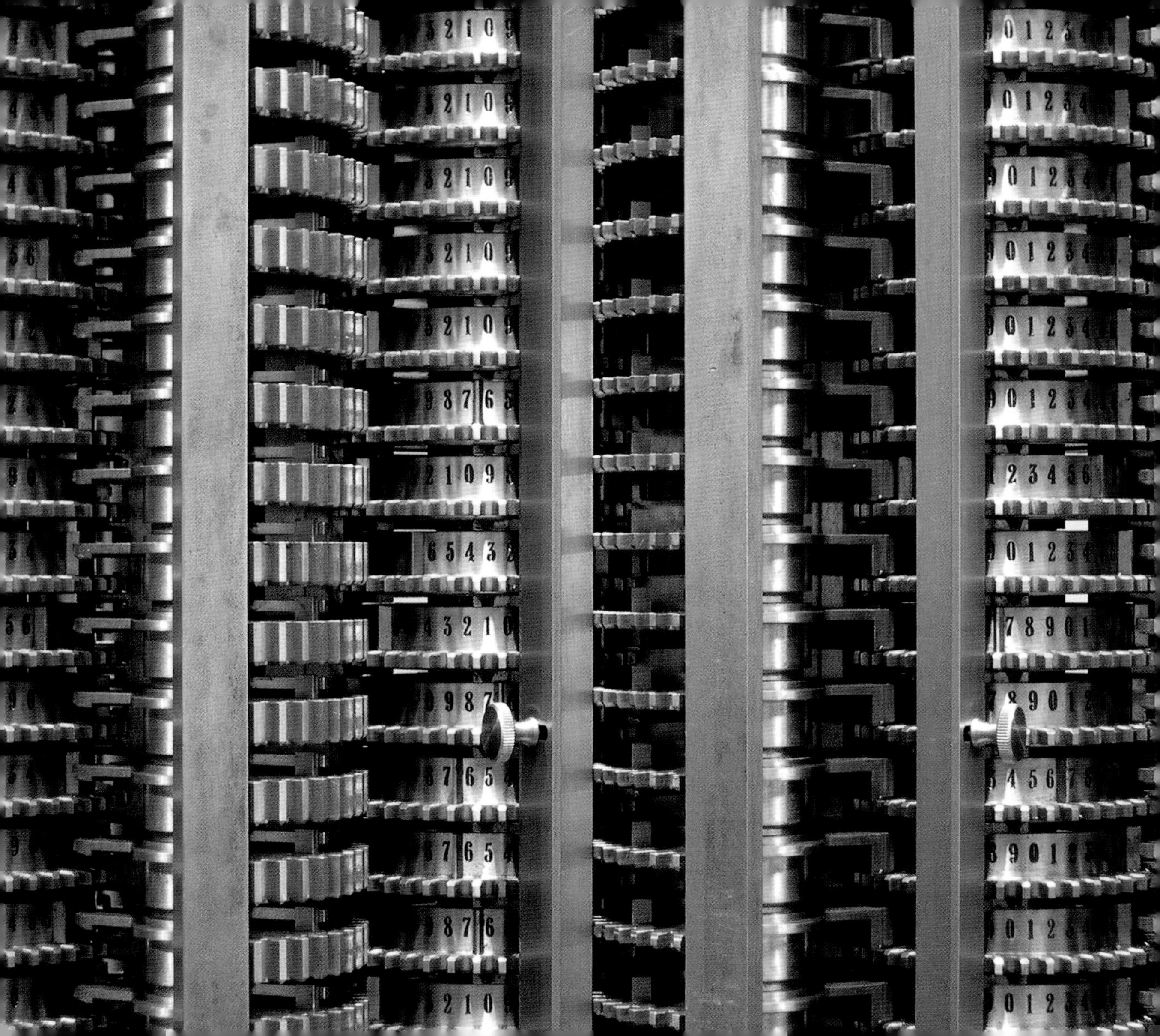

1823

# Cauchy's *Le Calcul Infinitésimal*

**Augustin Louis Cauchy** (1789–1857)

American mathematician William Waterhouse writes, "Calculus in 1800 was in a curious state. There was no doubt it was correct. Mathematicians of sufficient skill and insight had been successful with it for a century. Yet no one could explain clearly why it worked.... Then came Cauchy." In his 1823 *Résumé des leçons sur le calcul infinitésimal* (*Résumé of Lessons on Infinitesimal Calculus*), the prolific French mathematician Augustin Cauchy provides a rigorous development of calculus and a modern proof of the Fundamental Theorem of Calculus, which elegantly unites the two major branches of calculus (differential and integral) into a single framework.

Cauchy begins his treatise with a clear definition of the derivative. His mentor, French mathematician Joseph-Louis Lagrange (1736–1813), thought in terms of graphs of curves and considered the derivative a tangent to a curve. In order to determine a derivative, Lagrange would search for derivative formulas as necessary. Stephen Hawking writes, "Cauchy went far beyond Lagrange and defined the derivative of $f$ at $x$ as the limit of the difference quotient $\Delta y/\Delta x = [f(x + i) - f(x)]/i$" as $i$ approaches zero, which is our modern, nongeometric definition of the derivative.

Similarly, by clarifying the notion of the integral in calculus, Cauchy demonstrated the Fundamental Theorem of Calculus, which establishes a way in which we can compute the integral of $f(x)$ from $x = a$ to $x = b$ for any continuous function $f$. More particularly, the Fundamental Theorem of Calculus states that if $f$ is an integrable function in the interval $[a, b]$, and if $H(x)$ is the integral of $f(x)$ from $a$ to $x \leq b$, then the derivative of $H(x)$ is identical to $f(x)$. In other words, $H'(x) = f(x)$.

Waterhouse concludes, "Cauchy did not really establish new foundations; he swept away all the dust to reveal the whole edifice of calculus already standing on bedrock...."

**SEE ALSO** Zeno's Paradoxes (c. 445 B.C.), Discovery of Calculus (c. 1665), L'Hôpital's *Analysis of the Infinitely Small* (1696), Agnesi's *Instituzioni Analitiche* (1748), and Laplace's *Théorie Analytique des Probabilités* (1812).

*Augustin Louis Cauchy, lithograph by Gregoire et Deneux.*

B.on Augustin Cauchy

# Barycentric Calculus

## **August Ferdinand Möbius** (1790–1868)

German mathematician August Ferdinand Möbius, famous for his one-sided loop called the Möbius strip, also made a major contribution to mathematics with his barycentric calculus, a geometrical method for defining a point as the center of gravity of certain other points to which coefficients or weights are ascribed. We can think of Möbius's *barycentric coordinates* (or *barycentrics*) as coordinates with respect to a reference triangle. These coordinates are usually written as triples of numbers, which can be visualized as corresponding to masses placed at the vertices of the triangle. In this way, these masses determine a point, which is the geometric centroid of the three masses. The new algebraic tools, developed by Möbius in his 1827 book *Der Barycentrische Calcul* (*The Barycentric Calculus*), have since turned out to have wide application. This classic book also discusses related topics in analytical geometry such as projective transformations.

The word *barycentric* is derived from the Greek *barys* for "heavy" and refers to the center of mass. Möbius understood that several weights positioned along a straight stick can be replaced by a single weight at the stick's center of mass. From this simple principle, he constructed a mathematics system in which numerical coefficients are assigned to every point in space.

Today, barycentric coordinates are treated as a form of general coordinates that are used in many branches of mathematics and in computer graphics. Many of the advantages of barycentric coordinates occur in the field of **Projective Geometry**, which is concerned with *incidences*—that is, where elements such as lines, planes, and points either coincide or do not. Projective geometry is also concerned with the relationships between objects and the mappings that result from projecting the objects onto another surface, which can be visualized as shadows of solid objects.

**SEE ALSO** Descartes' *La Géométrie* (1637), Projective Geometry (1639), and The Möbius Strip (1858).

*Barycentric coordinates. Point* P *is the barycenter of* A, B, *and* C, *and we say that the barycentric coordinates of* P *are* [A, B, C]. *Triangle* ABC *would balance on a pin placed beneath the barycenter.*

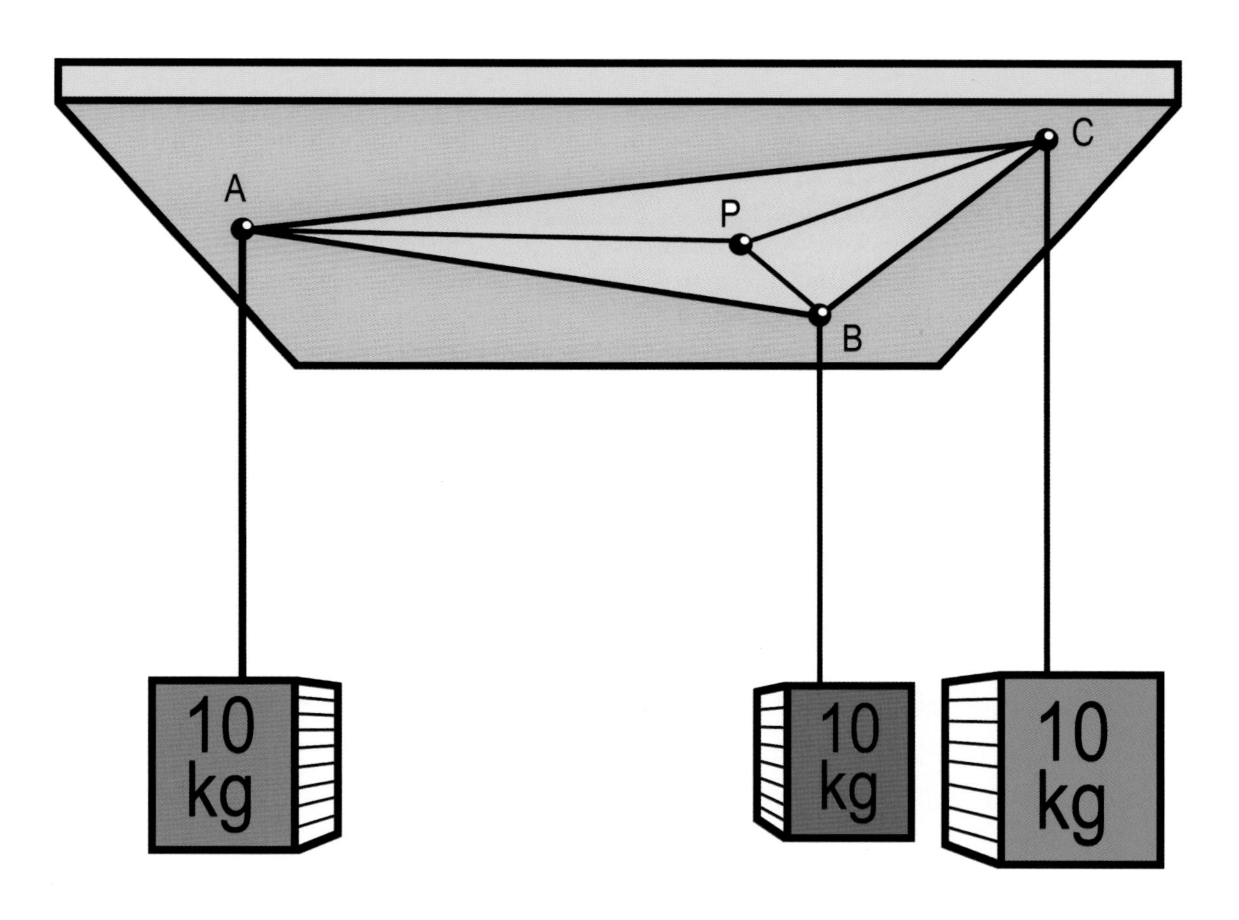


*BRIAN C. MANSFIELD*

# Non-Euclidean Geometry

**Nicolai Ivanovich Lobachevsky** (1792–1856), **János Bolyai** (1802–1860), **Georg Friedrich Bernhard Riemann** (1826–1866)

Since the time of Euclid (c. 325–270 B.C.), the so-called parallel postulate seemed to reasonably describe how our three-dimensional world works. According to this postulate, given a straight line and a point not on that line, in their plane only one straight line through the point exists that never intersects the original line.

Over time, the formulations of non-Euclidean geometry, in which this postulate does not hold, have had dramatic consequences. Einstein said about non-Euclidean geometry: "To this interpretation of geometry, I attach great importance, for should I have not been acquainted with it, I never would have been able to develop the theory of relativity." In fact, Einstein's General Theory of Relativity represents space-time as a non-Euclidean geometry in which space-time actually warps, or curves, near gravitating bodies such as the sun and planets. This can be visualized by imagining a bowling ball sinking into a rubber sheet. If you were to place a marble into the depression formed by the stretched rubber sheet, and give the marble a sideways push, it would orbit the bowling ball for a while, like a planet orbiting the sun.

In 1829, Russian mathematician Nicolai Lobachevsky published *On the Principles of Geometry*, in which he imagined a perfectly consistent geometry that results from assuming that the parallel postulate is false. Several years earlier, Hungarian mathematician János Bolyai had worked on a similar non-Euclidean geometry, but his publication was delayed until 1932. In 1854, German mathematician Bernhard Riemann generalized the findings of Bolyai and Lobachevsky by showing that various non-Euclidean geometries are possible, given the appropriative number of dimensions. Riemann once remarked, "The value of non-Euclidean geometry lies in its ability to liberate us from preconceived ideas in preparation for the time when exploration of physical laws might demand some geometry other than the Euclidean." His prediction was realized later with Einstein's General Theory of Relativity.

**SEE ALSO** Euclid's *Elements* (300 B.C.), Omar Khayyam's *Treatise* (1070), Descartes' *La Géométrie* (1637), Projective Geometry (1639), Riemann Hypothesis (1859), Beltrami's Pseudosphere (1868), and Weeks Manifold (1985).

*One form of non-Euclidean geometry is exemplified by Jos Leys's hyperbolic tiling. Artist M. C. Escher also experimented with non-Euclidean geometries in which the entire universe could be compressed and represented in a finite disk.*

1831

# Möbius Function

## **August Ferdinand Möbius** (1790–1868)

In 1831, August Möbius introduced his exotic Möbius function, today written as $\mu(n)$. To understand the function, imagine placing all the integers into just one of three large mailboxes. The first mailbox is painted with a big "0," the second with "+1," and a third with "−1." In mailbox 0, Möbius places multiples of square numbers (other than 1), including {4, 8, 9, 12, 16, 18,…}. A *square number* is a number such as 4, 9, or 16 that is the square of another integer. For example, $\mu(12) = 0$, because 12 is a multiple of the square number 4 and is thus placed in mailbox "0."

In the −1 mailbox, Möbius places any number that factors into an odd number of distinct prime numbers. For example, $5 \times 2 \times 3 = 30$, so 30 is in this list because it has these three prime factors. All prime numbers are also on this list because they only have one prime factor, themselves. Thus, $\mu(29) = -1$ and $\mu(30) = -1$. The probability that a number falls in the −1 mailbox turns out to be $3/\pi^2$ —the same probability as for falling in the +1 mailbox.

Let's further consider the +1 mailbox, in which Möbius places all the numbers, such as 6, that factor into an even number of distinct primes ($2 \times 3 = 6$). For completeness, Möbius put 1 into this bin. Numbers in this mailbox include {1, 6, 10, 14, 15, 21, 22,…}. The first 20 terms of the wonderful Möbius function are $\mu(n) = \{1, -1, -1, 0, -1, 1, -1, 0, 0, 1, -1, 0, -1, 1, 1, 0, -1, 0, -1, 0\}$.

Amazingly, scientists have found practical uses of the Möbius function in various physical interpretations of subatomic particle theory. The Möbius function is also fascinating because almost everything about its behavior is unsolved and because numerous elegant mathematical identities exist that involve $\mu(n)$.

**SEE ALSO** Cicada-Generated Prime Numbers (c. 1 Million B.C.), Sieve of Eratosthenes (c. 240 B.C.), and Andrica's Conjecture (1985).

*August Ferdinand Möbius, from the frontispiece in Möbius's* Werke.

Adolf Neumann gest.

1832

# Group Theory

**Évariste Galois** (1811–1832)

French mathematician Évariste Galois was responsible for Galois theory, an important branch of abstract algebra, and famous for his contributions to group theory, which concerns the mathematical study of symmetry. In particular, in 1832, he produced a method of determining when a general equation could be solved by radicals, thereby, in essence, giving a kick start to modern group theory.

Martin Gardner writes, "In 1832...he was killed by a pistol shot....He was not yet 21. Some early, fragmentary work had been done on groups, but it was Galois who laid the foundations of modern group theory and even named it, all in a long, sad letter that he wrote to a friend the night before his fatal duel." One key aspect of a group is that it is a set of elements with an operation that combines any two of its elements to form a third element within that set. For example, consider the set of integers and the operation of addition, which together form a group. Adding two integers always yields an integer. A geometrical object can be characterized by a group called a *symmetry group* that specifies the symmetry features of the object. This group contains a set of transformations that leave the object unchanged when applied. Today, important topics in group theory are often illustrated to students using the **Rubik's Cube**.

The circumstances that led to Galois's death have never been fully explained. Perhaps his death resulted from a quarrel over a woman or for political reasons. In any case, preparing for the end, he spent the night feverishly outlining his mathematical ideas and discoveries. The figure at right shows a page from his last night's writing on quintic equations (equations with the term $x^5$).

The next day, Galois was hit in the stomach. He lay helpless on the ground. There was no physician to help him, and the victor casually walked away, leaving Galois to writhe in agony. His mathematical reputation and legacy rests on fewer than 100 pages of a posthumously published work of genius.

**SEE ALSO** Wallpaper Groups (1891), Langlands Program (1967), Rubik's Cube (1974), Monster Group (1981), and The Quest for Lie Group $E_8$ (2007).

*The frantic mathematical scribbling Galois made during the night before his fatal duel. On this page, on the left below the center, are the words* Une femme, *with* femme *crossed out—a reference to the woman at the center of the duel.*

φx = (fx)² + (x − a)(Fx)²
(fx)² + (x − a)(Fx)² = 0
indivisible
Unité, indivisibilité de la république
Liberté, égalité, fraternité ou la mort
Logarithme
(fx + Fx √φy)ⁿ

1834

# Pigeonhole Principle

**Johann Peter Gustav Lejeune Dirichlet** (1805–1859)

The first statement of the Pigeonhole Principle was made by German mathematician Johann Dirichlet in 1834, although he referred to it as *Schubfachprinzip* ("Drawer Principle"). The phrase "Pigeonhole Principle" was first used in a serious mathematics journal by mathematician Raphael M. Robinson in 1940. Simply stated, if we have $m$ pigeon homes and $n$ pigeons, we can be sure that at least one home houses more than one pigeon if $n > m$.

This simple assertion has been used in applications that range from computer data compression to problems that involve infinite sets that cannot be put into one-to-one correspondence. The Pigeonhole Principle has also been generalized to probabilistic applications so that if $n$ pigeons are *randomly* placed in $m$ pigeonholes with uniform probability of $1/m$, then at least one pigeonhole will hold more than one pigeon with probability $1 - m!/[(m - n)!m^n]$. Let us consider some examples that may demonstrate unintuitive results.

Because of the Pigeonhole Principle, there must be at least two people in New York City with the same number of hairs on their heads. Let the hairs be represented as pigeonholes and the people as pigeons. New York City has more than 8 million people, and the human head has far less than a million hairs; thus, at least two people must exist with the same number of hairs on their heads.

A paper with the dimensions of a dollar bill is colored blue and red on one surface. Is it *always* possible to find two points of the same color exactly one inch apart on such a surface, no matter how intricately the surface is colored? To solve, draw an equilateral triangle with one-inch edges. Consider the colors as pigeonholes and the triangle vertices as pigeons. At least two of its vertices must be of the same color. This proves that two points of the same color must exist exactly one inch apart.

**SEE ALSO** Dice (c. 3000 B.C.), Laplace's *Théorie Analytique des Probabilités* (1812), and Ramsey Theory (1928).

*Given* m *pigeon homes and* n *pigeons, at least one home must house more than one pigeon if* n > m.

1843

# Quaternions

**Sir William Rowan Hamilton** (1805–1865)

Quaternions are four-dimensional numbers, conceived in 1843 by Irish mathematician William Hamilton. Quaternions have since been used to describe the dynamics of motion in three dimensions and applied in such fields as computer graphics for virtual realities, video game programming, signal processing, robotics, bioinformatics, and studies of the geometry of space-time. The flight software of the Space Shuttle makes use of quaternions for guidance computations, navigation, and flight control, for reasons of speed, compactness, and reliability.

Despite the potential usefulness of quaternions, some mathematicians were skeptical at first. Scottish physicist William Thomson (1824–1907) wrote, "Quaternions came from Hamilton after his really good work had been done, and though beautifully ingenious, have been an unmixed evil to those who have touched them in any way." On the other hand, engineer and mathematician Oliver Heaviside wrote in 1892, "The invention of quaternions must be regarded as a most remarkable feat of human ingenuity. Vector analysis, without quaternions, could have been found by any mathematician… but to find out quaternions required genius." Interestingly, Theodore Kaczynski (the "Unabomber") wrote intricate mathematical treatises on quaternions before he went on his killing spree.

Quaternions can be represented in four dimensions by $Q = a_0 + a_1i + a_2j + a_3k$ where $i$, $j$, and $k$ are (like the imaginary number $i$) unit vectors in three orthogonal (perpendicular) directions, and they are perpendicular to the real number axis. To add or multiply two quaternions, we treat them as polynomials in $i$, $j$, and $k$, but use the following rules to deal with products: $i^2 = j^2 = k^2 = -1$; $ij = -ji = k$; $jk = -kj = i$; and $ki = -ik = j$. Hamilton tells us that he cut these formulas into a stone of Brougham Bridge in Dublin while walking with his wife, after the ideas came to him in a flash.

**SEE ALSO** Imaginary Numbers (1572).

*Physicist Leo Fink rendered this 3-D section of a 4-D quaternion fractal. The intricate surface represents the complicated behavior of* $Q_{n+1} = Q_n^2 + c$, *where* Q *and* c *are quaternion numbers, and* $c = -0.35 + 0.7i + 0.15j + 0.3k$.

1844

# Transcendental Numbers

**Joseph Liouville** (1809–1882), **Charles Hermite** (1822–1901), **Ferdinand von Lindemann** (1852–1939)

In 1844, French mathematician Joseph Liouville considered the following interesting number: 0.110001000000000000000001000…, known today as the Liouville constant. Can you guess its significance or what rule he used to create it?

Liouville showed that his unusual number was transcendental, thus making this number among the first to be proven transcendental. Notice that the constant has 1 in each decimal place corresponding to a factorial, and zeros elsewhere. This means that the 1s occur only in the 1st, 2nd, 6th, 24th, 120th, 720th, etc. places.

Transcendental numbers are so exotic that they were only "discovered" relatively recently in history, and you may only be familiar with one of them, $\pi$, and perhaps **Euler's Number, *e***. These numbers cannot be expressed as the root of any algebraic equation with rational coefficients. This means, for example, that $\pi$ could not exactly satisfy equations like $2x^4 - 3x^2 + 7 = 0$.

Proving that a number is transcendental is difficult. French mathematician Charles Hermite proved *e* was transcendental in 1873, and German mathematician Ferdinand von Lindemann proved $\pi$ was transcendental in 1882. In 1874, German mathematician Georg Cantor surprised many mathematicians by demonstrating that "almost all" real numbers are transcendental. Thus, if you could somehow put all the numbers in a big jar, shake the jar, and pull one out, it would be virtually certain to be transcendental. Yet despite the fact that transcendental numbers are "everywhere," only a few are known and named. There are lots of stars in the sky, but how many can you name?

Aside from his mathematical pursuits, Liouville was interested in politics and was elected to the French Constituting Assembly in 1848. After a later election defeat, Liouville became depressed. His mathematical ramblings became interspersed with poetical quotes. Nonetheless, during the course of his life, Liouville wrote more than 400 serious mathematical papers.

**SEE ALSO** Quadrature of the Lune (c. 440 B.C.), $\pi$ (c. 250 B.C.), Euler's Number, *e* (1727), Stirling's Formula (1730), Cantor's Transfinite Numbers (1874), Normal Number (1909), and Champernowne's Number (1933).

*French mathematician Charles Hermite, c. 1887. Hermite proved in 1873 that Euler's number e was transcendental.*

# Catalan Conjecture

**Eugène Charles Catalan** (1814–1894), **Preda Mihăilescu** (b. 1955)

Deceptively simple-looking challenges involving whole numbers can confound even the most brilliant mathematicians. As in the case of Fermat's Last Theorem, centuries may pass before simple conjectures concerning integers are proved or refuted. Some problems may never be solved, even with the combined efforts of humans and computers.

To set the stage for understanding the Catalan conjecture, consider the squares of whole numbers (integers) greater than 1, that is, 4, 9, 16, 25,...and also consider the sequence of cubes, 8, 27, 64, 125....If we merge the two lists and place them in order, we obtain 4, 8, 9, 16, 25, 27, 36,....Notice that 8 (the cube of 2) and 9 (the square of 3) are consecutive integers. In 1844, Belgian mathematician Eugène Catalan conjectured that 8 and 9 are the *only* powers of integers that are consecutive! If other such pairs had existed, they might have been found by searching for integer values for which $x^p - y^q = 1$ is true and for values of $x$, $y$, $p$, and $q$ greater than 1. Catalan believed that only one solution exists: $3^2 - 2^3 = 1$.

This history of the Catalan conjecture has a colorful cast of characters. Hundreds of years before Catalan, Frenchman Levi ben Gerson (1288–1344)—better known as Gersonides or the Ralbag—had already demonstrated a more restricted version of the conjecture, namely that the only powers of 2 and 3 that differ by 1 are $3^2$ and $2^3$. The Ralbag was a famous rabbi, philosopher, mathematician, and Talmudist.

Let's skip forward to 1976 when Robert Tijdeman of the University of Leiden in the Netherlands showed that if examples of other consecutive powers existed, then they would have to be finite in number. Finally, in 2002, Preda Mihăilescu of the University of Paderborn in Germany proved Catalan's conjecture.

**SEE ALSO** Fermat's Last Theorem (1637) and Euler's Polygon Division Problem (1751).

*Belgian mathematician Eugène Charles Catalan. In 1844, Catalan conjectured that 8 and 9 are the* only *powers of integers that are consecutive.*

# The Matrices of Sylvester

**James Joseph Sylvester** (1814–1897), **Arthur Cayley** (1821–1895)

In 1850, in his paper "On a New Class of Theorems," British mathematician James Sylvester was the first to use the word *matrix* when referring to a rectangular arrangement, or array, of elements that can be added and multiplied. Matrices are often used to describe a system of linear equations or simply to represent information that depends on two or more parameters.

Credit for understating and identifying the complete significance of the algebraic properties of matrices is given to the English mathematician Arthur Cayley for his later work on matrices in 1855. Because Cayley and Sylvester enjoyed many years of close collaboration, they are often considered the joint founders of matrix theory.

Although matrix theory flourished in the mid-1800s, simple concepts of matrices date back to before the birth of Christ, when the Chinese knew of **Magic Squares** and also began to apply matrix methods to solve simultaneous equations. In the 1600s, Japanese mathematician Seki Kowa (in 1683) and German mathematician Gottfried Leibniz (in 1693) also explored the early use of matrices.

Both Sylvester and Cayley studied at Cambridge, but Sylvester was ineligible for a degree because he was a Jew, although he was ranked second in Cambridge's mathematical examinations. Before Cambridge, Sylvester had attended the Royal Institution in Liverpool, where students tormented him for his religion, thus causing him to escape to Dublin.

Cayley worked as a lawyer for more than a decade, while publishing about 250 mathematics papers. During his time at Cambridge, he published another 650 papers. Cayley was first to introduce matrix multiplication.

Today, matrices are used in numerous areas, including data encryption and decryption, object manipulation in computer graphics (including video games and medical imaging), solving systems of simultaneous linear equations, quantum mechanical studies of atomic structure, equilibrium of rigid bodies in physics, graph theory, game theory, economics models, and electrical networks.

**SEE ALSO** Magic Squares (c. 2200 B.C.), Thirty-Six Officers Problem (1779), and Sylvester's Line Problem (1893).

*Portrait of James Joseph Sylvester, the frontispiece to Volume 4 of* The Collected Mathematical Papers of James Joseph Sylvester, *edited by H. F. Baker (Cambridge University Press, 1912).*

*Yours faithfully*
*J. J. Sylvester*

# Four-Color Theorem

**Francis Guthrie** (1831–1899), **Kenneth Appel** (b. 1932), **Wolfgang Haken** (b. 1928)

Mapmakers have believed for centuries that just four colors were sufficient for coloring any map drawn on a plane, so that no two distinct regions sharing a common edge are the same color, although two regions can share a common vertex and have the same color. Today, we know for certain that while some planar maps require fewer colors, no map requires more than four. Four colors are sufficient for maps drawn on spheres and cylinders. Seven colors are sufficient to paint any map on a torus (the surface of a doughnut shape).

In 1852, mathematician and botanist Francis Guthrie was the first to conjecture that four colors must be sufficient when he attempted to color a map of counties of England. Since the time of Guthrie, mathematicians had tried in vain to *prove* the consequences of this seemingly simple four-color observation, and it remained one of the most famous unsolved problems in topology.

Finally, in 1976, mathematicians Kenneth Appel and Wolfgang Haken succeeded in proving the four-color theorem with the help of a computer testing thousands of cases, making it the first problem in pure mathematics to make use of a computer to produce an essential component for the proof. Today, computers are playing increasing roles in mathematics, helping mathematicians verify proofs so complex that they sometimes defy human comprehension. The four-color theorem is one example. Another is the classification of finite simple groups, embodied in a 10,000-page multi-author project. Alas, the traditional people-centered methods for ensuring that a proof is correct breaks down when a paper reaches thousands of pages.

Surprisingly, the four-color theorem has been of little practical importance for mapmakers and cartographers. For example, a study of atlases through time reveals no pressing desire to try to minimize the number of colors used, and books on cartography and mapmaking history often use more colors than needed.

**SEE ALSO** Kepler Conjecture (1611), Riemann Hypothesis (1859), Klein Bottle (1882), and The Quest for Lie Group $E_8$ (2007).

*This map of the state of Ohio, scanned from an 1881 original, makes use of four colors. Note that no two distinct regions sharing a common edge are the same color.*

A B C D E F G H I J K L M N O P Q R S T U V W X Y Z
MICHIGAN
HILLSDALE
LENAWEE
MONROE
LAKE ERIE
WILLIAMS
FULTON
LUCAS
OTTAWA
DEFIANCE
HENRY
WOOD
SANDUSKY
ERIE
LORAIN
CUYAHOGA
GEAUGA
LAKE
ASHTABULA
TRUMBULL
PAULDING
PUTNAM
HANCOCK
SENECA
HURON
MEDINA
SUMMIT
PORTAGE
MAHONING
VAN WERT
ALLEN
WYANDOT
CRAWFORD
RICHLAND
ASHLAND
WAYNE
STARK
COLUMBIANA
MERCER
AUGLAIZE
HARDIN
MARION
MORROW
KNOX
HOLMES
TUSCARAWAS
CARROLL
JEFFERSON
SHELBY
LOGAN
UNION
DELAWARE
COSHOCTON
HARRISON
DARKE
CHAMPAIGN
MIAMI
LICKING
MUSKINGUM
GUERNSEY
BELMONT
CLARKE
FRANKLIN
MADISON
PERRY
NOBLE
MONROE
PREBLE
MONTGOMERY
GREENE
FAIRFIELD
PICKAWAY
FAYETTE
MORGAN
BUTLER
WARREN
CLINTON
ROSS
HOCKING
ATHENS
WASHINGTON
HAMILTON
CLERMONT
HIGHLAND
VINTON
MEIGS
BROWN
ADAMS
PIKE
JACKSON
SCIOTO
GALLIA
LAWRENCE
KENTUCKY
WEST VIRGINIA
PENNSYLVANIA
WOOD
JACKSON
MASON
PUTNAM
CABELL
WETZEL
TYLER
PLEASANTS
BOONE
KENTON
CAMPBELL
PENDLETON
BRACKEN
MASON
LEWIS
GREENUP
SANDUSKY
CLEVELAND
TOLEDO
COLUMBUS
CINCINNATI
DAYTON
SPRINGFIELD
ZANESVILLE
NEWARK
MANSFIELD
LIMA
PORTSMOUTH
CHILLICOTHE
MARIETTA
PARKERSBURG
WHEELING
POMEROY
MAP OF HAMILTON CO., OHIO.
SCALE OF MILES.

1854

# Boolean Algebra

**George Boole** (1815–1864)

English mathematician George Boole's most important work was his 1854 *An Investigation into the Laws of Thought, on Which Are Founded the Mathematical Theories of Logic and Probabilities*. Boole was interested in reducing logic to a simple algebra involving just two quantities, 0 and 1, and three basic operations: *and*, *or*, and *not*. In modern times, Boolean algebra has had vast applications in telephone switching and the design of modern computers. Boole looked upon this work as "the most valuable...contribution that I have made or am likely to make to Science and the thing by which I would desire if at all to be remembered hereafter...."

Alas, Boole died at the age of 49 after he developed a bad fever. Unfortunately, his wife believed that a remedy should resemble the cause, and she dumped buckets of water over him while he was in his bed, because his illness had been precipitated by being out in the cold rain.

Mathematician Augustus De Morgan (1806–1871) praised his work, saying "Boole's system of logic is but one of many proofs of genius and patience combined....That the symbolic processes of algebra, invented as tools of numerical calculation, should be competent to express every act of thought, and to furnish the grammar and dictionary of an all-containing system of logic, would not have been believed until it was proved...."

Approximately seventy years after Boole's death, American mathematician Claude Shannon (1916–2001) was introduced to Boolean algebra while still a student, and he showed how Boolean algebra could be used to optimize the design of systems of telephone routing switches. He also demonstrated that circuits with relays could solve Boolean algebra problems. Thus, Boole, with Shannon's help, provided one of the foundations for our Digital Age.

**SEE ALSO** Aristotle's *Organon* (c. 350 B.C.), Gros's *Théorie du Baguenodier* (1872), Venn Diagrams (1880), Boole's *Philosophy and Fun of Algebra* (1909), *Principia Mathematica* (1910–1913), Gödel's Theorem (1931), Gray Code (1947), Information Theory (1948), and Fuzzy Logic (1965).

*Ukrainian artist and photographer Mikhail Tolstoy illustrates his creative conception of a binary stream composed of ones and zeros. The artwork reminds him of the binary information flowing through digital networks such as the Internet.*

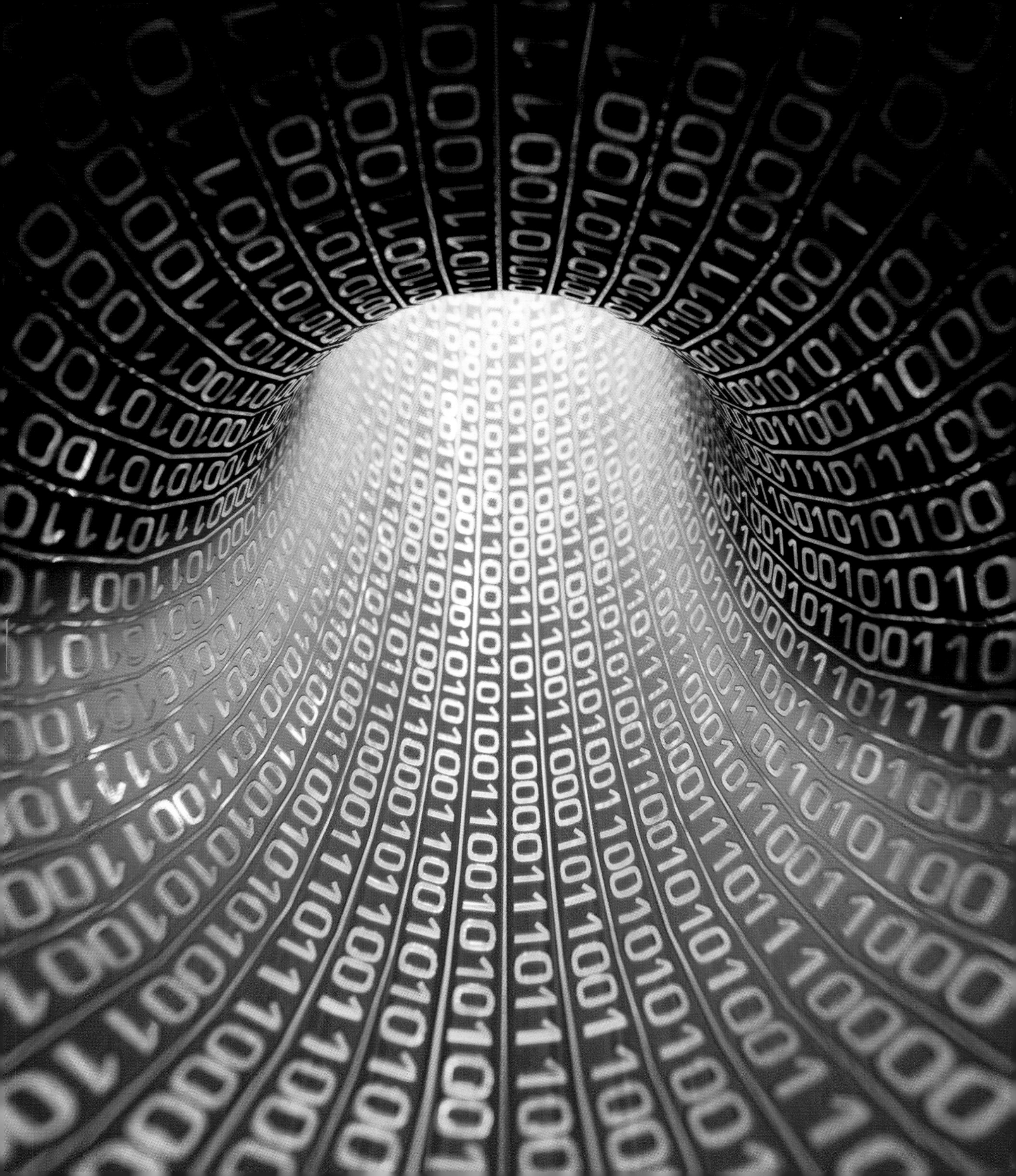

1857

# Icosian Game

**Sir William Rowan Hamilton** (1805–1865)

In 1857, Irish mathematician, physicist, and astronomer William Hamilton described the Icosian Game, the objective of which is to find a path along the edges of a dodecahedron (a polyhedron with 12 faces) so that every vertex (corner) is visited only once. Today, in the field of graph theory, mathematicians refer to a Hamiltonian path as one in which a path visits each graph vertex exactly once. A Hamiltonian cycle (or Hamiltonian circuit)—required for the Icosian Game—implies that the path returns to the starting point. British mathematician Thomas Kirkman (1806–1895) posed the Icosian Game problem more generally: Given a graph of a polyhedron, does a cycle exist that passes through every vertex?

The term *Icosian* comes from Hamilton's invention of a kind of algebra called *Icosian calculus*, based on the symmetry properties of the icosahedron. He solved his puzzle using this algebra and its associated *icosians* (special kinds of vectors). All Platonic solids are Hamiltonian. In 1974, mathematician Frank Rubin described an efficient search procedure that can find some or all Hamilton paths and circuits in graphs.

A London toy manufacturer bought the rights to the Icosian Game and created a puzzle that had nails at each vertex of the dodecahedron. Each nail stood for a major city. The player traced out his path by wrapping a string around each nail as he traveled. The toy was also sold in other forms—for example, as a flat pegboard with holes at the nodes of the dodecahedron. (A flat model of a dodecahedron can be created by puncturing one of its faces and stretching the object flat so that it lies on a plane.) Alas, the game did not sell well, partly because it was fairly easy to solve. Perhaps Hamilton's focus on deep theories made him overlook the fact that trial and error would soon lead to a solution!

**SEE ALSO** Platonic Solids (c. 350 B.C.), Archimedean Semi-Regular Polyhedra (c. 240 B.C.), Königsberg Bridges (1736), Euler's Formula for Polyhedra (1751), Pick's Theorem (1899), Geodesic Dome (1922), Császár Polyhedron (1949), Szilassi Polyhedron (1977), Spidrons (1979), and Solving of the Holyhedron (1999).

*Teja Krašek's creative rendition of the Icosian Game. The objective is to find a path along the edges of this dodecahedron so that every corner is visited only once. In 1859, a London toy manufacturer bought the rights to the game.*

1857

# Harmonograph

**Jules Antoine Lissajous** (1822–1880), **Hugh Blackburn** (1823–1909)

The harmonograph is a Victorian art device that usually employs just two pendulums to trace out paths that can be studied from both an artistic and mathematical perspective. In one version, a pendulum moves a pen. The other pendulum moves a table with a sheet of paper. The combined effect of the two pendulums produces a complicated motion that steadily decays to a single point due to friction. Each path of the pen, upon each revolution, is a short distance away from the path on the previous revolution, giving the patterns a wavy, spider-web-like appearance. By varying the frequency and phases of the pendulums relative to one another, a wide range of patterns is generated.

In the simplest version, the patterns may be characterized as Lissajous curves that describe complex harmonic motion and can be represented (assuming no friction) by the curves produced by $x(t) = \mathrm{A}\sin(at + d)$, $y(t) = B\sin(bt)$, where $t$ is time, and A and $B$ are amplitudes. The ratio of $a$ to $b$ controls the relative frequencies, and $d$ is a phase difference. From relatively few parameters, a huge panoply of ornamental curves is produced.

The first harmonographs were constructed in 1857, when French mathematician and physicist Jules Antoine Lissajous demonstrated patterns produced by two tuning forks, attached to small mirrors that vibrated at different frequencies. A beam of light reflected off the mirrors to produce the intricate curves that delighted a general public.

British mathematician and physicist Hugh Blackburn is credited with making the first more traditional pendulum versions of the harmonograph, and many variations of Blackburn's harmonograph have been created up to the present day. More complex harmonographs may employ additional pendulums that hang off one another. In my novel *The Heaven Virus*, we encounter a zany alien harmonograph with "a pen that oscillates on a platform that oscillated on another platform, which oscillated on another platform, and so on for ten different platforms."

**SEE ALSO** Differential Analyzer (1927), Chaos and the Butterfly Effect (1963), Ikeda Attractor (1979), and Butterfly Curve (1989).

*A harmonograph rendition produced by Ivan Moscovich. In the 1960s, Moscovich created mechanically efficient, large harmonographs by linking pendulums to a vertical surface. Moscovich, a famous puzzle designer, was in the Auschwitz concentration camp, and liberated by British troops in 1945.*

# The Möbius Strip

**August Ferdinand Möbius** (1790–1868)

German mathematician August Ferdinand Möbius was a shy, unsociable, absentminded professor whose most famous discovery, the Möbius strip, was made when he was almost seventy years old. To create the strip yourself, simply join the two ends of a ribbon after giving one end a 180-degree twist with respect to the other end. The result is a one-sided surface—a bug can crawl from any point on such a surface to any other point without ever crossing an edge. Try coloring a Möbius strip with a crayon. It's impossible to color one side red and the other green because the strip has only one side.

Years after Möbius's death, the popularity and applications of the strip grew, and it has become an integral part of mathematics, magic, science, art, engineering, literature, and music. The Möbius strip is the ubiquitous symbol for recycling where it represents the process of transforming waste materials into useful resources. Today, the Möbius strip is everywhere, from molecules and metal sculptures to postage stamps, literature, technology patents, architectural structures, and models of our entire universe.

August Möbius had simultaneously discovered his famous strip with a contemporary scholar, the German mathematician Johann Benedict Listing (1808–1882). However, Möbius seems to have taken the concept a little further than Listing, as Möbius more closely explored some of the remarkable properties of this strip.

The Möbius strip is the first one-sided surface discovered and investigated by humans. It seems far-fetched that no one had described the properties of one-sided surfaces until the mid-1800s, but history has recorded no such observations. Given that the Möbius strip is often the first and only exposure of a wide audience to the study of topology—the science of geometrical shapes and their relationships to one another—this elegant discovery deserves a place in this book.

**SEE ALSO** Königsberg Bridges (1736), Euler's Formula for Polyhedra (1751), Knight's Tours (1759), Barycentric Calculus (1827), Reuleaux Triangle (1875), Klein Bottle (1882), and Boy's Surface (1901).

*Multiple Möbius strips, an artwork created by Teja Krašek and Cliff Pickover. The Möbius strip is the first one-sided surface discovered and investigated by humans.*

1858

# Holditch's Theorem

**Hamnet Holditch** (1800–1867)

Draw a smooth, closed, convex curve $C_1$. Place a chord of constant length inside curve $C_1$, and let the chord slide around inside the curve so that the two ends of the chord touch $C_1$ at all times. (You can visualize this as moving a stick around on the surface of a puddle that has the shape of curve $C_1$.) Label a point on the stick so that it divides the stick into two parts of length $p$ and $q$. As you move the stick, the point traces out a new closed curve $C_2$ within the original curve. Assuming that $C_1$ is shaped in such a way that the stick can actually pass around $C_1$ once, Holditch's theorem states that the area between the curves $C_1$ and $C_2$ will be $\pi pq$. Interestingly, this area is totally independent of the shape of $C_1$.

Mathematicians have marveled at Holditch's theorem for more than a century. For example, in 1988, British mathematician Mark Cooker wrote, "Two things immediately struck me as astonishing. First, the formula for the area is independent of the size of the given curve $C_1$. Second, [the equation for the area] is the area of an ellipse of semi-axes $p$ and $q$, but there are no ellipses in the theorem!"

The theorem was published by Rev. Hamnet Holditch in 1858. Holditch was president of Caius College in Cambridge during the middle part of the 1800s. The Holditch curve $C_2$ for a circle $C_1$ of radius $R$ is another circle, which has radius $r = \sqrt{R^2 - pq}$.

**SEE ALSO** $\pi$ (c. 250 B.C.) and Jordan Curve Theorem (1905).

*As the stick slides around the outer curve, a point on the stick traces out the inner curve. Holditch's theorem states that the area between the curves will be* $\pi$pq *and is independent of the shape of the outer curve. (Figure by Brian Mansfield.)*

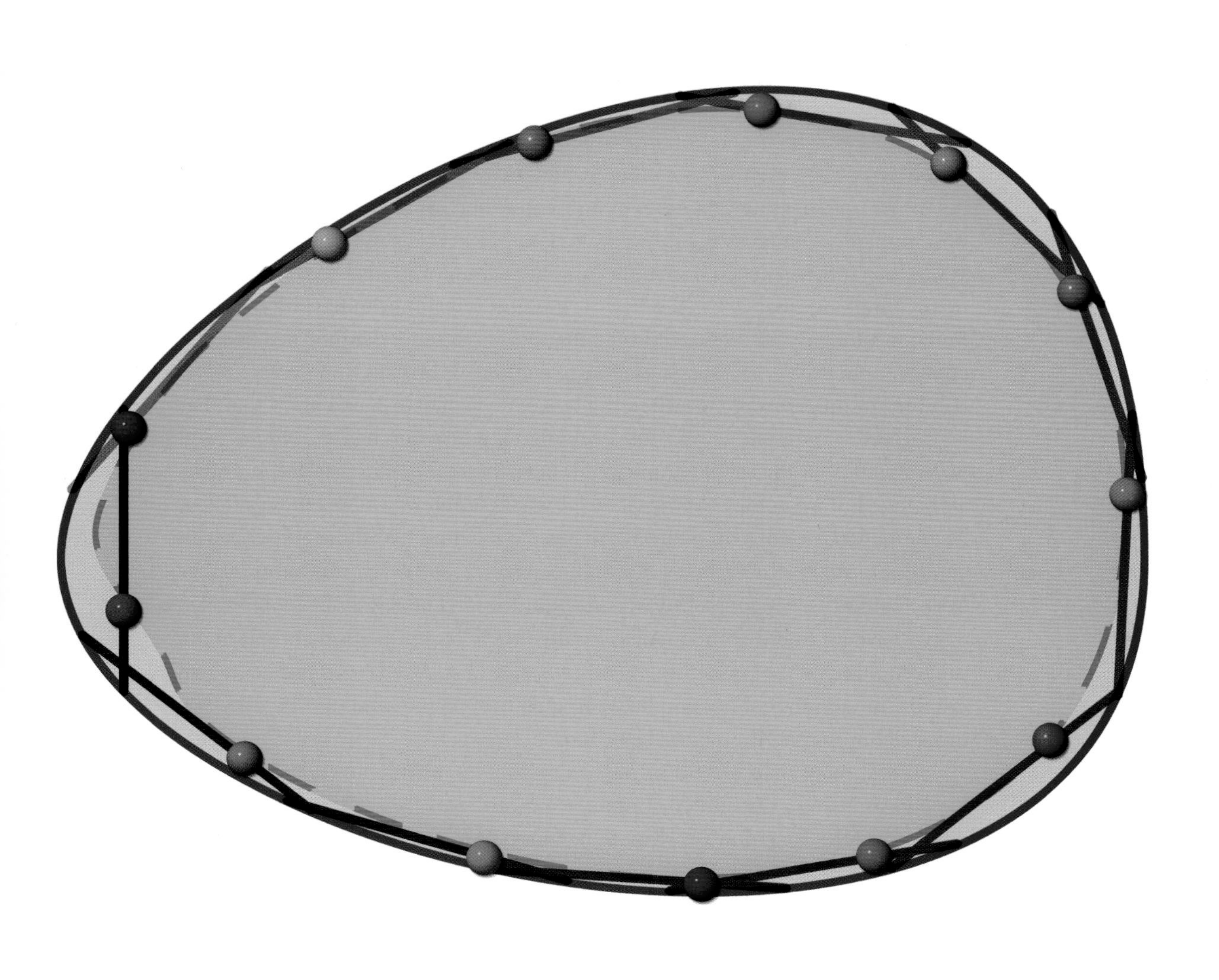

1859

# Riemann Hypothesis

**Georg Freidrich Bernhard Riemann** (1826–1866)

Many mathematical surveys indicate that the "proof of the Riemann hypothesis" is the most important open question in mathematics. The proof involves the *zeta function*, which can be represented by a complicated-looking curve that is useful in number theory for investigating properties of prime numbers. Written as $\zeta(x)$, the function was originally defined as the infinite sum $\zeta(x) = 1 + (1/2)^x + (1/3)^x + (1/4)^x + \ldots$etc. When $x = 1$, this series has no finite sum. For values of $x$ larger than 1, the series adds up to a finite number. If $x$ is less than 1, the sum is again infinite. The complete zeta function, studied and discussed in the literature, is a more complicated function that is equivalent to this series for values of $x$ greater than 1, but has finite values for any real or complex number, except for when the real part is equal to 1. We know that the function equals zero when $x$ is $-2, -4, -6, \ldots$ and that the function has an infinite number of zero values for the set of complex numbers, the real part of which is between zero and one—but we do not know exactly for what complex numbers these zeros occur. Mathematician Georg Bernhard Riemann conjectured that these zeros occur for those complex numbers the real part of which equals 1/2. Although vast numerical evidence exists that favors this conjecture, it is still unproven. The proof of Riemann's hypothesis would have profound consequences for the theory of prime numbers and in our understanding of the properties of complex numbers. Amazingly, physicists may have found a mysterious connection between quantum physics and number theory through investigations of the Riemann Hypothesis.

Today, more than 11,000 volunteers around the world are working on the Riemann hypothesis, using a distributed computer software package at *Zetagrid.net* to search for the zeros of the Riemann zeta function. More than 1 billion zeros for the zeta function are calculated every day.

**SEE ALSO** Cicada-Generated Prime Numbers (c. 1 Million B.C.), Sieve of Eratosthenes (c. 240 B.C.), Harmonic Series Diverges (c. 1350), Imaginary Numbers (1572), Four-Color Theorem (1852), and Hilbert's 23 Problems (1900).

*Tibor Majlath's rendition of the Riemann zeta function $\zeta(s)$ in the complex plane. The four small bulls-eye patterns at top and bottom correspond to zeros at $Re(s) = ½$. The plot extends from $-32$ to $+32$ in the real and imaginary directions.*

# Beltrami's Pseudosphere

**Eugenio Beltrami** (1835–1899)

The pseudosphere is a geometrical object that resembles two musical horns glued together at their rims. The "mouthpieces" of the two horns are located at the ends of two infinitely long tails, as if to be blown only by the omnipotent gods. The peculiar shape was first discussed in depth in the 1868 paper "Essay on an Interpretation of Non-Euclidean Geometry" by Italian mathematician Eugenio Beltrami, famous for his work in geometry and physics. To produce the surface, a curve called a *tractrix* is rotated about its asymptote.

Whereas an ordinary sphere has a property called *positive curvature* everywhere on its surface, a pseudosphere has a constant negative curvature, which means that it can be thought of as maintaining a constant concavity over its entire surface (except at its central cusp). Thus, a sphere is a closed surface with a finite area, while a pseudosphere is an open surface with infinite area. British science writer David Darling writes, "In fact, although both the two-dimensional plane and a pseudosphere are infinite, the pseudosphere manages to have more room! One way to think of this is that a pseudosphere is more intensely infinite than the plane." The negative curvature of a pseudosphere requires that the angles of a triangle drawn on its surface add up to less than 180°. The geometry of the pseudosphere is called *hyperbolic*, and some astronomers in the past have suggested that our entire universe might be described by hyperbolic geometry with properties of a pseudosphere. The pseudosphere is of historical importance because it was one of the first models for a **Non-Euclidean** space.

Beltrami's interests ranged far beyond mathematics. His four-volume work, *Opere Matematiche*, discusses optics, thermodynamics, elasticity, magnetism, and electricity. Beltrami was a member of the Accademia dei Lincei, serving as president of this scientific academy in 1898. He was elected to the Italian Senate a year before his death.

**SEE ALSO** Torricelli's Trumpet (1641), Minimal Surface (1774), and Non-Euclidean Geometry (1829).

*A variant of the classic Beltrami pseudosphere, this depiction of a breather pseudosphere, rendered by Paul Nylander, also has a constant negative curvature.*

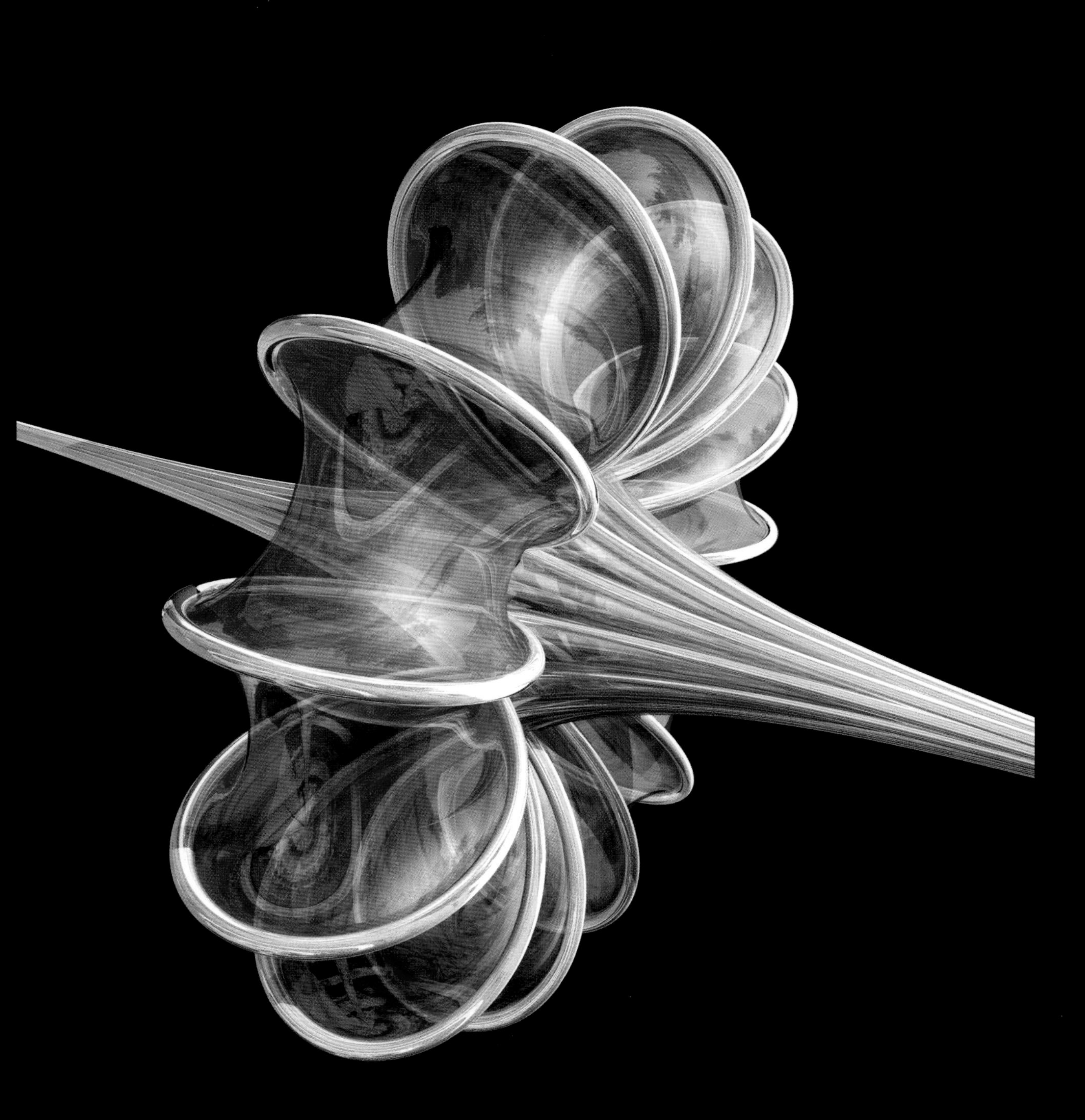

1872

# Weierstrass Function

**Karl Theodor Wilhelm Weierstrass** (1815–1897)

In the early 1800s, mathematicians often thought of a continuous function $f(x)$ as having a derivative (a unique tangent line) that could be specified along most points in the curve. In 1872, German mathematician Karl Weierstrass stunned mathematical colleagues at the Berlin Academy by proving this thinking to be false. His function, which was continuous everywhere but differentiable (possessing a derivative) nowhere, was defined by $f(x) = \Sigma a^k \cos(b^k \pi x)$, where the sum is from $k = 0$ to $\infty$. Here, $a$ is a real number with $0 < a < 1$, $b$ is an odd positive integer, and $ab > (1 + 3\pi/2)$. The summation symbol $\Sigma$ indicates that the function is constructed from an infinite number of trigonometric functions to produce a densely nested oscillating structure.

Of course, mathematicians were well aware that functions might not be differentiable at a few troublesome points, such as the bottom of the inverted wedge shape specified by $f(x) = |x|$, which has no derivative at $x = 0$. However, after Weierstrass's demonstration of a nowhere-differentiable curve, mathematicians were in a quandary. Mathematician Charles Hermite wrote to Thomas Stieltjes in 1893, "I turn away with fear and horror from the lamentable plague of continuous functions that do not have derivatives. . . ."

In 1875, Paul du Bois-Reymond published the Weierstrass function, making it the first published function of its kind. Two years earlier, he had given a draft of the paper to Weierstrass to read. (The draft contained a different function $f(x) = \Sigma \sin(a^n x)/b^n$, with $(a/b) > 1$ for $k = 0$ to $\infty$, which was changed before the paper was published.)

Like other **fractal** shapes, the Weierstrass function displays increasing detail with progressive magnification. Other mathematicians, such as Czech mathematician Bernard Bolzano and German mathematician Bernhard Riemann, had worked on similar (unpublished) constructions in 1830 and 1861, respectively. Another example of an everywhere-continuous but nowhere-differentiable curve is the fractal Koch curve.

**SEE ALSO** Peano Curve (1890), Koch Snowflake (1904), Hausdorff Dimension (1918), Coastline Paradox (c. 1950), and Fractals (1975).

*This Weierstrass surface, assembled from numerous related Weierstrass curves, was approximated and rendered by Paul Nylander using* $f_a(x) = \Sigma[\sin(\pi k^a x)/\pi k^a]$ *(*$0 < x < 1$*;* $2 < a < 3$*; and the sum was from* $k = 1$ *to 15).*

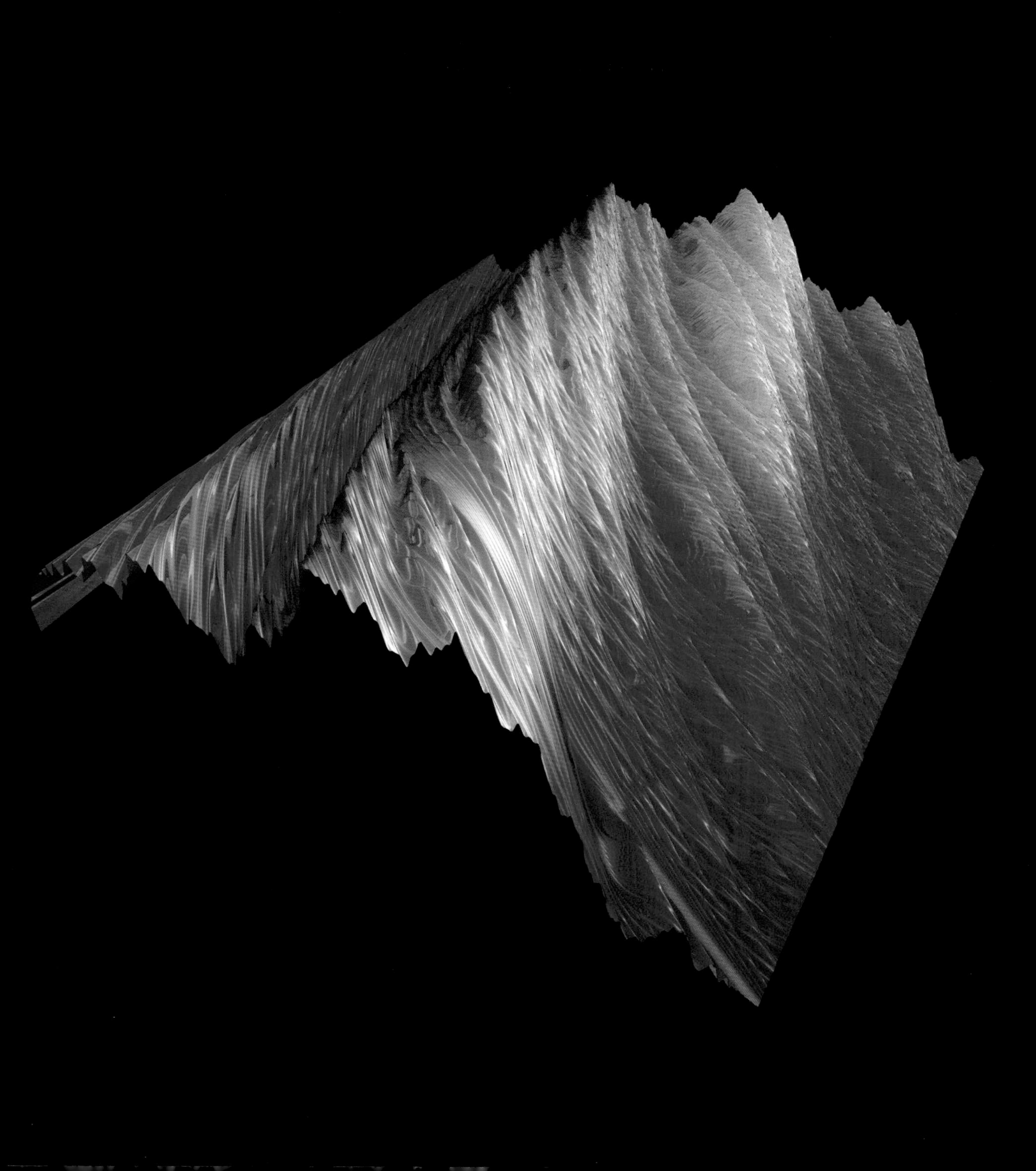

1872

# Gros's *Théorie du Baguenodier*

**Louis Gros** (c. 1837–c. 1907)

Baguenaudier is one of the oldest-known mechanical puzzles. In 1901, English mathematician Henry E. Dudeney remarked, "Certainly no home should be without this fascinating, historic, and instructive puzzle."

The objective of Baguenaudier is to remove all of the rings from a stiff horizontal loop. On the first move, it is possible to remove one or two rings from one end of the wire. The entire procedure is complicated because rings must be put back onto the wire loop in order to remove other rings, and the procedure is repeated many times. It turns out that the minimum number of moves needed is $(2^{n+1} - 2)/3$ if the number of rings $n$ is even and $(2^{n+1} - 1)/3$ if $n$ is odd. Martin Gardner writes, "Twenty-five rings require 22,369,621 steps. Assuming that a skilled operator can do 50 steps a minute, he could solve the puzzle... in a little more than two years."

According to legend, the puzzle was invented by the Chinese general Chu-ko Liang (A.D. 181–234) to keep his wife busy when he was away at the wars. In 1872, Louis Gros, a French magistrate, demonstrated an explicit connection between these rings and binary numbers in his booklet *Théorie du Baguenodier* (a spelling he preferred). Each ring can be represented by a binary digit: 1 for *on*, and 0 for *off*. Specifically, Gros showed that if the rings were in a set of known states, it was possible to compute a binary number that indicated exactly how many more steps were necessary and sufficient to solve the puzzle. Gros's work involved one of the first examples of what is now called the **Gray Code**, in which two successive binary numbers differ in only one digit. In fact, computer scientist Donald Knuth wrote that Gros was the "true inventor of the Gray binary code," which is today widely used to facilitate error correction in digital communications.

**SEE ALSO** Boolean Algebra (1854), Fifteen Puzzle (1874), Tower of Hanoi (1883), Gray Code (1947), and Instant Insanity (1966).

*The ancient Baguenaudier puzzle had led to various U.S. patents from the 1970s that describe similar puzzles. For example, one version can be easily disassembled even if not solved. Another allows the number of rings to be varied to change the difficulty level. (Figures are from U.S. patents 4,000,901 and 3,706,458.)*

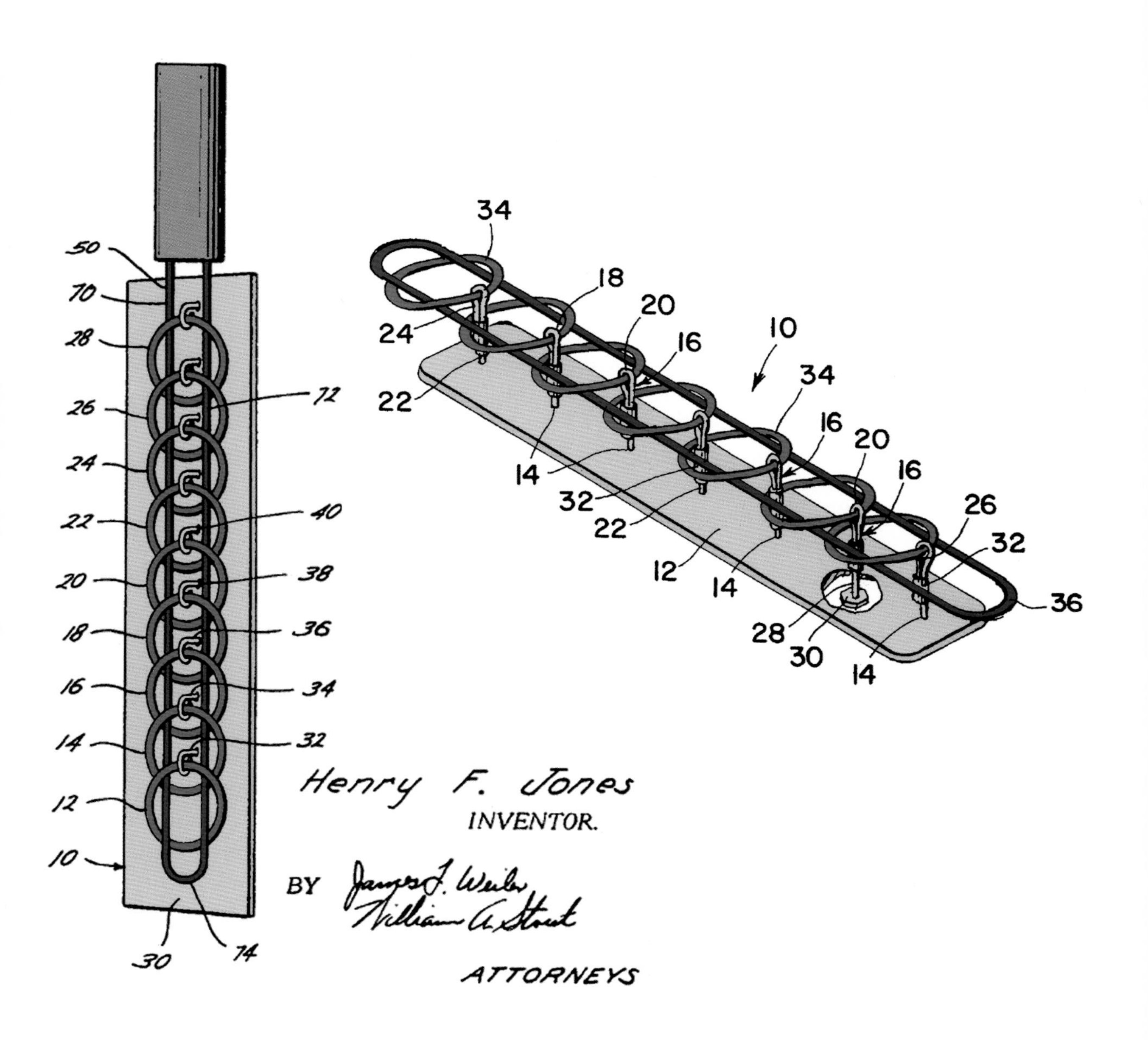
50
70
28
26
24
22
20
18
16
14
12
10
30
74
72
40
38
36
34
32
34
18
20
24
16
10
22
34
16
20
14
16
32
22
26
12
32
14
36
28
30
14
Henry F. Jones
INVENTOR.
BY
ATTORNEYS

1874

# The Doctorate of Kovalevskaya

**Sofia Kovalevskaya** (1850–1891)

The Russian mathematician Sofia Kovalevskaya made valuable contributions to the theory of differential equations and was the first woman in history to receive a doctorate in mathematics. Like most other mathematical geniuses, Sofia fell in love with mathematics at a very young age. She wrote in her autobiography: "The meaning of these concepts I naturally could not yet grasp, but they acted on my imagination, instilling in me a reverence for mathematics as an exalted and mysterious science which opens up to its initiates a new world of wonders, inaccessible to ordinary mortals." When Sofia was 11 years old, the walls of her bedroom were papered with mathematician Mikhail Ostrogradski's lecture notes on differential and integral analysis.

In 1874, Kovalevskaya received her doctorate, *summa cum laude*, from Göttingen University for her work on partial differential equations, Abelian integrals, and the structure of the rings of Saturn. However, despite this doctorate and enthusiastic letters of recommendation from mathematician Karl Weierstrass, for years Kovalevskaya was unable to obtain an academic position because she was a woman. However, she finally began to lecture at the University of Stockholm in Sweden in 1884, and she was appointed to a five-year professorship the same year. In 1888, the Paris Academy of Science awarded her a special prize for her theoretical treatments of rotating solids.

Kovalevskaya also deserves a place in the history of mathematics as she was the first Russian female mathematician of extreme notability, the third woman in history to become a professor in Europe—behind Laura Bassi (1711–1778) and Maria Agnesi (1718–1799)—and the first woman to hold a university chair in mathematics anywhere. She achieved these triumphs despite harsh resistance. For example, her father forbade her to study mathematics, but she secretly studied at night when the family slept. Russian women could not live apart from their families without written permission of the father; thus, she was forced to marry so that she could go abroad to further her education. Later in life, she wrote, "It is impossible to be a mathematician without being a poet in soul."

**SEE ALSO** The Death of Hypatia (415), Agnesi's *Instituzioni Analitiche* (1748), Boole's *Philosophy and Fun of Algebra* (1909), and Noether's *Idealtheorie* (1921).

*Sofia Kovalevskaya was the first woman to earn a doctorate in mathematics in Europe.*

1874

# Fifteen Puzzle

**Noyes Palmer Chapman** (1811–1889)

Although not a serious mathematical milestone like many of the entries in this book, the Fifteen Puzzle caused such a stir among the public that it is worthy of mention for historical reasons. Today, you can purchase a variant of the puzzle with 15 squares (tiles) and one vacant spot in a $4 \times 4$ frame or box. At the start, the squares sequentially contain the numbers 1 through 15 and then a gap. In a version of the puzzle in Sam Loyd's 1914 *Cyclopedia*, the starting configuration had the 14 and 15 reversed.

| 1 | 2 | 3 | 4 |
|---|---|---|---|
| 5 | 6 | 7 | 8 |
| 9 | 10 | 11 | 12 |
| 13 | 15 | 14 | |

*Unsolvable Fifteen Puzzle (Starting Position)*

For Loyd, the goal was to "slide" the squares up, down, right, and left to arrive at the sequence 1 through 15 (with the 14 and 15 having swapped positions). In *Cyclopedia*, Loyd claims that a prize of $1,000 was offered for the solution; alas, it is impossible to solve the puzzle from this starting position.

The original version of the game, developed in 1874 by New York postmaster Noyes Palmer Chapman, became an instant success in 1880 much like the Rubik's cube 100 years later. Originally, the tiles were loose, and the player placed them randomly, and then attempted to solve. Starting with random configurations, the puzzle can only be solved 50 percent of the time!

Mathematicians have since determined precisely which initial arrangements of the tiles can lead to solutions. German mathematician W. Ahrens noted, "The Fifteen Puzzle bobbed up in the United States; it spread quickly, and owing to the uncountable number of devoted players it had conquered, it became a plague." Interestingly, the chess superstar Bobby Fischer was an expert at solving the puzzle in less than 30 seconds if it started with any solvable configuration.

**SEE ALSO** Instant Insanity (1966) and Rubik's Cube (1974).

*In the 1880s, the Fifteen Puzzle took the world by storm, much like the Rubik's Cube did in modern times. Mathematicians have since precisely determined which initial arrangements of the tiles can lead to solutions.*

1
12
11
2
13
9
10
3
14
15
8
4
5
6
7

1874

# Cantor's Transfinite Numbers

**Georg Cantor** (1845–1918)

German mathematician Georg Cantor founded modern set theory and introduced the mind-boggling concept of transfinite numbers that can be used to denote the relative "sizes" of an infinite collection of objects. The smallest transfinite number is called *aleph-nought*, written as $\aleph_0$, which counts the number of integers. If the number of integers is infinite (with $\aleph_0$ members), are there yet higher levels of infinity? It turns out that even though there are an infinite number of integers, rational numbers (numbers that can be expressed as fractions), and irrational numbers (like the square root of 2 that cannot be expressed as a fraction), the infinite number of irrationals is in some sense greater than the infinite number of rationals or integers. Similarly, there are more *real* numbers (which include rational and irrational numbers) than there are integers.

Cantor's shocking concepts about infinity drew widespread criticism—which likely contributed to Cantor's bouts of severe depression and multiple institutionalizations—before being accepted as a fundamental theory. Cantor also equated his concept of the Absolute Infinite, which transcended the transfinite numbers, with God. He wrote, "I entertain no doubts as to the truths of the transfinites, which I recognized with God's help and which, in their diversity, I have studied for more than twenty years." In 1884, Cantor wrote to Swedish mathematician Gösta Mittag-Leffler explaining that he was not the creator of his new work, but merely a reporter. God had provided the inspiration, leaving Cantor only responsible for the organization and style of his papers. Cantor said that he knew that transfinites were real because "God had told me so," and it would have diminished God's power had God only created *finite* numbers. Mathematician David Hilbert described Cantor's work as "the finest product of mathematical genius and one of the supreme achievements of purely intellectual human activity."

**SEE ALSO** Aristotle's Wheel Paradox (c. 320 B.C.), Transcendental Numbers (1844), Hilbert's Grand Hotel (1925), and Continuum Hypothesis Undecidability (1963).

*Photo of Georg Cantor and his wife, taken around 1880. Cantor's startling ideas about infinity initially drew widespread criticism, which may have exacerbated his severe and chronic battles with depression.*

1875

# Reuleaux Triangle

**Franz Reuleaux** (1829–1905)

The Reuleaux triangle (RT) is one example of a wide class of geometrical discoveries like the **Möbius Strip** that did not find many practical applications until relatively late in humankind's intellectual development. Not until around 1875, when the distinguished German mechanical engineer Franz Reuleaux discussed the famous curvy triangle, did the RT begin to find numerous uses. Although Reuleaux wasn't the first to draw and consider the shape formed from the intersection of three circles at the corners of an equilateral triangle, he was the first to demonstrate its constant-width properties and the first to use the triangle in numerous real-world mechanisms. The construction of the triangle is so simple that modern researchers have wondered why no one before Reuleaux had exploited its use. The shape is a close relative of a circle because of its constant width, meaning that the distance between two opposite points is always the same.

Various technology patents have focused on drill bits that cut square holes using the RT. At first, the notion of a drill that creates nearly square holes defies common sense. How can a revolving drill bit cut anything but a circular hole? But such drill bits exist. For example, the illustration shown here is from the 1978 patent U.S. 4,074,778 for a "Square Hole Drill" and is based on the RT. The RT also appears in patents for other drill bits as well as for novel bottles, rollers, beverage cans, candles, rotatable shelves, gearboxes, rotary engines, and cabinets.

Many mathematicians have studied the Reuleaux triangle, so we know a lot about its properties. For example, its area is $A = \frac{1}{2}(\pi - \sqrt{3})r^2$, and the area drilled by a RT drill bit covers 0.9877003907... of the area of an actual square. The small difference occurs because the Reuleaux drill bit produces a square with very slightly rounded corners.

**SEE ALSO** Astroid (1674) and The Möbius Strip (1858).

*A figure from a 1978 patent (U.S. patent 4,074,778) which describes a drill bit for drilling a square hole based on the Reuleaux triangle.*

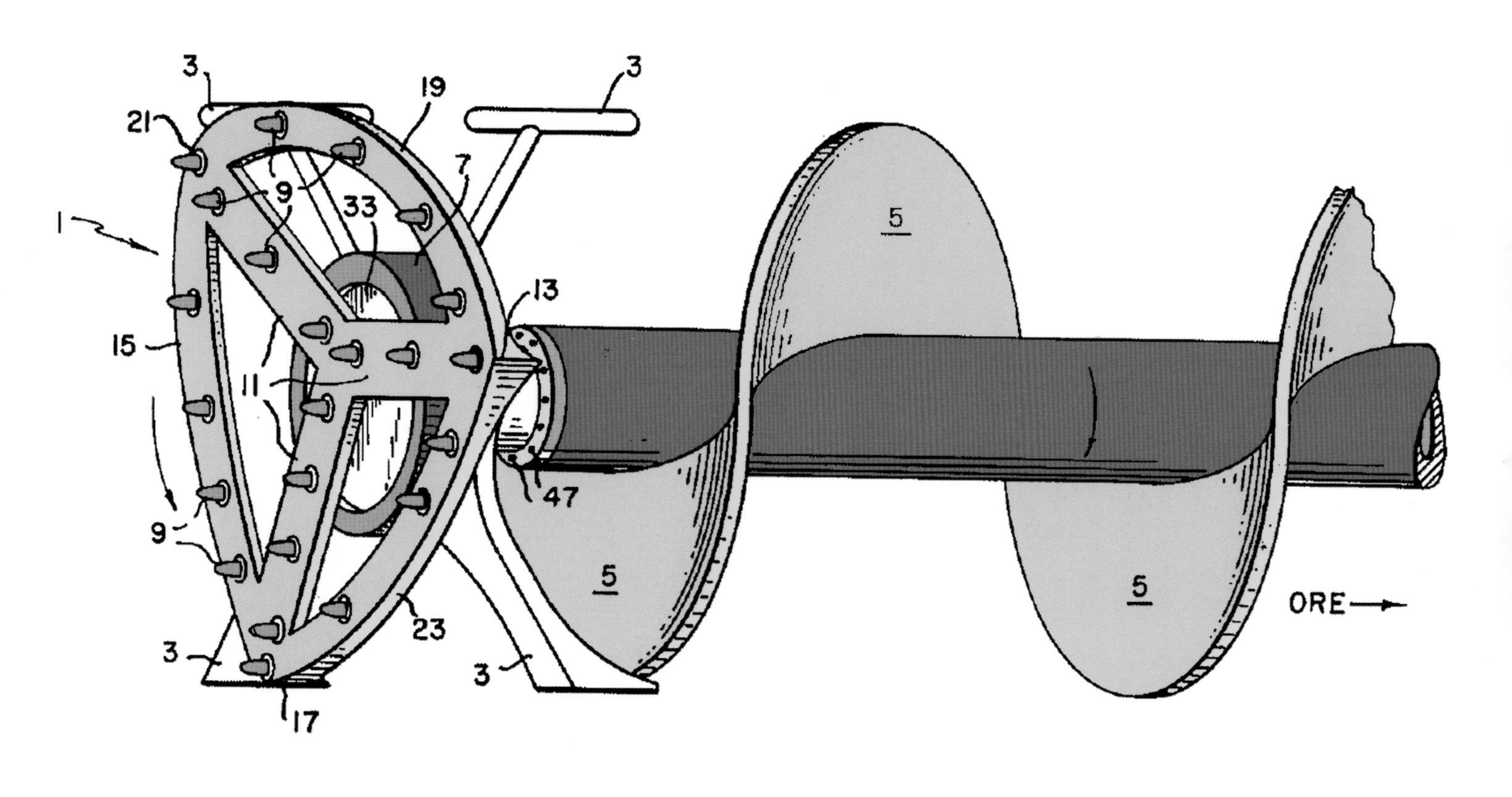

3
3
19
21
9
9
33
7
1
13
15
11
5
47
9
9
5
5
ORE
23
3
3
17

1876

# Harmonic Analyzer

**Jean Baptiste Joseph Fourier** (1768–1830), **William Thomson, Baron Kelvin of Largs** (1824–1907)

In the early 1800s, French mathematician Joseph Fourier found that any differentiable function can be represented to arbitrary accuracy by a sum of sine and cosine functions, no matter how complicated the function. For example, a periodic function $f(x)$ can be represented by the sum of $A_n \cdot \sin(nx) + B_n \cdot \cos(nx)$ for amplitudes $A_n$ and $B_n$.

A harmonic analyzer is a physical device for determining the coefficients $A_n$ and $B_n$. In 1876, Lord Kelvin, a British mathematical physicist, was first to invent the harmonic analyzer, for the analysis of curve traces related to ocean tidal observations. A paper with the curve of interest is wrapped around a main cylinder. The device is made to follow the curve, and then the positions of various subcomponents are determined to give the desired coefficients. Kelvin writes that the "kinematic machine" predicts not merely "the times and heights of high water, but the depths of water at any instant, showing them by a continuous curve, for...years in advance." Because the tides depend on the positions of the sun, moon, rotation of the earth, shape of the coastline, and sea floor profile, they can be quite complex.

In 1894, German mathematician Olaus Henrici (1840–1918) designed a harmonic analyzer for determining the harmonic components of complex sound waves such as those from musical instruments. The device employed several pulleys and glass spheres connected to measuring dials that gave the phase and amplitudes of 10 Fourier harmonic components.

In 1909, German engineer Otto Mader invented a harmonic analyzer that used gears and a pointer to trace a curve; the different gears corresponded to harmonics. The Montgomery harmonic analyzer of 1938 used optical and photoelectric means for determining the harmonic content of a curve. H. C. Montgomery of Bell Laboratories wrote that the device "is especially adapted to the analysis of speech and music, since it operates directly from a conventional type of sound track on film."

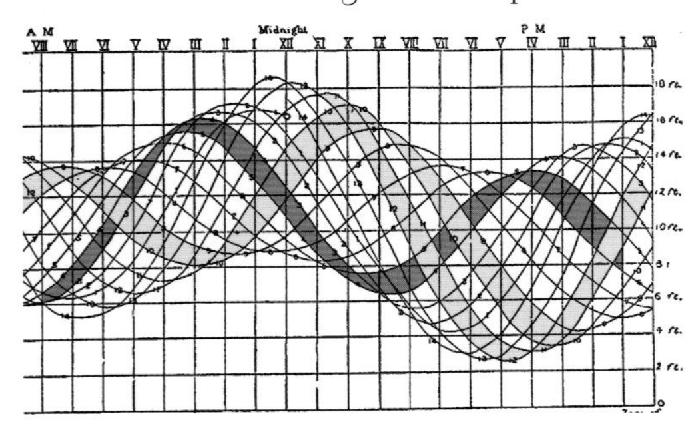


**SEE ALSO** Fourier Series (1807) and Differential Analyzer (1927).

LEFT: *Tidal record for two weeks (January 1–14, 1884) at Bombay. The tide was recorded on a cylindrical sheet that turned once every 24 hours.* RIGHT: *The harmonic analyzer of German mathematician Olaus Henrici.*

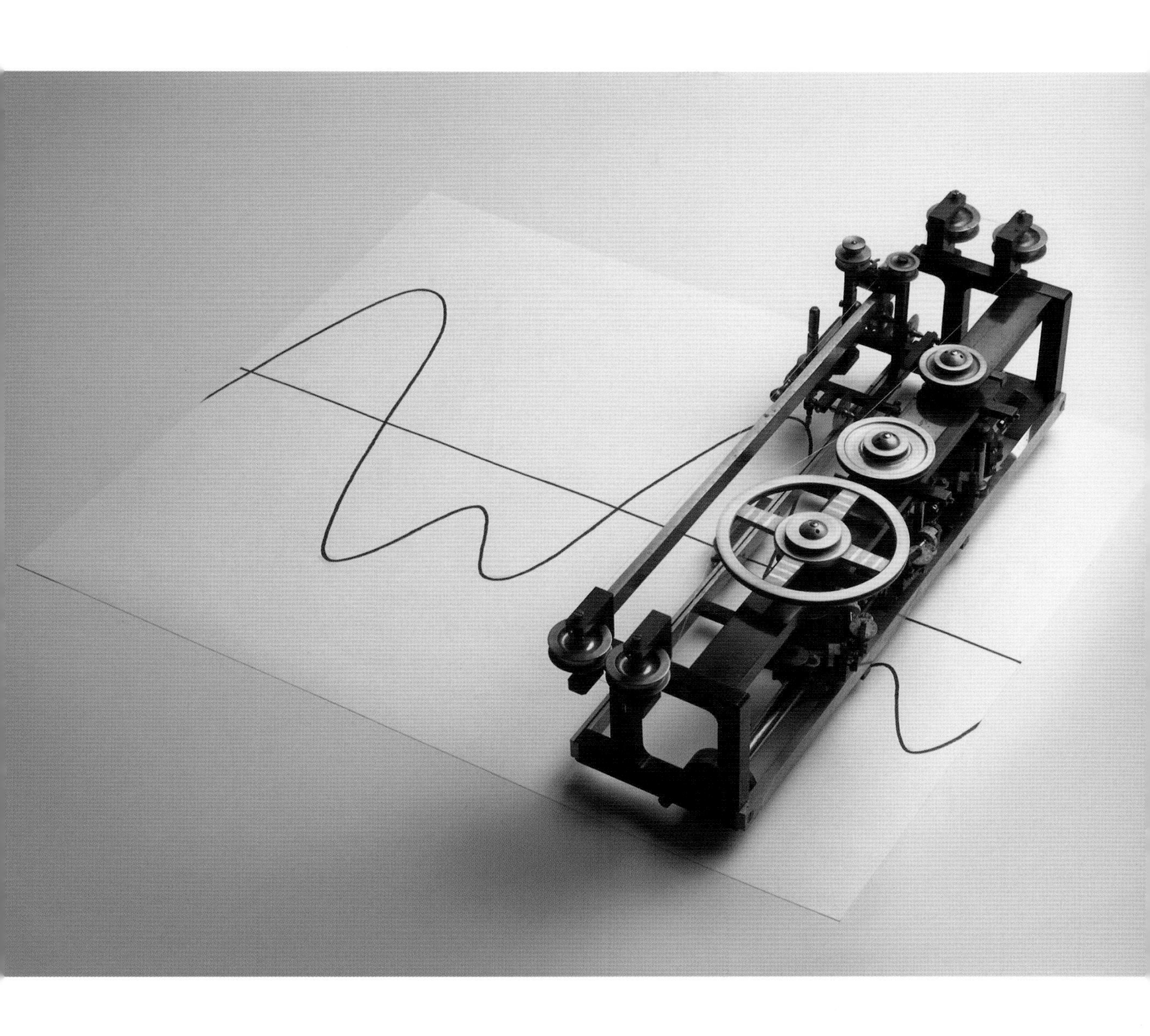

1879

# Ritty Model I Cash Register

**James Ritty** (1836–1918)

It is difficult to imagine how retail stores could operate efficiently before cash registers existed. Through the decades, cash registers became increasingly sophisticated and also functioned as theft deterrents. It is not an exaggeration to say that they became one of the major transforming mechanizations of the Industrial Age.

The first cash register was invented in 1879 by James Ritty. Ritty opened his first saloon in Dayton, Ohio, in 1871, referring to himself as a "Dealer in Pure Whiskies, Fine Wines, and Cigars." Ritty's biggest challenge came from his employees who sometimes secretly pocketed money received from customers.

While on a steamboat trip, Ritty studied a mechanism that counted the number of revolutions of the ship's propeller, and he began to imagine similar mechanisms that might be used to record cash transactions. Ritty's early machines had two rows of keys, each key corresponding to a money denomination from five cents through one dollar. Pressing the keys turned a shaft that moved an internal counter. He patented his design in 1879 as "Ritty's Incorruptible Cashier." Ritty soon sold his cash-register business to a salesman named Jacob H. Eckert, and in 1884, Eckert sold the company to John H. Patterson, who renamed the company the National Cash Register Company.

From Ritty's small seed, the modern cash register grew. Patterson added paper rolls to record transactions using hole punchers. When a transaction was finished, a bell rang on the cash register and the monetary amount was represented on a large dial. In 1906, inventor Charles F. Kettering designed a cash register with an electric motor. In 1974, the National Cash Register Company became NCR Corp. Today, cash register functionality goes beyond Ritty's wildest dream, as these number-crunching machines time-stamp transactions, retrieve prices from databases, and calculate appropriate tax amounts, various rates for preferred customers, and deductions for sale items.

**SEE ALSO** Curta Calculator (1948).

*A 1904 replica of the Ritty Model 1 cash register.*

00 5 10 15 20 25 30 35 40 45 50 55 60 65 70 75 80 85 90 95
1 2 3 6 7 8 9
L43702

# Venn Diagrams

**John Venn** (1834–1923)

In 1880, John Venn, a British philosopher and cleric in the Anglican Church, devised a scheme for visualizing elements, sets, and logical relationships. A *Venn diagram* usually contains circular areas representing groups of items sharing common properties. For instance, within the universe of all real and legendary creatures (the bounding rectangle in the first illustration), region *H* represents the humans, region *W* the winged creatures, and region A the angels. A glance at the diagram reveals that: (1) All angels are winged creatures (region A lies entirely within region *W*); (2) No humans are winged creatures (regions *H* and *W* are nonintersecting); and (3) No humans are angels (regions *H* and A are nonintersecting).

This is a depiction of a basic rule of logic—namely, that from the statements "all A is *W*" and "no *H* is *W*," it follows that "no *H* is A." The conclusion is evident when we look at the circles in the diagram.

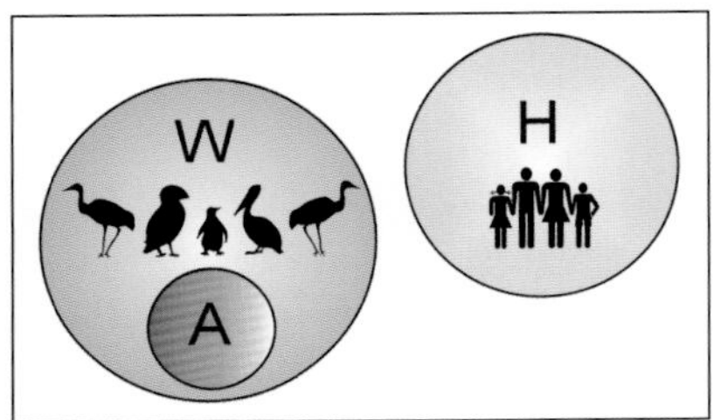


1

The uses of these kinds of diagrams in logic were used before Venn—for example, by mathematicians Gottfried Leibniz and Leonhard Euler—but Venn was the first to comprehensively study them and formalize and generalize their usage. In fact, Venn struggled with generalizing *symmetrical* diagrams for visualizing more sets with intersecting areas, but he only got as far as 4 sets using ellipses.

A century passed before Branko Grünbaum, a mathematician at the University of Washington, showed that rotationally symmetric Venn diagrams can be made from 5 congruent ellipses. The second illustration shows one of many different symmetrical diagrams for 5 sets.

Mathematicians gradually realized that rotationally symmetric diagrams can be drawn with prime numbers of petals only. However, symmetrical diagrams with 7 petals were so hard to find that mathematicians initially doubted their existence. In 2001, mathematician Peter Hamburger and artist Edit Hepp constructed an example for 11 petals, shown on the opposite page.

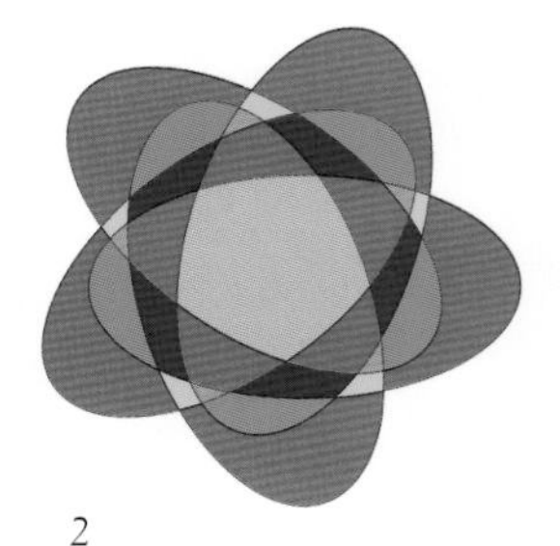

2

**SEE ALSO** Aristotle's *Organon* (c. 350 B.C.), Boolean Algebra (1854), *Principia Mathematica* (1910–1913), Gödel's Theorem (1931), and Fuzzy Logic (1965).

*Symmetric 11-Venn diagram, courtesy of Dr. Peter Hamburger and Edit Hepp.*

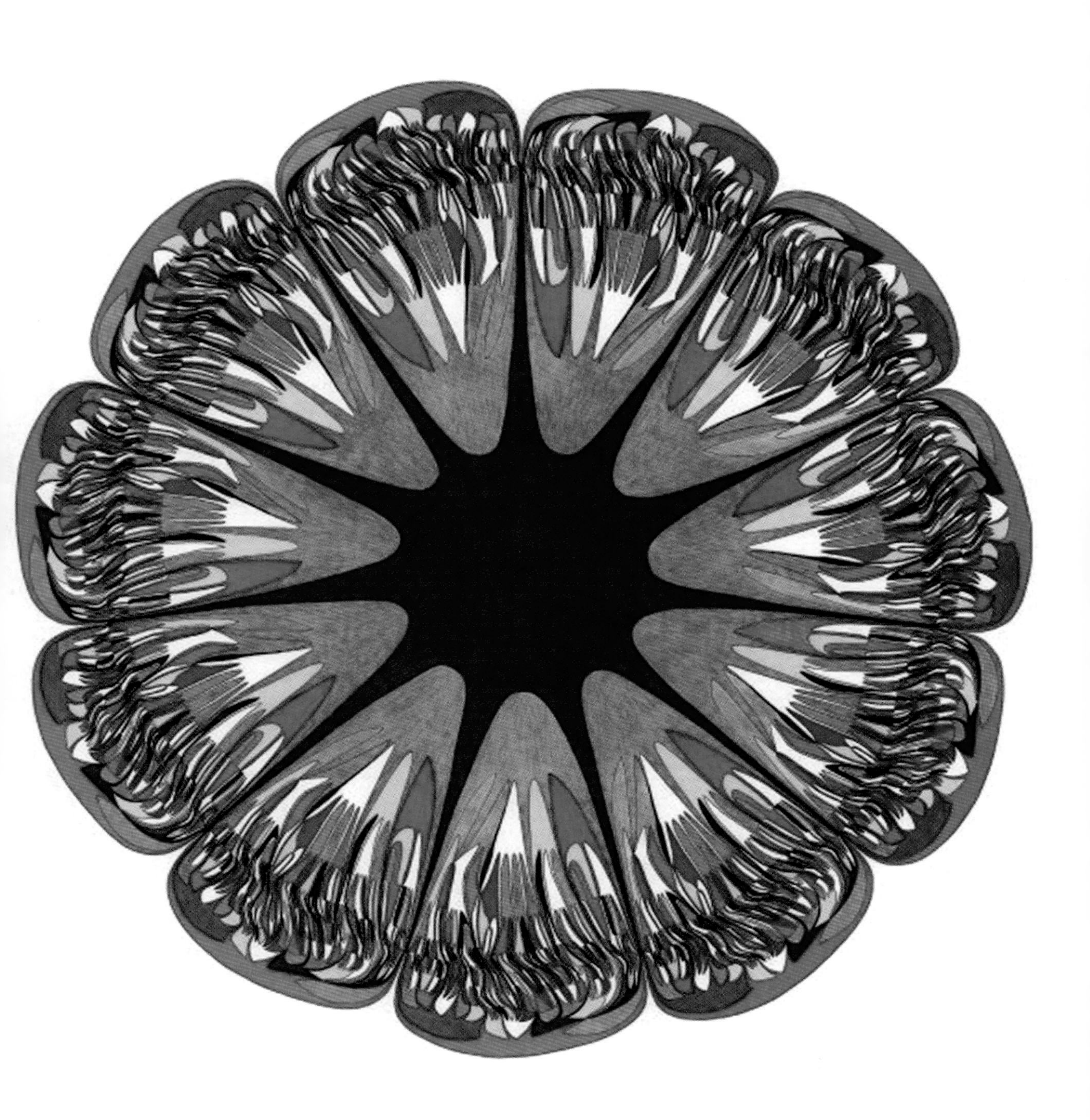

# Benford's Law

**Simon Newcomb** (1835–1909), **Frank Benford** (1883–1948)

Benford's law, also called the first-digit law or leading-digit phenomenon, asserts that in various number lists, the digit 1 tends to occur in the *leftmost position* with probability of roughly 30 percent, much greater than the expected 11.1 percent that would result if each digit occurred with a 1 out of 9 probability. Benford's law can be observed, for instance, in tables that list populations, death rates, stock prices, baseball statistics, and the area of rivers and lakes. Explanations for this phenomenon are very recent.

Benford's law is named after Dr. Frank Benford, a physicist at the General Electric Company who publicized his work in 1938, although it had been previously discovered by mathematician and astronomer Simon Newcomb in 1881. Pages of logarithms with numbers starting with the numeral 1 are said to be dirtier and more worn than other pages, because the number 1 occurs as the first digit about 30 percent more often than any other. In numerous kinds of data, Benford determined that the probability of any number $n$ from 1 through 9 being the first digit is $\log_{10}(1 + 1/n)$. Even the Fibonacci sequence—1, 1, 2, 3, 5, 8, 13... —follows Benford's law. Fibonacci numbers are far more likely to start with "1" than any other digit. It appears that Benford's law applies to any data that follows a "power law." For example, large lakes are rare, medium-size lakes are more common, and small lakes are even more common. Similarly, 11 Fibonacci numbers exist in the range 1–100, but only one in the next three ranges of 100 (101–200, 201–300, 301–400).

Benford's law has often been used to detect fraud. For example, accounting consultants can sometimes use the law to detect fraudulent tax returns in which the occurrence of digits does not follow what would naturally be expected according to Benford's law.

**SEE ALSO** Fibonacci's *Liber Abaci* (1202) and Laplace's *Théorie Analytique des Probabilités* (1812).

*Benford's Law can be observed in stock prices and other financial data, as well as in electricity bills and street addresses.*

KER INSUR. 33.50 70.53 44.53
TINTO 19.13 3.86 5.00
EORGE 15.63 33.51 70.52
ET 73.85 19.20 3.85
RA 29.93 15.65 33.51
21.10 29.55 74.94 19.19
MER 3.98 29.59 74.00
ELDG 37.40 21.13 29.55
17.70 3.99 21.10
22.79 37.48 3.98
42.52 17.72 37.45
HS 16.82 22.81 17.7
42.54 22.79

# Klein Bottle

**Felix Klein** (1849–1925)

The Klein bottle, first described in 1882 by German mathematician Felix Klein, is an object in which the flexible neck of a bottle wraps back *into* the bottle to form a shape with no inside and outside. This bottle is related to the **Möbius Strip**, and can theoretically be created by gluing two Möbius strips together along their edges. One way to build an imperfect physical model of a Klein bottle in our 3-D universe is to have it meet itself in a small, circular curve. Four dimensions are needed to create a true Klein bottle without self-intersections.

Imagine your frustration if you tried to paint just the outside of a Klein bottle. You start painting on what seems to be the bulbous "outside" surface and work your way down the slim neck. The 4-D object does not self-intersect, allowing you to continue to follow the neck, which is now "inside" the bottle. As the neck opens up to rejoin the bulbous surface, you find you are now painting inside the bottle. If our universe were shaped like a Klein bottle, we could find paths that would cause our bodies to reverse when we returned after a journey, so that, for example, our hearts would be on the right sides of our bodies.

Together with Toronto's Kingbridge Centre and Killdee Scientific Glass, astronomer Cliff Stoll has created the world's largest glass Klein bottle model. The Kingbridge Klein Bottle is about 43 inches (1.1 meters) tall and 20 inches (50 centimeters) in diameter, and is made of 33 pounds (15 kilograms) of clear Pyrex glass.

Because of the peculiar properties of the Klein bottle, mathematicians and puzzle enthusiasts study chess games and mazes played on Klein bottle surfaces. If a map were drawn on a Klein bottle, six different colors would be needed to ensure that no bordering areas would be colored the same.

**SEE ALSO** Minimal Surface (1774), Four-Color Theorem (1852), The Möbius Strip (1858), Boy's Surface (1901), and Turning a Sphere Inside Out (1958).

*The Klein bottle has a flexible neck that wraps back* into *the bottle to form a shape with no separate inside and outside. Four dimensions are needed to create a true Klein bottle without self-intersections.*

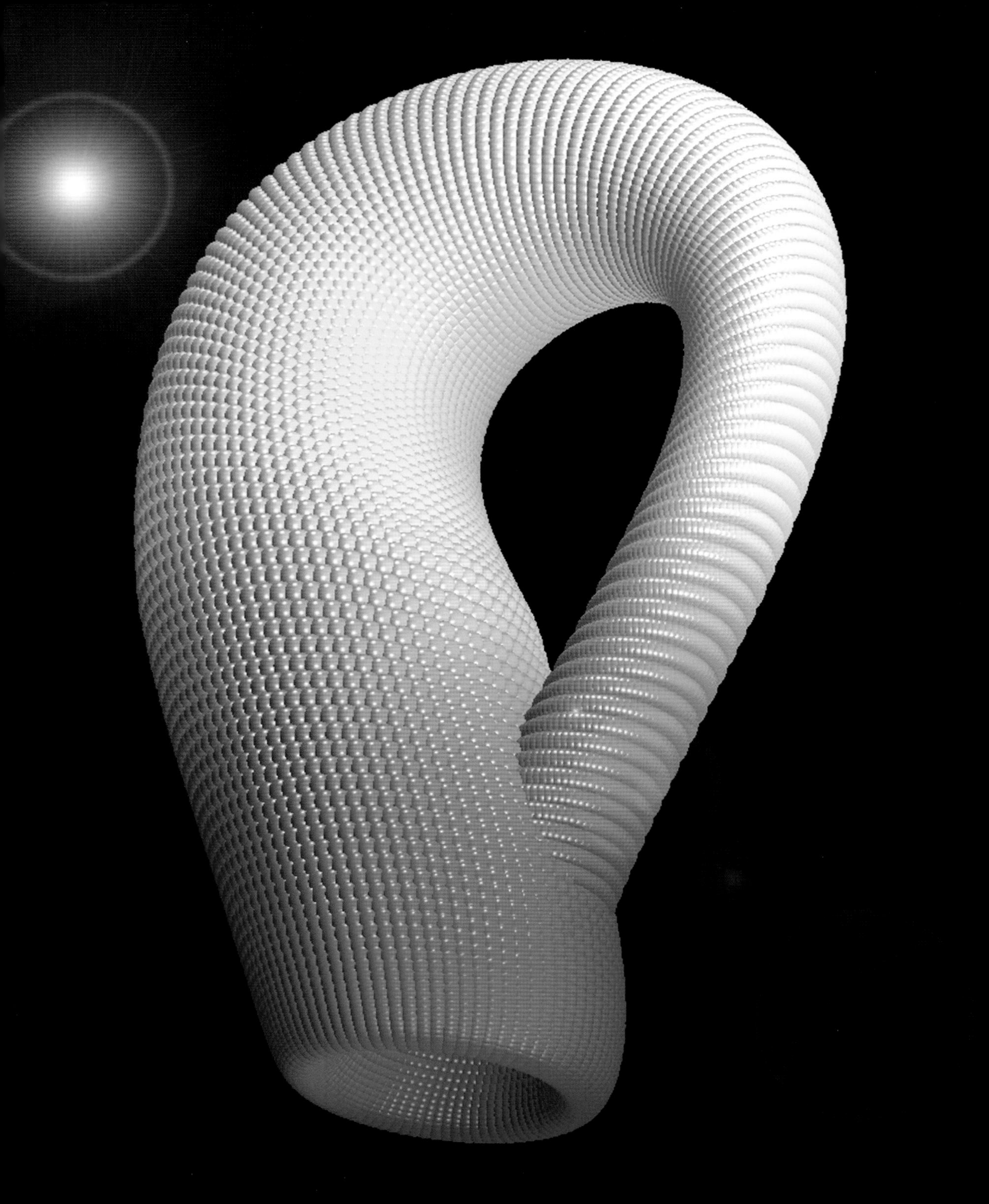

# Tower of Hanoi

**François Édouard Anatole Lucas** (1842–1891)

The Tower of Hanoi has intrigued the world since it was invented by French mathematician Édouard Lucas in 1883 and sold as a toy. This mathematical puzzle consists of several disks of different sizes that slide onto any of three pegs. The disks are initially stacked on one peg in order of size, with the smallest disk at the top. When playing the game, one can move one disk at a time to another peg by removing the top disk in any stack and placing it on the top of any other stack. A disk cannot be placed on top of a smaller disk. The goal is to move the entire starting stack (often with eight disks) to another peg. The minimum number of moves turns out to be $2^n - 1$, where $n$ is the number of disks.

The original game was said to be inspired by a legendary Indian Tower of Brahma, which employed 64 golden disks. The priests of Brahma continually moved these disks, using the same rules as in the Tower of Hanoi. When the last move of the puzzle is completed, the world will end. Note that if the priests were able to move disks at a rate of 1 per second, then $2^{64} - 1$ or 18,446,744,073,709,551,615 moves would require roughly 585 billion years—many times the current estimated age of our universe.

Simple algorithms exist for solutions involving three pegs, and the game is often used in computer programming classes to teach recursive algorithms. However, the optimal solution for the Tower of Hanoi problem with four or more pegs is still unknown. Mathematicians find the puzzle intriguing due to its relationships to other areas of math, including **Gray Codes** and finding Hamiltonian paths on an $n$-hypercube.

**SEE ALSO** Boolean Algebra (1854), Icosian Game (1857), Gros's *Théorie du Baguenodier* (1872), Tesseract (1888), Gray Code (1947), Instant Insanity (1966), and Rubik's Cube (1974).

*The Flag Tower of Hanoi, built in 1812, is located in Hanoi, Vietnam. It has a height of about 109.5 feet (33.4 meters), or 134.5 feet (41 meters) with the flag, and, according to some legends, was the likely inspiration for the name of the puzzle.*

1884

# *Flatland*

**Edwin Abbott Abbott** (1838–1926)

More than a century ago, Edwin Abbott Abbott—a clergyman and the headmaster of a school in Victorian England—wrote an influential book describing interactions between creatures with access to different spatial dimensions. The book is still popular among mathematics students and considered useful reading for anyone studying relationships between such dimensions.

Abbott encouraged readers to open their minds to new ways of perceiving. *Flatland* described a race of two-dimensional creatures, living in a flat plane, totally unaware of the existence of a higher dimension all around them. If we were able to look down on a two-dimensional world, we would be able to see inside every structure at once. A creature with access to a fourth spatial dimension could see inside our own bodies and remove a tumor without penetrating the skin. Flatlanders could be unaware that you were poised inches above their planar world, recording all the events of their lives. If you wanted to remove a Flatlander from jail, you could lift him "up" and deposit him elsewhere in Flatland. This act would appear miraculous to a Flatlander who would not even have the word *up* in his vocabulary.

Today, computer graphic projections of 4-D objects bring us a step closer to higher-dimensional phenomena, but even the most brilliant of mathematicians are often unable to grasp the fourth dimension, just as the square protagonist of *Flatland* had trouble understanding the third dimension. In one of *Flatland*'s most emotional scenes, the two-dimensional hero is confronted by the changing shapes of a three-dimensional being as it passes through Flatland. The square can only see the creature's cross sections. Abbott believed that study of the fourth spatial dimension is important in expanding our imagination, increasing our reverence for the universe, and increasing our humility—perhaps the first steps in any attempt to better understand the nature of reality or to glimpse the divine.

**SEE ALSO** Euclid's *Elements* (300 B.C.), Klein Bottle (1882), and Tesseract (1888).

*The cover of* Flatland, *6th Edition, by Edwin Abbott Abbott. Notice that "My Wife" is portrayed as a line within the pentagonal house. In* Flatland, *women could be particularly dangerous, due to their sharp ends.*

*"O day and night, but this is wondrous strange"*

# FLATLAND

## A ROMANCE OF MANY DIMENSIONS

### *By* A Square

(Edwin A. Abbott)

*No Dimensions*

*POINTLAND*

*One Dimension*

*LINELAND*

*Two Dimensions*

*FLATLAND*

*Three Dimensions*

*SPACELAND*

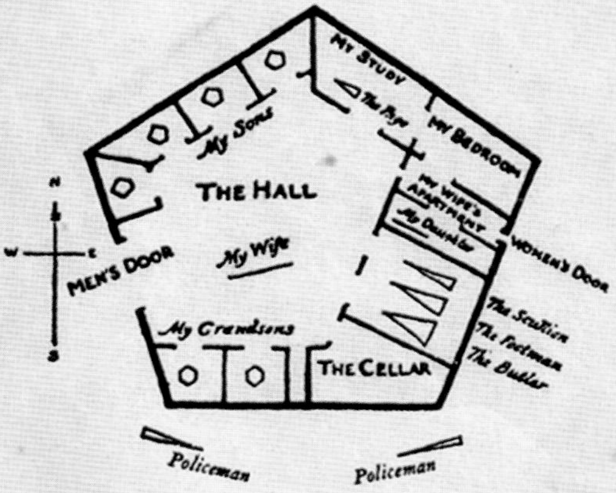

*"And therefore as a stranger give it welcome."*

BASIL BLACKWELL · OXFORD

*Price Seven Shillings and Sixpence net*

# Tesseract

## **Charles Howard Hinton** (1853–1907)

I know of no subject in mathematics that has intrigued both children and adults as much as the idea of a fourth dimension, a spatial direction different from all the directions of our everyday three-dimensional space. Theologians have speculated that the afterlife, heaven, hell, angels, and our souls could reside in a fourth dimension. Mathematicians and physicists frequently use the fourth dimension in their calculations. It's part of important theories that describe the very fabric of our universe.

The tesseract is the four-dimensional analog of the ordinary cube. The term *hypercube* is used more generally when referring to cube analogues in other dimensions. Just as a cube can be visualized by dragging a square into the third dimension and watching the shape that the square traces through space, a tesseract is produced by the trail of a cube moving into the fourth dimension. Although it is difficult to visualize a cube being shifted a distance in a direction perpendicular to all three of its axes, computer graphics often help mathematicians develop a better intuition for higher-dimensional objects. Note that a cube is bounded by square faces and a tesseract by cubical faces. We can write down the number of corners, edges, faces, and solids for these kinds of higher-dimensional objects:

| | Corners | Edges | Faces | Solids | Hypervolumes |
|---|---|---|---|---|---|
| Point | 1 | 0 | 0 | 0 | |
| Line segment | 2 | 1 | 0 | 0 | 0 |
| Square | 4 | 4 | 1 | 0 | 0 |
| Cube | 8 | 12 | 6 | 1 | 0 |
| Hypercube | 16 | 32 | 24 | 8 | 1 |
| Hyperhypercube | 32 | 80 | 80 | 40 | 10 |

The word *tesseract* was coined and first used in 1888 by British mathematician Charles Howard Hinton in his book *A New Era of Thought*. Hinton, a bigamist, was also famous for his set of colored cubes that he claimed could be used to help people visualize the fourth dimension. When used at séances, the Hinton cubes were thought to help people glimpse ghosts of dead family members.

**SEE ALSO** Euclid's *Elements* (300 B.C.), Prince Rupert's Problem (1816), Klein Bottle (1882), *Flatland* (1884), Boole's *Philosophy and Fun of Algebra* (1909), Rubik's Cube (1974), and Perfect Magic Tesseract (1999).

*Rendering of a tesseract by Robert Webb using Stella4D software. The tesseract is the four-dimensional analog of the ordinary cube.*

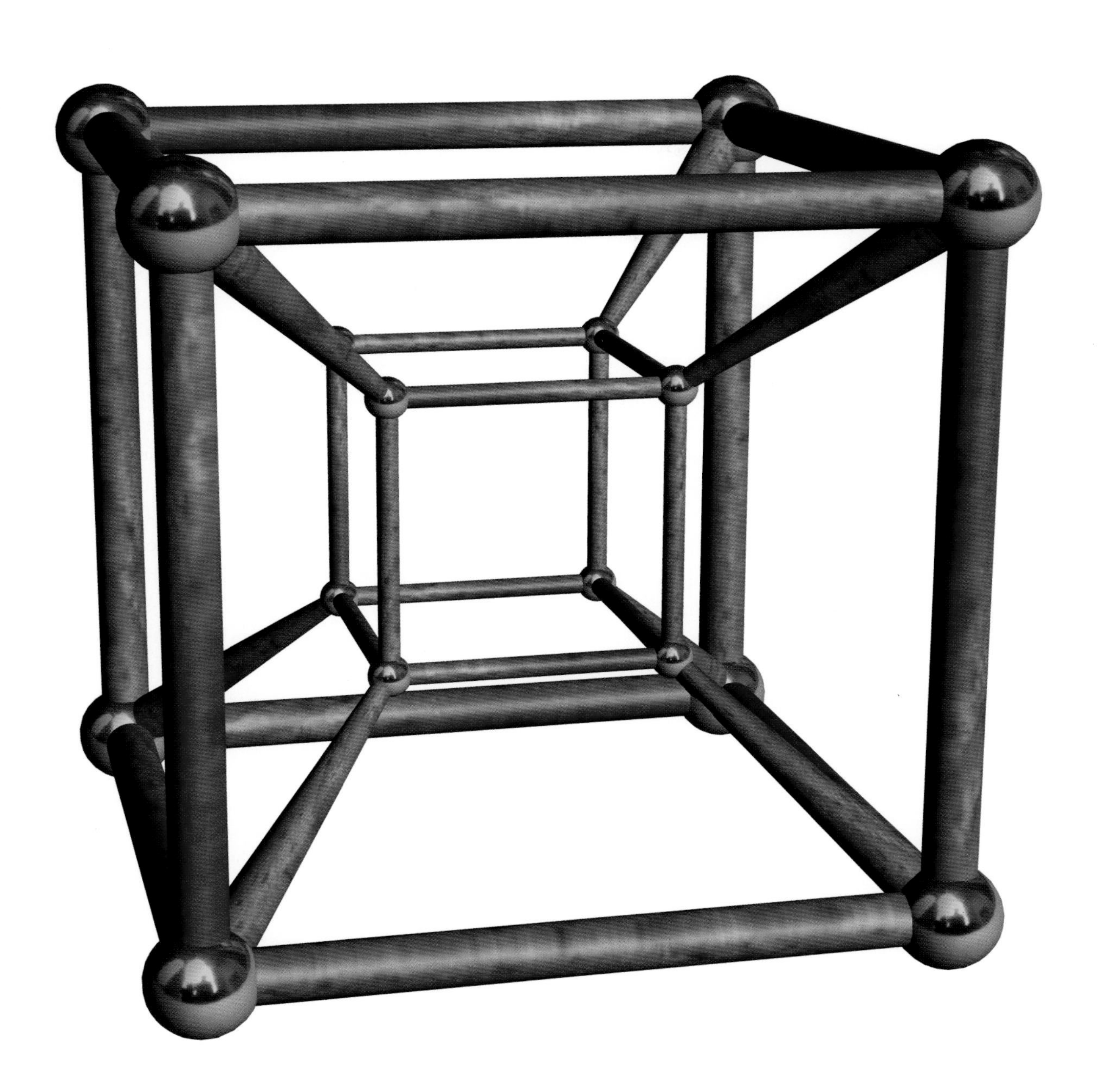

# Peano Axioms

**Giuseppe Peano** (1858–1932)

Schoolchildren know the simple arithmetic rules of counting, addition, and multiplication, but where do these simplest of rules come from, and how do we know that they are correct? Italian mathematician Giuseppe Peano was familiar with Euclid's five axioms, or assumptions, that laid the foundation for geometry, and Peano was interested in creating the same kind of foundation for arithmetic and number theory. The five Peano axioms involve non-negative integers and can be stated as: 1) 0 is a number; 2) The successor of any number is a number; 3) If $n$ and $m$ are numbers and if their successors are equal, then $n$ and $m$ are equal; 4) 0 is not the successor of any number; and 5) If S is a set of numbers containing 0 and if the successor of any number in $S$ is also in $S$, then $S$ contains all the numbers.

Peano's fifth axiom allows mathematicians to determine whether a property is true of all non-negative numbers. To accomplish this, we first must show that 0 has the property. Next, we must show that, for any number $i$, if $i$ has the property, then this implies that $i + 1$ also has the property. A metaphor helps. Imagine an infinite line of matches, nearly touching. If we want them all to burn, the first match must light, and each match in the line must be sufficiently close that it will catch fire. If one match along the line has too great a separation, the fire stops. With Peano's axioms, we can build a system of arithmetic that involves an infinite set of numbers. The axioms provide a foundation for our number system and help mathematicians construct other number systems used in modern mathematics. Peano first presented his axioms in his 1889 *Arithmetices principia, nova methodo exposita* (*The Principles of Arithmetic, Presented by a New Method*).

**SEE ALSO** Euclid's *Elements* (300 B.C.), Aristotle's *Organon* (c. 350 B.C.), Boolean Algebra (1854), Venn Diagrams (1880), Hilbert's Grand Hotel (1925), Gödel's Theorem (1931), and Fuzzy Logic (1965).

*The work of Italian mathematician Giuseppe Peano touches upon philosophy, mathematical logic, and set theory. He taught mathematics at the University of Turin until the day before he died from a heart attack.*

1890

# Peano Curve

## Giuseppe Peano (1858–1932)

In 1890, Italian mathematician Giuseppe Peano presented one of the first examples of a space-filling curve. British science writer David Darling calls the discovery an "earthquake on the traditional structure of mathematics." When discussing this new class of curves, Russian mathematician Naum Vilenkin wrote that "everything was in ruins, that all the basic mathematical concepts had lost their meaning."

The term *Peano curve* is often used synonymously with *space-filling curve*, and such curves can often be created by an iterative process that eventually builds a zigzagging line that covers the entire space in which it resides. Martin Gardner writes, "Peano curves were a profound shock to mathematicians. Their paths seem to be one-dimensional, yet at the limit they occupy a two-dimensional area. Should they be called *curves*? To make things worse, Peano curves can be drawn just as easily to fill cubes and hypercubes. . . ." Peano curves are continuous, yet like the boundary of a **Koch Snowflake** or the **Weierstrass Function**, no point on the curve has a unique tangent. Space-filling curves have a **Hausdorff Dimension** of 2.

Space-filling curves have had practical applications and have suggested efficient routes to take when visiting a number of towns. For example, John J. Bartholdi III, a professor in the School of Industrial and Systems Engineering at the Georgia Institute of Technology, has used Peano curves to build a routing system for an organization that delivers hundreds of meals to the poor and to route blood delivery by the American Red Cross to hospitals. Because delivery locations tend to be clustered around urban areas, Bartholdi's use of space-filling curves produces very good routing suggestions because the curves tend to visit all the locations on a region of a map before moving on to another region. Scientists have also experimented with space-filling curves for weapons targeting because the mathematical technique can be implemented so that it runs very efficiently on a computer sent into orbit around the Earth.

**SEE ALSO** Knight's Tours (1759), Weierstrass Function (1872), Tesseract (1888), Koch Snowflake (1904), Hausdorff Dimension (1918), and Fractals (1975).

*The Hilbert cube is a three-dimensional extension to a traditional two-dimensional Peano curve. This 4-inch (10.2-centimeter) bronze and stainless-steel sculpture was designed by Carlo H. Sequin at the University of California at Berkeley.*

# Wallpaper Groups

**Evgraf Stepanovich Fedorov** (1853–1919), **Arthur Moritz Schönflies** (1853–1928), **William Barlow** (1845–1934)

The phrase "wallpaper groups" refers to ways in which a plane may be tiled so that the resulting pattern repeats indefinitely in two dimensions. Seventeen wallpaper patterns exist, each identified by symmetries such as those involving translations (for example, shifting or sliding) and rotations.

The eminent Russian crystallographer E. S. Fedorov discovered and classified these patterns in 1891, and these patterns were also studied, independently, by the German mathematician A. M. Schönflies and the English crystallographer William Barlow. Thirteen of these patterns (formally known as *isometries*) include some kind of rotational symmetry, while four do not. Five show hexagonal symmetries. Twelve show rectangular symmetries. Martin Gardner writes, "Seventeen different symmetry groups [exist] that exhibit all the fundamentally different ways in which patterns can be repeated endlessly in two dimensions. The elements of these groups are simply operations performed on one basic pattern: sliding along the plane, rotating it, or giving it a mirror reversal. The seventeen symmetry groups are of great importance in the study of crystal structure."

The geometer H. S. M. Coxeter has noted that the art of filling a plane with a repeated pattern reached its peak in thirteenth-century Spain, where the Islamic Moors used all 17 groups in their beautiful decorations of the Alhambra, a palace and fortress. Because some Islamic traditions discouraged the use of images of people in artworks, symmetrical wallpaper patterns became particularly attractive as decorations. The Alhambra palace in Granada contains intricate arabesque designs that decorate the tiles, plasterwork, and woodcarvings.

Dutch artist M. C. Escher's (1898–1972) visit to the Alhambra palace influenced his own art, which is often replete with symmetries. Escher once said his trips to the Alhambra were "the richest source of inspiration I have ever tapped." Escher tried to "enhance" the artworks of the Moors by using geometric grids as the basis for his sketches, which he then superimposed with animal designs.

**SEE ALSO** Group Theory (1832), Squaring a Rectangle (1925), Voderberg Tilings (1936), Penrose Tiles (1973), Monster Group (1981), and The Quest for Lie Group $E_8$ (2007).

*The Alhambra palace and fortress complex. The Islamic Moors used numerous different wallpaper groups in their beautiful decorations of the Alhambra.*

# Sylvester's Line Problem

**James Joseph Sylvester** (1814–1897), **Tibor Gallai** (1912–1992)

Sylvester's line problem—also known as Sylvester's problem of collinear points or the Sylvester-Gallai theorem—stumped the entire mathematical community for forty years. It states that given a finite number of points in the plane, either: 1) A line exists that passes through exactly two of the points, or 2) All the points are collinear, or lying in the same straight line. English mathematician James Sylvester made the conjecture in 1893 but could not provide a proof. The Hungarian-born mathematician Paul Erdös studied the problem in 1943, and it was correctly solved by Hungarian mathematician Tibor Gallai in 1944.

Sylvester had actually asked readers to "Prove that it is not possible to arrange any finite number of real points so that a right line through every two of them shall pass through a third, unless they all lie in the same right line." (Sylvester used the term *right line* to indicate a straight line.)

Stimulated by Sylvester's conjecture, in 1951 mathematician Gabriel Andrew Dirac (1925–1984)—stepson of Paul Dirac and nephew of Eugene Wigner—conjectured that for any collection of $n$ points, not all collinear, there exist at least $n/2$ lines containing exactly two points. Today, only two counterexamples to Dirac's conjecture are known.

Mathematician Joseph Malkevitch writes about Sylvester's problem, "Some easy-to-state problems in mathematics stand out, despite their seeming simplicity, because initially they eluded solution.... Erdös expressed his surprise that the Sylvester problem went unanswered for so many years....One seminal problem can open up many avenues of ideas, which even to this day are still being explored." Sylvester said in his 1877 address to Johns Hopkins University, "Mathematics is not a book confined within a cover.... It is not a mine, whose treasures... fill only a limited number of veins.... It is limitless.... Its possibilities are as infinite as the worlds which are forever crowding in and multiplying upon the astronomer's gaze."

**SEE ALSO** Euclid's *Elements* (300 B.C.), Pappus's Hexagon Theorem (c. 340), The Matrices of Sylvester (1850), and Jung's Theorem (1901).

*Given a scattering of a finite number of points—not all along a single line (and represented here by colored spheres)—the Sylvester-Gallai theorem tells us that there must exist at least one line containing exactly two points.*

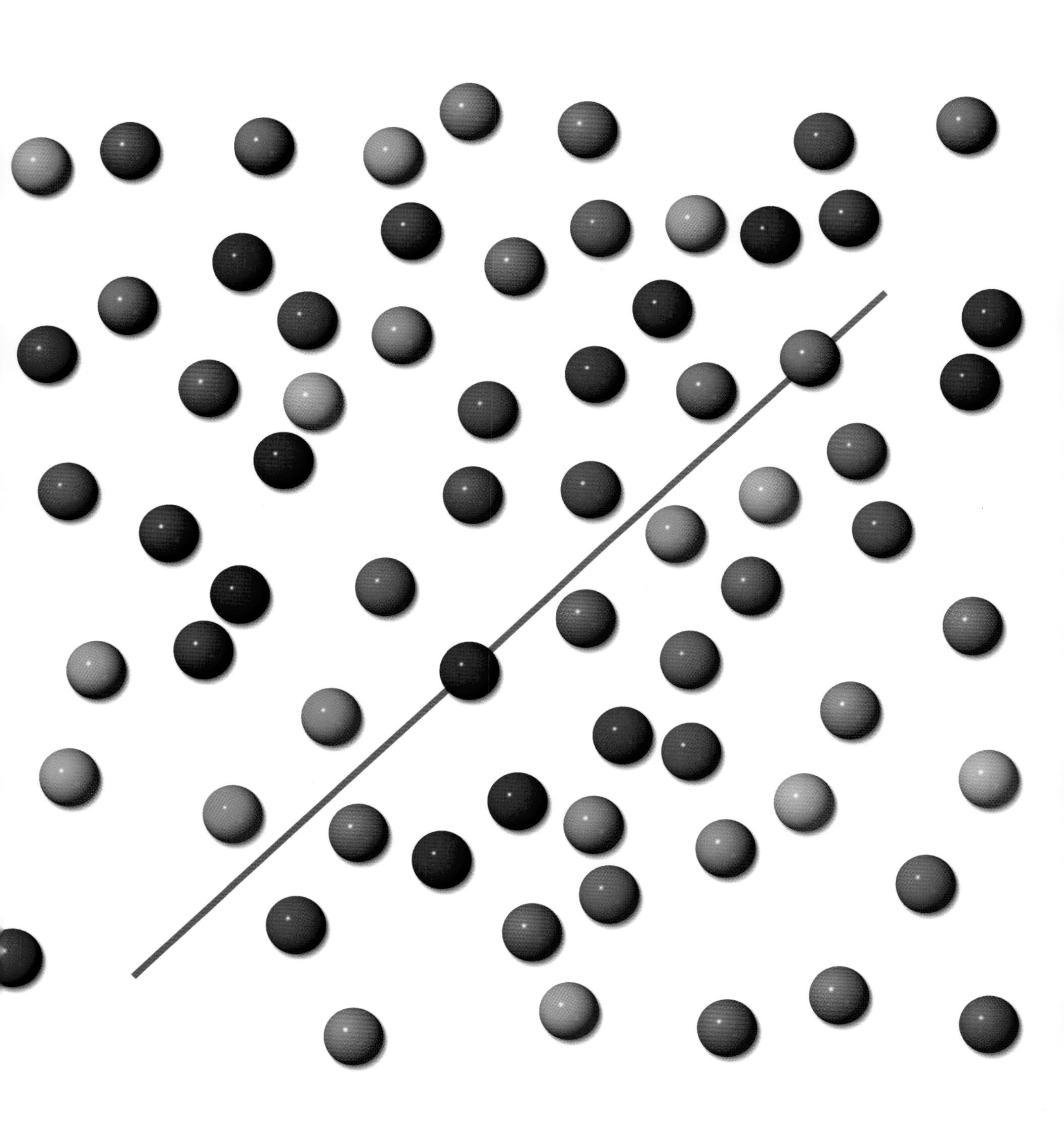

1896

# Proof of the Prime Number Theorem

**Johann Carl Friedrich Gauss** (1777–1855), **Jacques Salomon Hadamard** (1865–1963), **Charles-Jean de la Vallée-Poussin** (1866–1962), **John Edensor Littlewood** (1885–1977)

Mathematician Don Zagier has commented that "despite their simple definition and role as the building blocks of the natural numbers, the prime numbers grow like weeds among the natural numbers…and nobody can predict where the next one will sprout…. Even more astonishing…the prime numbers exhibit stunning regularity, there are laws governing their behavior, and they obey these laws with almost military precision."

Consider $\pi(n)$, which is the number of primes less than or equal to a given number $n$. In 1792, when only 15 years old, Carl Gauss became fascinated by the occurrence of prime numbers, and he proposed that $\pi(n)$ was approximately equal to $n/\ln(n)$, where ln is the natural logarithm. One consequence of the prime number theorem is that the $n$th prime number is approximately equal to $n\ln(n)$, with the relative error of this approximation approaching 0 as $n$ approaches infinity. Gauss later refined his estimate to $\pi(n) \sim \mathrm{Li}(n)$ where $\mathrm{Li}(n)$ is the integral from 2 to $n$ of $dx/\ln(x)$.

Finally, in 1896, French mathematician Jacques Hadamard and Belgian mathematician Charles-Jean de la Vallée-Poussin independently proved Gauss's theorem. Based on numerical experiments, mathematicians had conjectured that $\pi(n)$ was always somewhat less than $\mathrm{Li}(n)$. However, in 1914, Littlewood proved that $\pi(n) < \mathrm{Li}(n)$ reverses *infinitely* often if one were able to search though huge values of $n$. In 1933, South African mathematician Stanley Skewes showed that the first crossing of $\pi(n) - \mathrm{Li}(n) = 0$ occurs before 10^10^10^34, a number referred to as Skewes' number, where ^ indicates the raising to a power. Since 1933, this value has been reduced to around $10^{316}$.

English mathematician G. H. Hardy (1877–1947) once described Skewes' number as "the largest number which has ever served any definite purpose in mathematics," although the Skewes' number has since lost this lofty accolade. Around 1950, Paul Erdös and Atle Selberg discovered an elementary proof of the prime number theorem—that is, a proof that uses only real numbers.

**SEE ALSO** Cicada-Generated Prime Numbers (c. 1 Million B.C.), Sieve of Eratosthenes (240 B.C.), Goldbach Conjecture (1742), Constructing a Regular Heptadecagon (1796), Gauss's *Disquisitiones Arithmeticae* (1801), Riemann Hypothesis (1859), Brun's Constant (1919), Gilbreath's Conjecture (1958), Ulam Spiral (1963), Erdös and Extreme Collaboration (1971), Public-Key Cryptography (1977), and Andrica's Conjecture (1985).

*Prime numbers, represented in boldface, "grow like weeds among the natural numbers…and nobody can predict where the next one will sprout…." Although the number 1 used to be considered a prime, today mathematicians generally consider 2 to be the first prime.*

**1**, **2**, **3**, 4, **5**, 6, **7**, 8, 9, 10, **11**, 12, **13**, 14, 15, 16, **17**, 18, **19**, 20, 21, 22, **23**, 24, 25, 26, 27, 28, **29**, 30, **31**, 32, 33, 34, 35, 36, **37**, 38, 39,
40, **41**, 42, **43**, 44, 45, 46, **47**, 48, 49, 50, 51, 52, **53**, 54, 55, 56, 57, 58, **59**, 60, **61**, 62, 63, 64, 65, 66, **67**, 68, 69, 70, **71**, 72, **73**, 74, 75,
76, 77, 78, **79**, 80, 81, 82, **83**, 84, 85, 86, 87, 88, **89**, 90, 91, 92, 93, 94, 95, 96, **97**, 98, 99, 100, **101**, 102, **103**, 104, 105, 106, **107**, 108,
**109**, 110, 111, 112, **113**, 114, 115, 116, 117, 118, 119, 120, 121, 122, 123, 124, 125, 126, **127**, 128, 129, 130, **131**, 132, 133, 134, 135,
136, **137**, 138, **139**, 140, 141, 142, 143, 144, 145, 146, 147, 148, **149**, 150, **151**, 152, 153, 154, 155, 156, **157**, 158, 159, 160, 161, 162,
**163**, 164, 165, 166, **167**, 168, 169, 170, 171, 172, **173**, 174, 175, 176, 177, 178, **179**, 180, **181**, 182, 183, 184, 185, 186, 187, 188, 189,
190, **191**, 192, **193**, 194, 195, 196, **197**, 198, **199**, 200, 201, 202, 203, 204, 205, 206, 207, 208, 209, 210, **211**, 212, 213, 214, 215, 216,
217, 218, 219, 220, 221, 222, **223**, 224, 225, 226, **227**, 228, **229**, 230, 231, 232, **233**, 234, 235, 236, 237, 238, **239**, 240, **241**, 242, 243,
244, 245, 246, 247, 248, 249, 250, **251**, 252, 253, 254, 255, 256, **257**, 258, 259, 260, 261, 262, **263**, 264, 265, 266, 267, 268, **269**, 270,
**271**, 272, 273, 274, 275, 276, **277**, 278, 279, 280, **281**, 282, **283**, 284, 285, 286, 287, 288, 289, 290, 291, 292, **293**, 294, 295, 296, 297,
298, 299, 300, 301, 302, 303, 304, 305, 306, **307**, 308, 309, 310, **311**, 312, **313**, 314, 315, 316, **317**, 318, 319, 320, 321, 322, 323, 324,
325, 326, 327, 328, 329, 330, **331**, 332, 333, 334, 335, 336, **337**, 338, 339, 340, 341, 342, 343, 344, 345, 346, **347**, 348, **349**, 350, 351,
352, **353**, 354, 355, 356, 357, 358, **359**, 360, 361, 362, 363, 364, 365, 366, **367**, 368, 369, 370, 371, 372, **373**, 374, 375, 376, 377, 378,
**379**, 380, 381, 382, **383**, 384, 385, 386, 387, 388, **389**, 390, 391, 392, 393, 394, 395, 396, **397**, 398, 399, 400, **401**, 402, 403, 404, 405,
406, 407, 408, **409**, 410, 411, 412, 413, 414, 415, 416, 417, 418, **419**, 420, **421**, 422, 423, 424, 425, 426, 427, 428, 429, 430, **431**, 432,
**433**, 434, 435, 436, 437, 438, **439**, 440, 441, 442, **443**, 444, 445, 446, 447, 448, **449**, 450, 451, 452, 453, 454, 455, 456, **457**, 458, 459,
460, **461**, 462, **463**, 464, 465, 466, **467**, 468, 469, 470, 471, 472, 473, 474, 475, 476, 477, 478, **479**, 480, 481, 482, 483, 484, 485, 486,
**487**, 488, 489, 490, **491**, 492, 493, 494, 495, 496, 497, 498, **499**, 500, 501, 502, **503**, 504, 505, 506, 507, 508, **509**, 510, 511, 512, 513,
514, 515, 516, 517, 518, 519, 520, **521**, 522, **523**, 524, 525, 526, 527, 528, 529, 530, 531, 532, 533, 534, 535, 536, 537, 538, 539, 540,
**541**, 542, 543, 544, 545, 546, **547**, 548, 549, 550, 551, 552, 553, 554, 555, 556, **557**, 558, 559, 560, 561, 562, **563**, 564, 565, 566, 567,
568, **569**, 570, **571**, 572, 573, 574, 575, 576, **577**, 578, 579, 580, 581, 582, 583, 584, 585, 586, **587**, 588, 589, 590, 591, 592, **593**, 594,
595, 596, 597, 598, **599**, 600, **601**, 602, 603, 604, 605, 606, **607**, 608, 609, 610, 611, 612, **613**, 614, 615, 616, **617**, 618, **619**, 620, 621,
622, 623, 624, 625, 626, 627, 628, 629, 630, **631**, 632, 633, 634, 635, 636, 637, 638, 639, 640, **641**, 642, **643**, 644, 645, 646, **647**, 648,
649, 650, 651, 652, **653**, 654, 655, 656, 657, 658, **659**, 660, **661**, 662, 663, 664, 665, 666, 667, 668, 669, 670, 671, 672, **673**, 674, 675,
676, **677**, 678, 679, 680, 681, 682, **683**, 684, 685, 686, 687, 688, 689, 690, **691**, 692, 693, 694, 695, 696, 697, 698, 699, 700, **701**, 702,
703, 704, 705, 706, 707, 708, **709**, 710, 711, 712, 713, 714, 715, 716, 717, 718, **719**, 720, 721, 722, 723, 724, 725, 726, **727**, 728, 729,
730, 731, 732, **733**, 734, 735, 736, 737, 738, **739**, 740, 741, 742, **743**, 744, 745, 746, 747, 748, 749, 750, **751**, 752, 753, 754, 755, 756,
**757**, 758, 759, 760, **761**, 762, 763, 764, 765, 766, 767, 768, **769**, 770, 771, 772, **773**, 774, 775, 776, 777, 778, 779, 780, 781, 782, 783,
784, 785, 786, **787**, 788, 789, 790, 791, 792, 793, 794, 795, 796, **797**, 798, 799, 800, 801, 802, 803, 804, 805, 806, 807, 808, **809**, 810,
**811**, 812, 813, 814, 815, 816, 817, 818, 819, 820, **821**, 822, **823**, 824, 825, 826, **827**, 828, **829**, 830, 831, 832, 833, 834, 835, 836, 837,
838, **839**, 840, 841, 842, 843, 844, 845, 846, 847, 848, 849, 850, 851, 852, **853**, 854, 855, 856, **857**, 858, **859**, 860, 861, 862, **863**, 864,
865, 866, 867, 868, 869, 870, 871, 872, 873, 874, 875, 876, **877**, 878, 879, 880, **881**, 882, **883**, 884, 885, 886, **887**, 888, 889, 890, 891,
892, 893, 894, 895, 896, 897, 898, 899, 900, 901, 902, 903, 904, 905, 906, **907**, 908, 909, 910, **911**, 912, 913, 914, 915, 916, 917, 918,
**919**, 920, 921, 922, 923, 924, 925, 926, 927, 928, **929**, 930, 931, 932, 933, 934, 935, 936, **937**, 938, 939, 940, **941**, 942, 943, 944, 945,
946, **947**, 948, 949, 950, 951, 952, **953**, 954, 955, 956, 957, 958, 959, 960, 961, 962, 963, 964, 965, 966, **967**, 968, 969, 970, **971**, 972,
973, 974, 975, 976, **977**, 978, 979, 980, 981, 982, **983**, 984, 985, 986, 987, 988, 989, 990, **991**, 992, 993, 994, 995, 996, **997**, 998, 999

1899

# Pick's Theorem

**Georg Alexander Pick** (1859–1942)

Pick's theorem is delightful for its simplicity, and it can be experimented with using a pencil and graph paper. Draw a simple polygon on an equally spaced grid so that all of the vertices (corners) of the polygon fall on grid points. Pick's theorem tells us that the area A of this polygon, in units squared, can be determined by counting the number $i$ of points located within the polygon and the number $b$ of boundary points located on the boundary of the polygon, according to $A = i + b/2 - 1$. Pick's theorem does not apply to polygons that have holes in them.

Austrian mathematician Georg Pick presented his theorem in 1899. In 1911, Pick introduced Albert Einstein to the work of relevant and key mathematicians, which helped Einstein to develop his general theory of relativity. When Hitler's troops invaded Austria in 1938, Pick, a Jew, fled to Prague. Sadly, his flight was not sufficient to save his life. The Nazis invaded Czechoslovakia and sent him to the Theresienstadt concentration camp in 1942, where he died. Of the approximately 144,000 Jews sent to Theresienstadt, about a quarter died on site and around 60 percent were shipped to Auschwitz or other death camps.

Mathematicians have since discovered that no direct analogue of Pick's theorem exists in three dimensions that allows us to calculate the volume of a polytope (for example, a polyhedron) by counting its interior and boundary points.

Using tracing paper that has a grid, we can use Pick's theorem to estimate the areas of regions on a map, if we approximate the region with a polygon. British science writer David Darling writes: "Over the past few decades,...various generalizations of Pick's theorem have been made to more general polygons, to higher-dimensional polyhedra, and to lattices other than square lattices....The theorem provides a link between traditional Euclidean geometry and the modern subject of digital (discrete) geometry."

**SEE ALSO** Platonic Solids (350 B.C.), Euclid's *Elements* (300 B.C.), and Archimedean Semi-Regular Polyhedra (c. 240 B.C.).

*According to Pick's theorem, the area of this polygon is* i + b/2 − 1, *where* i *is the number of points located within the polygon and* b *is the number of boundary points located on the boundary of the polygon.*

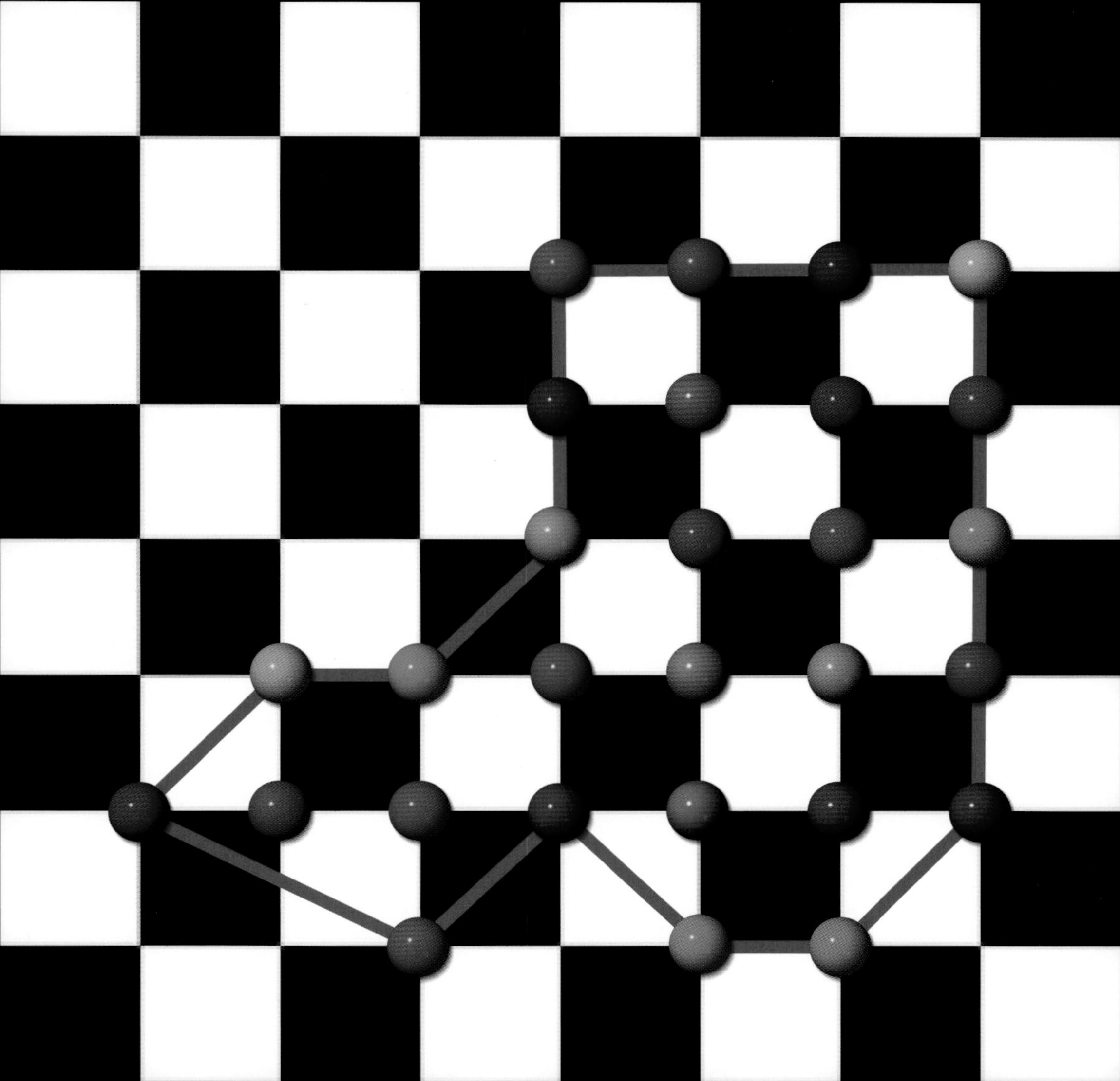

# Morley's Trisector Theorem

**Frank Morley** (1860–1937)

In 1899, Anglo-American mathematician and accomplished chess player Frank Morley proposed Morley's theorem that states that in *any* triangle, the three points of intersection of adjacent angle trisectors *always* form an equilateral triangle. *Trisectors* refers to the straight lines that divide the interior angles into three equal parts, and these lines intersect in six points, of which three are vertices of an equilateral triangle. Various proofs exist, and some of the earliest proofs were quite complicated.

Morley's colleagues found the result so beautiful and surprising that it came to be known as "Morley's Miracle." Richard Francis writes, "Apparently overlooked by ancient geometers or hastily abandoned because of trisection and constructability uncertainties, the problem came to light only a century ago. Though conjectured around 1900 by Frank Morley, resolution or rigorous proof was to await even more recent advances. This beautiful and elegant Euclidean theorem, mysteriously unnoticed across the ages, thus belongs to the twentieth century."

Morley taught at both the Quaker College in Haverford, Pennsylvania, and at Johns Hopkins University. In 1933, he published *Inversive Geometry* co-written with his son, the mathematician Frank V. Morley. The son wrote about his father in *One Contribution to Chess*: "He would begin to fiddle in his waistcoat pocket for a stub of pencil perhaps two inches long, and there would be a certain amount of scrabbling in a side pocket for an old envelope... until he would get up a little stealthily and make his way toward his study... and my mother would call out, 'Frank, you're not going to work!'—and the answer would always be, 'A little, not much!'—and the study door would close."

Morley's theorem continues to fascinate mathematicians. In 1998, Alain Connes, a French Fields Medalist, presented a new proof of Morley's theorem.

**SEE ALSO** Euclid's *Elements* (300 B.C.), Law of Cosines (c. 1427), Viviani's Theorem (1659), Euler's Polygon Division Problem (1751), and Ball Triangle Picking (1982).

*According to Morley's theorem—also known as* Morley's Miracle*—for* any *triangle, the three points of intersection of adjacent angle trisectors* always *form an equilateral triangle.*

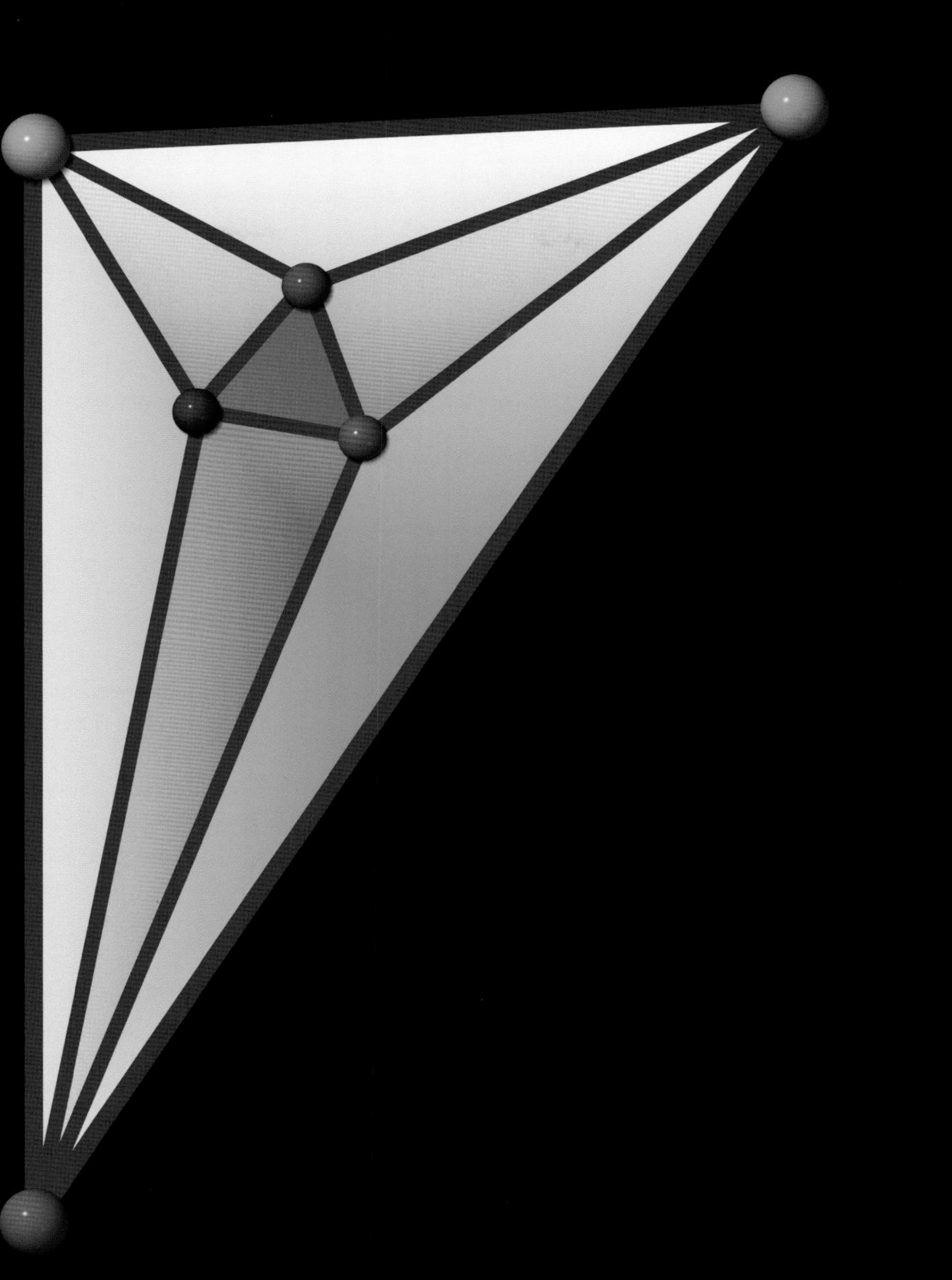

1900

# Hilbert's 23 Problems

**David Hilbert** (1862–1943)

German mathematician David Hilbert wrote, "A branch of science is full of life as long as it offers an abundance of problems; a lack of problems is a sign of death." In 1900, he presented 23 important mathematical problems to be targeted for solution in the twentieth century. Because of Hilbert's prestige, mathematicians spent a great deal of time tackling the problems through the years. His extremely influential speech on the subject started, "Who of us would not be glad to lift the veil behind which the future lies hidden; to cast a glance at the next advances of our science and at the secrets of its development during future centuries? What particular goals will there be toward which the leading mathematical spirits of coming generations will strive?"

About ten of the problems have since been cleanly solved, and many others have solutions that are accepted by some mathematicians but for which some controversy still remains. For example, the **Kepler Conjecture** (part of Problem 18), which raised questions about the efficiency of sphere packing, involved a computer-assisted proof, which may be difficult for people to verify.

One of the most famous problems still unresolved today is the **Riemann Hypothesis**, which concerns the distribution of the zeros of the Riemann zeta function (a very wiggly function). David Hilbert remarked, "If I were to awaken after having slept for a thousand years, my first question would be: Has the Riemann hypothesis been proven?"

Ben Yandell writes, "Solving one of Hilbert's Problems has been the romantic dream of many a mathematician.... In the last hundred years, solutions and significant partial results have come from all over the world. Hilbert's list is a thing of beauty, and aided by their romantic and historical appeal, these well-chosen problems have been an organizing force in mathematics."

**SEE ALSO** Kepler Conjecture (1611), Riemann Hypothesis (1859), and Hilbert's Grand Hotel (1925).

*Photograph of David Hilbert (1912), which appeared on postcards of faculty members at the University of Göttingen. Students often purchased such postcards.*

1900

# Chi-Square

**Karl Pearson** (1857–1936)

Scientists often obtain experimental results that do not agree with those anticipated according to the rules of probability. For example, when tossing a die, if the deviation from expectation is very large, we would say that the die is probably biased, such as would be the case for a die with unequally weighted sides.

The chi-square test was first published in 1900 by British mathematician Karl Pearson, and his method has since been used in countless fields ranging from cryptography and reliability engineering to the analysis of hitting records in baseball. When applying the test, events are assumed to be independent (as in our die-tossing example). The chi-square value can be calculated once we know each observed frequency, $O_i$, and each theoretical (i.e., expected) frequency, $E_i$. The formula can be expressed as $\chi^2 = \Sigma(O_i - E_i)^2/E_i$. If the frequency of expected and observed events agrees exactly, then $\chi^2 = 0$. The greater the differences, the larger the value for $\chi^2$. In practice, the significance of this difference is determined with reference to a chi-square table that helps researchers determine the degree of significance of the difference. Of course, researchers may also be suspicious if $\chi^2$ is too close to zero, and thus may look for values of $\chi^2$ that are either too low or too high.

As an example, let us test the hypothesis that a random sample of 100 insects has been drawn from a population in which butterflies and beetles are equal in frequency. If we observe 10 beetles and 90 butterflies, we obtain a $\chi^2$ value of $(10 - 50)^2/50 + (90 - 50)^2/50 = 64$, a huge value that suggests that our initial hypothesis—that we have randomly drawn from a population with the same number of butterflies as beetles—is probably incorrect.

Pearson received many awards for his work, although outside the field of mathematics he was a racist and advocated a "war" against "inferior races."

**SEE ALSO** Dice (3000 B.C.), Law of Large Numbers (1713), Normal Distribution Curve (1733), Least Squares (1795), and Laplace's *Théorie Analytique des Probabilités* (1812).

*Chi-square values help us test the hypothesis that a random sample of 100 insects has been drawn from a population in which butterflies and beetles are equal in frequency. For this figure, a value of 64 suggests that our hypothesis is probably incorrect.*

1901

# Boy's Surface

**Werner Boy** (1879–1914), **Bernard Morin** (b. 1931)

Boy's surface was discovered in 1901 by German mathematician Werner Boy. Like the **Klein Bottle**, this object is a single-sided surface with no edges. Boy's surface is also a non-orientable surface, which means that a two-dimensional creature can travel within the surface and find paths that will reverse the creature's handedness when it returns to its starting point. The **Möbius Strip** and Klein bottle also have non-orientable surfaces.

Formally speaking, Boy's surface is an immersion of a projective plane in three-dimensional space with no singularities (pinch points). Geometric recipes exist for its creation, and some of them involve the stretching of a disk and the gluing of the disk edge to the edge of a Möbius strip. During the process, the surface is allowed to pass through itself, but it may not be torn or have any pinch points. The Boy's surface is very difficult to visualize, although computer graphics help researchers have a better feel for the shape.

Boy's surface has three-fold symmetry. In other words, there exists an axis about which the shape can be rotated by 120° and look identical. Interestingly, Boy was able to sketch several models of the surface, but he could not determine the equations (that is, a parametric model) to describe the surface. Finally, in 1978, French mathematician Bernard Morin used computers to find the first parameterization. Morin, who was blind since childhood, had a successful career in mathematics.

Mathematics journalist Allyn Jackson writes, "Far from detracting from his extraordinary visualization ability, Morin's blindness may have enhanced it.... One thing that is difficult about visualizing geometric objects is that one tends to see only the outside of the objects, not the inside, which might be very complicated.... Morin has developed the ability to pass from outside to inside.... Because he is so accustomed to tactile information, Morin can, after manipulating a hand-held model for a couple of hours, retain the memory of its shape for years afterward."

**SEE ALSO** Minimal Surface (1774), The Möbius Strip (1858), Klein Bottle (1882), Turning a Sphere Inside Out (1958), and Weeks Manifold (1985).

*Boy's surface, rendered by Paul Nylander. This object is a single-sided surface with no edges.*

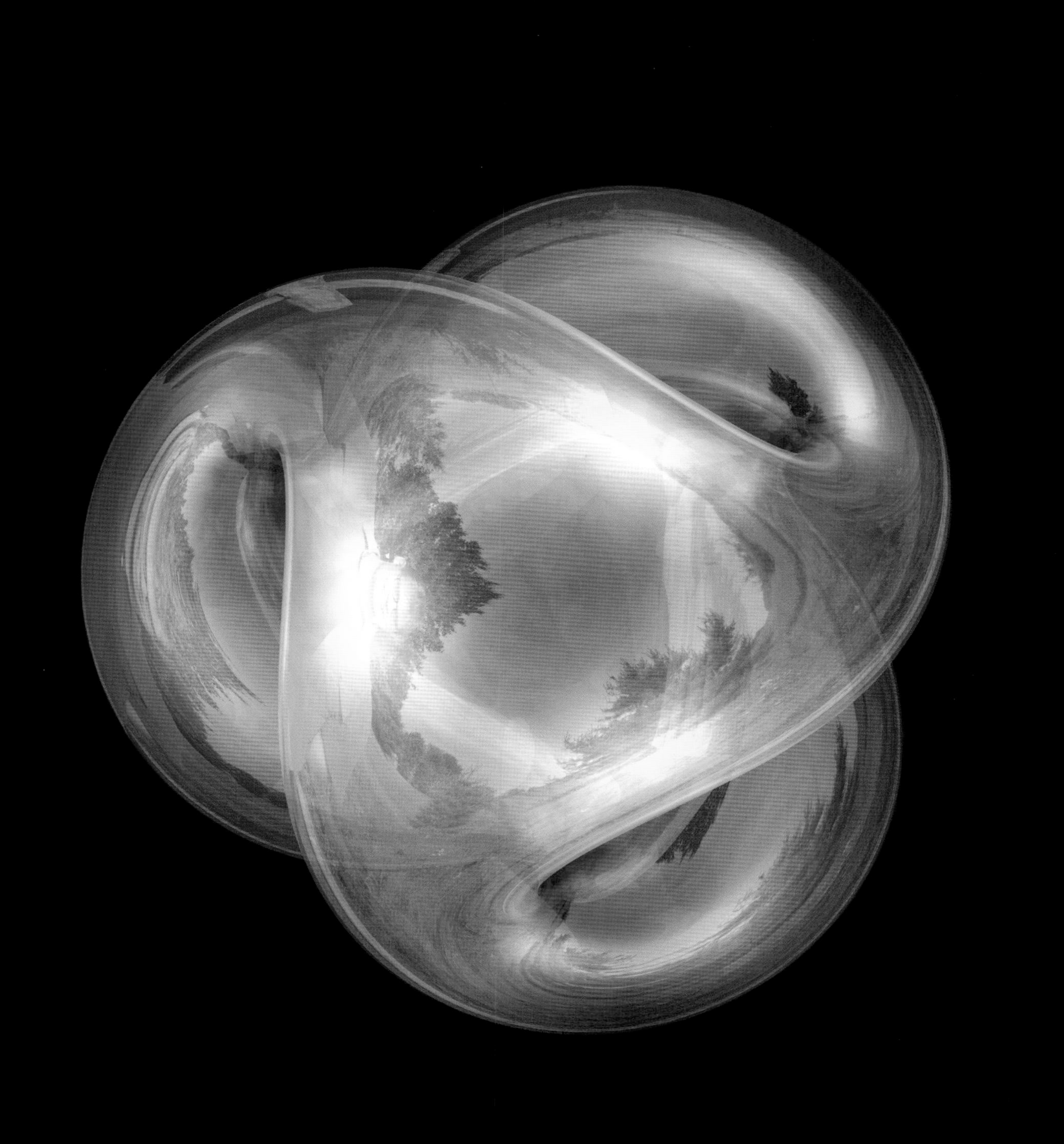

1901

# Barber Paradox

**Bertrand Russell** (1872–1970)

In 1901, British philosopher and mathematician Bertrand Russell uncovered a possible paradox or apparent contradiction that forced a modification to set theory. One version of Russell's paradox, also known as the Barber paradox, involves a town with one male barber who, every day, shaves every man who doesn't shave himself, and no one else. Does the barber shave himself?

The scenario seemed to demand that the barber shave himself if and only if he does not shave himself! Helen Joyce writes, "The paradox raises the frightening prospect that the whole of mathematics is based on shaky foundations, and that no proof can be trusted."

Russell's paradox, in its original form, involves the set of all sets that aren't members of themselves. Many sets $R$ are not members of themselves—for example, the set of cubes is not a cube. Examples of sets $T$ that do contain themselves as members are the set of all sets, or the set of all things except cubes. Every set would seem to be either of type $R$ or of type $T$, and no set can be both. However, Russell wondered about the set $S$ of all sets that aren't members of themselves. Somehow, $S$ is neither a member of itself nor not a member of itself. Russell realized that he had to alter set theory in a way to avoid such confusions and possible contradictions.

One possible refutation of the Barber paradox seems to be that we can simply say that such a barber does not exist. Nevertheless, Russell's paradox led to a cleaner form of set theory. German mathematician Kurt Gödel made use of similar observations when forming his incompleteness theorem. British mathematician Alan Turing also found Russell's work useful when studying the undecidability of the halting problem, which concerns the assessment of whether or not a computer program will finish in a finite number of steps.

**SEE ALSO** Zeno's Paradoxes (c. 445 B.C.), Aristotle's Wheel Paradox (c. 320 B.C.), St. Petersburg Paradox (1738), Zermelo's Axiom of Choice (1904), *Principia Mathematica* (1910–1913), Banach-Tarski Paradox (1924), Hilbert's Grand Hotel (1925), Gödel's Theorem (1931), Turing Machines (1936), Birthday Paradox (1939), Newcomb's Paradox (1960), Chaitin's Omega (1974), and Parrondo's Paradox (1999).

*The Barber paradox involves a town with one male barber who, every day, shaves every man who doesn't shave himself, and no one else. Does the barber shave himself?*

sterilizer

1901

# Jung's Theorem

## **Heinrich Wilhelm Ewald Jung** (1876–1953)

Imagine a finite set of scattered points, as you might see in a map of a constellation of stars or randomly placed drops of ink on a page. Draw a line between the two points that have the greatest separation. This largest possible distance $d$ between two points is called the *geometric span* of the set of points. Jung's theorem says that no matter how strangely scattered the points are, they are guaranteed to be enclosed by a circle with a radius no greater than $d/\sqrt{3}$. In the case of points arranged along the sides of an equilateral triangle with a side length of 1 unit, the enclosing circle touches all three vertices (corners) of the triangle and has a radius of $1/\sqrt{3}$.

Jung's theorem may be generalized to three dimensions in which the set of points can be enclosed by a sphere with a radius no greater than $\sqrt{6}d/4$. This means, for example, that if we have a collection of point-like objects in space, such as a flock of birds or a school of fish, then these objects are guaranteed to be enclosable by such a sphere. Jung's theorem has since been extended to various **Non-Euclidean Geometries** and spaces.

If we want to move the theorem into more mind-boggling territories, such as encapsulating birds in higher-dimensional hyperspheres of dimension $n$, we can resort to the wonderfully compact formula

$$r \leq d\sqrt{\frac{n}{2(n+1)}}$$

which means that a four-dimensional hypersphere of radius $d\sqrt{2/5}$ is guaranteed to trap a flock of starlings that are flying with access to the fourth dimension. German mathematician Heinrich Jung studied mathematics, physics, and chemistry at the University of Marburg and at the University of Berlin from 1895 to 1899, and he published his theorem in 1901.

**SEE ALSO** Euclid's *Elements* (300 B.C.), Non-Euclidean Geometry (1829), and Sylvester's Line Problem (1893).

*A flock of birds, no matter how complicated, can be enclosed by a sphere with a radius no greater than $\sqrt{6}d/4$ if we consider each bird to be a point in space. What can we say about a flock of starlings in a four-dimensional space?*

1904

# Poincaré Conjecture

**Henri Poincaré** (1854–1912), **Grigori Perelman** (b. 1966)

The Poincaré conjecture, posed in 1904 by French mathematician Henri Poincaré, concerns topology, the branch of mathematics involving the study of shapes and their interrelations. In 2000, the Clay Mathematics Institute offered $1 million for a proof of this conjecture, which can be conceptually visualized at a high level in terms of oranges and doughnuts. Imagine a loop of string wrapped around an orange. In theory, we can slowly shrink the loop to a point without tearing the string or the orange, and without the string leaving the surface of the orange. However, if a string is wrapped around a doughnut through its hole, the string can't be shrunk to a point without breaking the string or the doughnut. The surface of the orange is called *simply connected*, and the doughnut surface is not. Poincaré understood that a two-dimensional spherical shell (for example, modeled by the orange *surface*) is simply connected, and he asked if a three-dimensional sphere (the set of points in four-dimensional space that are the same distance away from a single point) has the same properties.

Finally, in 2002 and 2003, Russian mathematician Grigori Perelman proved the conjecture. Oddly enough, Perelman showed little interest in collecting the prize and simply placed his solution on the Internet rather than publishing it in a mainstream journal. In 2006, Perelman was awarded the prestigious **Fields Medal** for his solution, but he rejected the award, saying that it was "completely irrelevant" for him. For Perelman, if the proof was correct "then no other recognition is needed."

*Science* magazine reported in 2006, "Perelman's proof has fundamentally altered two distinct branches of mathematics. First, it solved a problem that for more than a century was the indigestible seed at the core of topology.... [Second], the work will lead to a much broader result...a 'periodic table' that brings clarity to the study of three-dimensional spaces, much as Mendeleev's table did for chemistry."

**SEE ALSO** Königsberg Bridges (1736), Klein Bottle (1882), Fields Medal (1936), and Weeks Manifold (1985).

*French mathematician Henri Poincaré, who posed the Poincaré conjecture in 1904. The conjecture remained unproven until 2002 and 2003 when Russian mathematician Grigori Perelman finally offered a valid proof.*

1904

# Koch Snowflake

## **Niels Fabian Helge von Koch** (1870–1924)

The Koch snowflake is often one of the first fractal objects to which students are exposed, and it is also among the earliest fractal objects described in the history of mathematics. The intricate shape appears in Swedish mathematician Helge von Koch's 1904 paper "On a Continuous Curve without Tangents, Constructible from Elementary Geometry." A related object, the Koch curve, starts with a line segment instead of an equilateral triangle for the process used to generate the curve.

To create the crinkly Koch curve, we may recursively alter a line segment, watching it sprout an infinite amount of edges in the process. Imagine splitting a line into three equal parts. Next, replace the middle portion with two lines, both of the same length as the first three, so that they form a V-shaped wedge (the top edges of an equilateral triangle). The shape now consists of four straight lines. For each of these lines, repeat the process of splitting and forming wedges.

Starting with a line 1 inch in length, the length of the growing curve at step $n$ in the procedure is $(4/3)^n$ inches. After a few hundred iterations, the length of the curve becomes longer than the diameter of the visible universe. In fact, the "final" Koch curve has infinite length and a fractal dimension of about 1.26, because it partially fills the 2-D plane in which it is drawn.

Even though the edge of a Koch snowflake has an infinite length, it encloses a finite area $(2\sqrt{3}s^2)/5$, where $s$ is the original side length, or equivalently, the area is simply 8/5 times the area of the original triangle. Note that a function has no definite tangent at a corner, which means that a function is not differentiable (has no unique derivatives) at corners. The Koch curve is non-differentiable everywhere (because it is so pointy!), even though the curve is continuous.

**SEE ALSO** Weierstrass Function (1872), Peano Curve (1890), Hausdorff Dimension (1918), Menger Sponge (1926), Coastline Paradox (c. 1950), and Fractals (1975).

*Koch snowflake tiling. To create this pattern, mathematician and artist Robert Fathauer uses different snowflake sizes.*

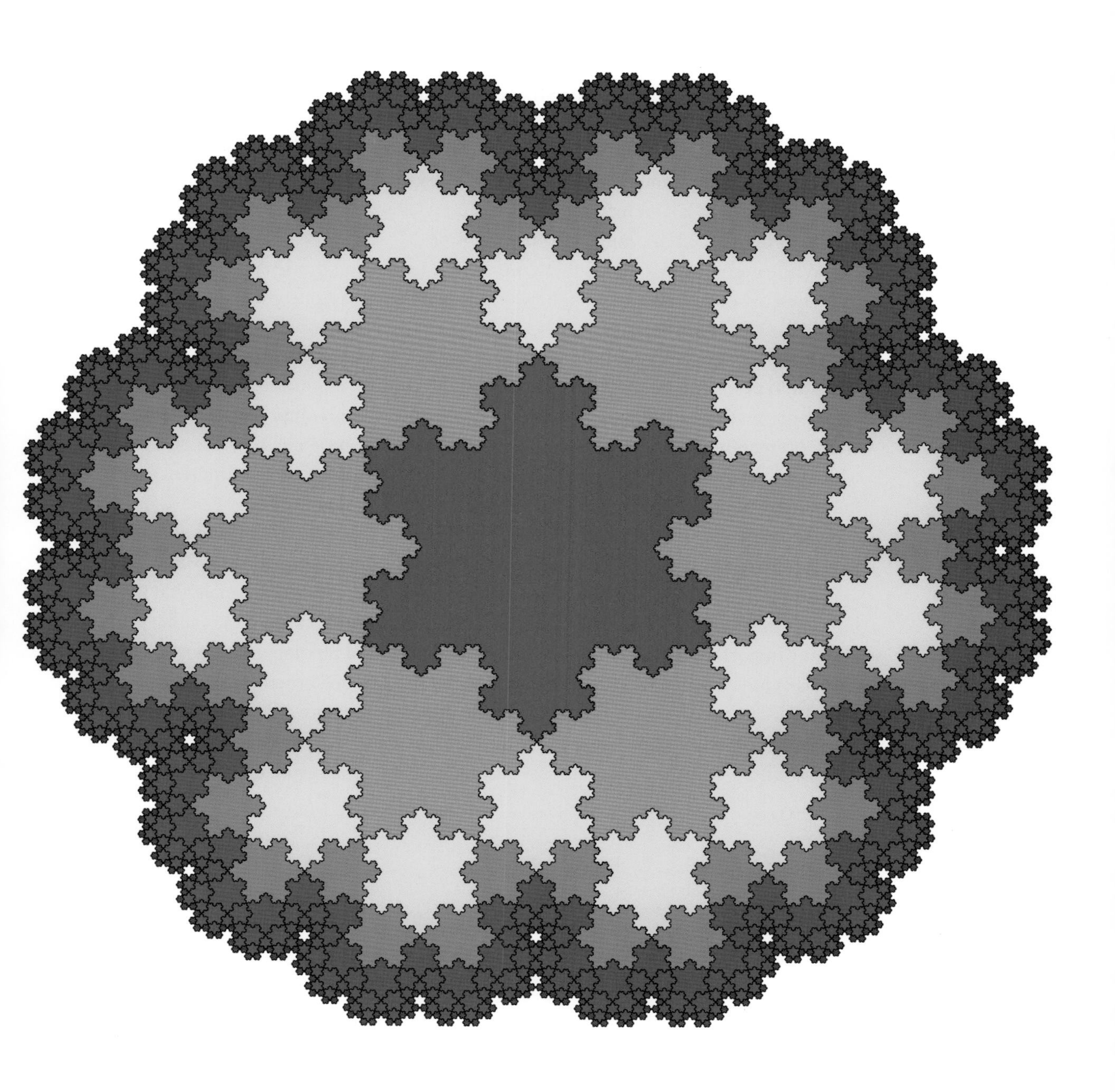

1904

# Zermelo's Axiom of Choice

**Ernst Friedrich Ferdinand Zermelo** (1871–1953)

David Darling calls this axiom in set theory "one of the most controversial axioms in mathematics." The axiom was formulated in 1904 by the German mathematician Ernst Zermelo, who was later appointed to an honorary chair at the University of Freiburg, which he renounced in protest of Hitler's regime.

While complex to write mathematically, the axiom can be visualized using a long shelf of goldfish bowls. Each bowl must contain at least one goldfish. The axiom of choice (AC) simply says that you can always choose, in theory, one goldfish from each bowl, even if there are *infinitely* many bowls, even if we have no "rule" for which goldfish to pluck from each bowl, and even if the goldfish are indistinguishable.

Using mathematical language, if S is a collection of non-empty sets having no element in common, then a set must exist that has exactly one element in common with every set $s$ of S. Looking at this another way, there exists a choice function $f$ with the property that, for each set $s$ in the collection, $f(s)$ is a member of $s$.

Before the AC, there was no reason to believe that we could always find a mathematical rationale for which fish to pick from the bowls if some of the bowls had infinitely many fish, or at least no reason to think that we could always find a rationale that would take less than an infinite amount of time to use. It turns out that the AC is at the core of many important mathematical theorems in algebra and topology, and most mathematicians today accept AC because it is *so useful*. Eric Schecter writes, "When we accept AC, this means we are agreeing to the convention that we shall permit ourselves to use a hypothetical choice function $f$ in proofs, as though it 'exists' in some sense, even in cases where we cannot give an explicit example of it or an explicit algorithm for it."

**SEE ALSO** Peano Axioms (1889), Barber Paradox (1901), and Hilbert's Grand Hotel (1925).

*In theory, even if we have* infinitely *many goldfish bowls, we can always choose one goldfish from each bowl, even if we have no "rule" for which goldfish to pluck from each bowl, and even if the goldfish are indistinguishable.*

1905

# Jordan Curve Theorem

## **Marie Ennemond Camille Jordan** (1838–1922), **Oswald Veblen** (1880–1960)

Find a loop of wire, twist it in a very convoluted fashion that does not include any self-intersections, and lay it flat on a table to create a maze of sorts. You place an ant in the structure. If the maze is sufficiently complex, it is visually difficult to determine if the ant is inside or outside the loop. One way to determine if the ant is within the loop is to count the number of times an imaginary straight line drawn from the ant to the outside world crosses the wire. If the line crosses the curve an even number of times, the ant is outside the maze; if an odd number of times, the ant is inside.

French mathematician Camille Jordan researched these kinds of rules for determining the inside and outside of curves, and he is most famous for his theorem showing that a simply closed curve divides a plane into an inside and outside, now called the Jordan curve theorem (JCT). Although this may seem obvious, Jordan realized that a rigorous proof was necessary and difficult. Jordan's work with curves appeared in his *Cours d'analyse de l'École Polytechnique* (*Analysis Course from the École Polytechnique*), first published in three volumes between 1882 and 1887. The JCT appeared in the third edition of the text, published between 1909 and 1915. American mathematician Oswald Veblen is usually credited with being the first to provide a precise proof of the JCT, in 1905.

Note that a Jordan curve is a plane curve that is a deformed circle, and it must be simple (the curve cannot cross itself) and closed (has no endpoints and also completely encloses an area). On a plane or sphere, Jordan curves have an inside and outside—and to get from one to the other, at least one line must be crossed. However, on a torus (the surface of a doughnut shape), Jordan curves do not necessarily exhibit this property.

**SEE ALSO** Königsberg Bridges (1736), Holditch's Theorem (1858), Poincaré Conjecture (1904), Alexander's Horned Sphere (1924), and Sprouts (1967).

*Jordan curves by mathematician and artist Robert Bosch.* TOP: *Is the red dot inside or outside of the Jordan curve?* BOTTOM: *The white line is a Jordan curve; the green and blue regions are its interior and exterior, respectively.*

1906

# Thue-Morse Sequence

**Axel Thue** (1863–1922), **Marston Morse** (1892–1977)

The Thue-Morse (TM) sequence is a binary sequence that begins 01101001....In my book *Mazes for the Mind*, when the sequence is converted to sounds, one character remarks, "It's the strangest thing ye ever heard. It ain't exactly irregular and it ain't exactly regular, either." The sequence is named in honor of the Norwegian mathematician Axel Thue and American mathematician Marston Morse. In 1906, Thue introduced the sequence as an example of an aperiodic, recursively computable string of symbols. In 1921, Morse applied it to his studies of differential geometry, and numerous fascinating properties and applications have since been discovered.

One way to generate the sequence is to start with a zero and then repeatedly do the following replacements: $0 \rightarrow 01$ and $1 \rightarrow 10$ to produce the following successive generations: 0, 01, 0110, 01101001, 0110100110010110. . . . Notice that some terms, such as the third term 0110, are palindromes (sequences that read the same backward or forward).

You can generate the sequence in another way: Each generation is obtained from the preceding one by appending its complement. For example, if you see a 0110, you append to it a 1001. You can also generate the sequence by starting with the numbers 0, 1, 2, 3,. . . and writing them in binary notation: 0, 1, 10, 11, 100, 101, 110, 111,. . . .Next, calculate the sum of the digits modulo 2 for each binary number—that is, divide the sum by 2 and use the remainder. This also yields the TM sequence: 0, 1, 1, 0, 1, 0, 0, 1,. . .

The sequence is self-similar. For example, retaining every other term of the infinite sequence reproduces the sequence. Retaining every other pair also reproduces the sequence. In other words, you take the first two numbers, skip the next two numbers, and so forth. Although aperiodic, the sequence is anything but random. It has strong short-range and long-range structures. For example, there can never be more than two adjacent terms that are identical.

**SEE ALSO** Boolean Algebra (1854), Penrose Tiles (1973), Fractals (1975), and Audioactive Sequence (1986).

*Mark Dow's artwork composed of square tiles that contain a set of symmetrical spirals. The 1s and 0s of the Thue-Morse sequence control the two orientations of the tiles as they fill a checkerboard array.*

1909

# Brouwer Fixed-Point Theorem

**Luitzen Egbertus Jan Brouwer** (1881–1966)

David Darling refers to the Brouwer fixed-point theorem as "an amazing result in topology and one of the most useful theorems in mathematics." Max Beran says the theorem "takes his breath away." In order to help visualize the theorem, imagine we have two sheets of graph paper of the same size, one atop the other. Your messy roommate takes one piece, crumples it into a messy blob, and tosses it onto the other sheet so that no piece of the blob extends beyond the edge of the bottom paper. The theorem states that at least one point exists in the blob that lies exactly above the same position on the bottom sheet where it was originally. (We assume that the roommate does not tear the paper.)

The same theorem works in other dimensions. Imagine a ball-shaped lemonade bowl with an opening on top. Your messy roommate stirs the lemonade. Even if all the points in the liquid move, Brouwer's theorem insists that there must be some point in the lemonade that is in exactly the same spot as it was before your roommate started stirring.

In the more precise language of mathematics, the theorem states that a continuous function from an $n$-ball into an $n$-ball (where $n > 0$ is the dimension) must have a fixed point.

Dutch mathematician Luitzen Brouwer proved the theorem for the case for $n = 3$ in 1909. French mathematician Jacques Hadamard proved the general case in 1910. According to Martin Davis, Brouwer was often combative, and toward the end of his life Brouwer became isolated and "under the spell of totally unfounded financial worries and a paranoid fear of bankruptcy, persecution, and illness." He was hit by a car and killed in 1966 while crossing the street.

**SEE ALSO** Projective Geometry (1639), Königsberg Bridges (1736), Hairy Ball Theorem (1912), Hex (1942), and Ikeda Attractor (1979).

*Randomly tossed crumpled papers help visualize Dutch mathematician Luitzen Brouwer's fixed-point theorem—"an amazing result in topology and one of the most useful theorems in mathematics."*

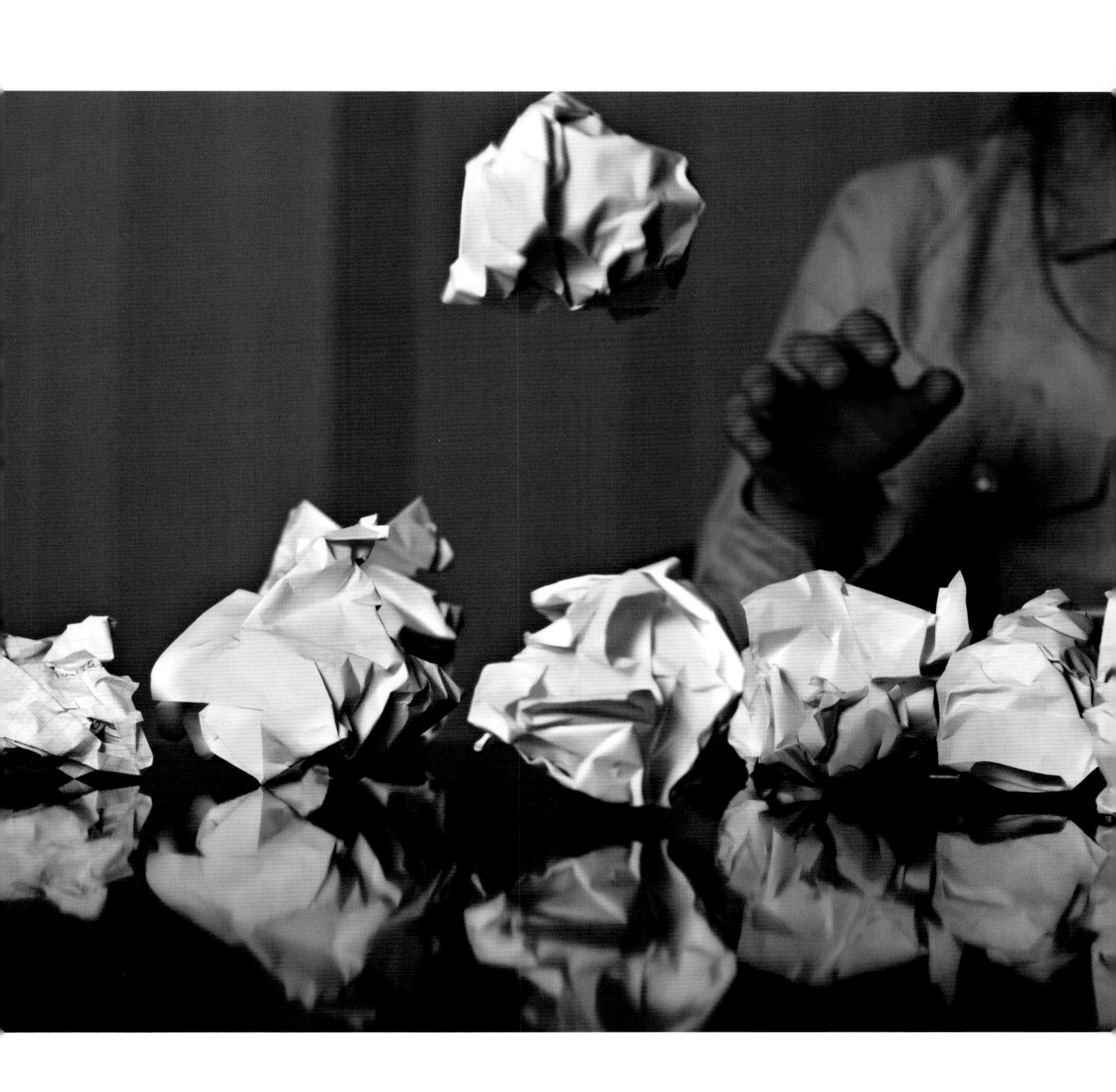

1909

# Normal Number

**Félix Édouard Justin Émile Borel** (1871–1956)

The search for patterns in the endless stream of digits in numbers like π is an ongoing quest for mathematicians. Mathematicians conjecture that π is "normal," which means that any finite pattern of digits occurs with the same frequency as would be found for a completely random sequence.

The quest for possible patterns in π played a key role in Carl Sagan's novel *Contact,* in which aliens coded a picture of a circle in the digits of π. The theological implications are intriguing, making the reader wonder if the universe could have been carefully constructed to reveal messages in the constants of nature. In fact, if π is a normal number, somewhere inside its endless digits is almost surely a very close representation for all of us—the atomic coordinates of all our atoms, our genetic code, all our thoughts, all our memories. Be happy: π makes us immortal!

Sometimes mathematicians use the phrase "absolutely normal" to denote normality in every base and "simply normal" if the number is normal in a particular base. (For example, our decimal system is "base 10" because it uses the 10 digits, 0 through 9.) Normality means that all digits are equally likely, all pairs of digits equally likely, all triplets of digits equally likely, and so forth. For example, for π, the digit 7 is expected to appear roughly 1 million times among the first 10 million digits of its decimal expansion. It actually occurs 1,000,207 times, which is very close to the expected value.

French mathematician and politician Émile Borel introduced the concept of normal numbers in 1909 as a way to characterize the digits of π, which seemed to have the properties of a random string of digits. In 1933, the artificially constructed **Champernowne's Number** was one of the first numbers found to be normal in base 10. The first absolutely normal number was constructed by Wacław Sierpiński in 1916. As is the case for π, it is conjectured, but not yet proved, that the numbers $\sqrt{2}$, *e*, and ln(2) are also normal.

**SEE ALSO** π (c. 250 B.C.), Euler's Number, *e* (1727), Transcendental Numbers (1844), and Champernowne's Number (1933).

Piece of π, *an artwork created by considering a portion of the endless digits of* π *and representing each digit by a color. The number* π *is conjectured to be "normal" and to have characteristics of a completely random sequence.*

# Boole's *Philosophy and Fun of Algebra*

**Mary Everest Boole** (1832–1916)

Mary Everest Boole was a self-taught mathematician, known for her intriguing 1909 book *Philosophy and Fun of Algebra*. She was the wife of George Boole (1815–1864), the British mathematician and philosopher who invented **Boolean Algebra**, which became the foundation for modern computer arithmetic. She was also responsible for editing his monumental 1854 book *Laws of Thought*. Her *Philosophy and Fun of Algebra* gives modern historians a glimpse of math education during the early 1900s.

At one point in her life, Mary worked at Queens College, the first women's college in England. Alas, she lived in an era when women were not allowed to receive degrees or teach at the college. Although she desperately wanted to teach, she accepted a job working at a library, where she advised many students. Her perseverance and zeal for mathematics and education make her a hero in the eyes of some modern-day feminists.

Toward the end of her book, she discusses imaginary numbers, like $\sqrt{-1}$, which she treats with mystic reverence: "[A top Cambridge mathematics student] got thinking about the square root of minus one as if it were a reality, till he lost his sleep and dreamed that he was the square root of minus one and could not extract himself; and he became so ill that he could not go to his examination at all." She also writes that "Angels, and the square roots of negative quantities...are messengers from the As-Yet-Unknown; and come to tell us where we are to go next; and the shortest road to get there; and where we ought not to go just at present."

Mathematics seemed to be in the Boole blood. Mary's oldest daughter married Charles Howard Hinton (1853–1907), who also provided mystical interpretations of **tesseracts** and tools for visualizing the fourth dimension. Another daughter, Alicia, is famous for work with *polytopes*, a term that she coined and that refers to generalizations of polygons to higher dimensions.

**SEE ALSO** Imaginary Numbers (1572), Boolean Algebra (1854), Tesseract (1888), and The Doctorate of Kovalevskaya (1874).

*Mary Everest Boole, author of* Philosophy and Fun of Algebra *and wife of mathematician George Boole, who invented Boolean algebra.*

1911–1913

# *Principia Mathematica*

**Alfred North Whitehead** (1861–1947), **Bertrand Russell** (1872–1970)

British philosophers and mathematicians Bertrand Russell and Alfred North Whitehead collaborated for eight years to produce their landmark work *Principia Mathematica* (three volumes, nearly 2,000 pages, 1910–1913), which aimed to demonstrate that mathematics can be stated using concepts of logic such as class and membership in a class. The *Principia* attempted to derive mathematical truths from axioms and inference rules in symbolic logic.

The Modern Library ranks *Principia* as the twenty-third most important nonfiction book of the twentieth century, in a list that includes such books as James Watson's *The Double Helix* and William James's *The Varieties of Religious Experience*. According to *The Stanford Encyclopedia of Philosophy*, "Written as a defense of logicism (i.e., the view that mathematics is in some significant sense reducible to logic), the book was instrumental in developing and popularizing modern mathematical logic. It also served as a major impetus for research in the foundations of mathematics throughout the twentieth century. Next to **Aristotle's *Organon***, it remains the most influential book on logic ever written."

Although *Principia* succeeded in providing derivations of many major theorems in mathematics, some critics were nervous about some the book's assumptions, such as the axiom of infinity (that is, there exists an infinite number of objects), which seemed to be an empirical assumption rather than a logical one. Therefore, it is still an open question as to whether mathematics can be reduced to logic. Nevertheless, *Principia* was extremely influential in emphasizing the connections between logicism and traditional philosophy, thus catalyzing new research in diverse areas of philosophy, mathematics, economics, linguistics, and computer science.

In *Principia*, after a few hundred pages, the authors prove that 1 + 1 = 2. Cambridge University Press, the publisher of the book, had decided that publishing *Principia* would result in an estimated loss of 600 pounds. Only after the authors agreed to give some money to Cambridge was the book published.

**SEE ALSO** Aristotle's *Organon* (c. 350 B.C.), Peano Axioms (1889), Barber Paradox (1901), and Gödel's Theorem (1931).

*After a few hundred pages of* Principia, *volume 1, the authors note that 1 + 1 = 2. The proof is actually completed in Volume II, accompanied by the comment, "The above proposition is occasionally useful."*

$*54 \cdot 43.\quad \vdash :. \alpha, \beta \in 1 . \supset : \alpha \cap \beta = \Lambda . \equiv . \alpha \cup \beta \in 2$

*Dem.*

$$\vdash . *54 \cdot 26 . \supset \vdash :. \alpha = \iota`x . \beta = \iota`y . \supset : \alpha \cup \beta \in 2 . \equiv . x \neq y .$$

$$[*51 \cdot 231] \qquad \equiv . \iota`x \cap \iota`y = \Lambda .$$

$$[*13 \cdot 12] \qquad \equiv . \alpha \cap \beta = \Lambda \qquad (1)$$

$$\vdash . (1) . *11 \cdot 11 \cdot 35 . \supset$$

$$\vdash :. (\exists x, y) . \alpha = \iota`x . \beta = \iota`y . \supset : \alpha \cup \beta \in 2 . \equiv . \alpha \cap \beta = \Lambda \qquad (2)$$

$$\vdash . (2) . *11 \cdot 54 . *52 \cdot 1 . \supset \vdash . \text{Prop}$$

From this proposition it will follow, when arithmetical addition has been defined, that $\mathbf{1} + \mathbf{1} = \mathbf{2}$.

1912

# Hairy Ball Theorem

**Luitzen Egbertus Jan Brouwer** (1881–1966)

In 2007, materials scientist Francesco Stellacci of the Massachusetts Institute of Technology made use of the hairy ball theorem (HBT) in mathematics to force nanoparticles to stick together to form long chainlike structures. According to a very high-level view of the theorem, first proved in 1912 by Dutch mathematician Luitzen Brouwer, if a sphere is covered in hair and we try to smoothly brush those hairs to make them all lie flat, we will always leave behind at least one hair standing up straight or a hole (for example, a bald spot).

Stellacci's team covered gold nanoparticles with sulfurous molecular hairs. Because of the HBT, hairs were likely to protrude in one or more locations, and these points became unstable defects on the particle surfaces, making it easy to substitute these standouts with chemicals that behaved as handles so that the particles could stick to one another, and perhaps someday be used to form nanowires in electronics devices.

Using mathematical language, the HBT states that any continuous tangent vector field on the sphere must have at least one point where the vector field is zero. Consider a continuous function $f$ that assigns a vector in 3-D space to every point $p$ on a sphere such that $f(p)$ is always tangent to the sphere at $p$. This means that at least one $p$ exists such that $f(p) = 0$. In other words, "the hair on a fur ball can't be brushed so that it lies flat at every point."

The implications of the theorem are intriguing. For example, since wind may be thought of as vectors with magnitudes and directions, the theorem states that somewhere on the Earth's surface, the horizontal wind speed must be zero, no matter how windy it is at every other location. Interestingly, the hairy ball theorem does not apply to the surface of a torus (for example, a doughnut surface), and thus it is theoretically possible to create an admittedly unappetizing hairy doughnut where all the hairs lie flat.

**SEE ALSO** Brouwer Fixed-Point Theorem (1909).

*If we try to smoothly brush the hairs on a hairy sphere to make them all lie flat, we will always leave behind at least one hair standing up straight or a hole (for example, a bald spot).*

1913

# Infinite Monkey Theorem

## **Félix Édouard Justin Émile Borel** (1871–1956)

The infinite monkey theorem states that a monkey pressing keys at random on a typewriter keyboard for an infinite amount of time will almost surely type a particular finite text, such as the Bible. Let us consider a single biblical phrase, "In the beginning, God created the heavens and the earth." How long would it take a monkey to type this phrase? Assume that there are 93 symbols on a keyboard. The phrase contains 56 letters (counting spaces and the period at the end). If the probability of hitting the correct key on the typewriter is $1/n$, where $n$ is the number of possible keys, then the probability of the monkey correctly typing 56 consecutive characters in the target phrase is, on average, $1/93^{56}$, which means that the monkey would have to try more than $10^{100}$ times, on average, before getting it right! If the monkey pressed one key per second, he'd be typing for well over the current age of the universe.

Interestingly, if we were to save characters that are typed correctly, the monkey would obviously require many fewer keystrokes. Mathematical analysis reveals that the monkey, after only 407 trials, would have a 50/50 chance that the correct sentence was typed! This crudely illustrates how evolution can produce remarkable results when harnessing nonrandom changes by preserving useful features and eliminating non-adaptive ones.

French mathematician Émile Borel mentioned the "dactylographic" (that is, typewriting) monkeys in a 1913 article, in which he commented on the likelihood of one million monkeys typing 10 hours a day to produce books in a library. The physicist Arthur Eddington wrote in 1928, "If an army of monkeys were strumming on typewriters, they *might* write all the books in the British Museum. The chance of their doing so is decidedly more favorable than the chance of [all gas molecules in a vessel suddenly moving to] one half of the vessel."

**SEE ALSO** Law of Large Numbers (1713), Laplace's *Théorie Analytique des Probabilités* (1812), Chi-Square (1900), and The Rise of Randomizing Machines (1938).

*According to the infinite monkey theorem, a monkey pressing keys at random on a typewriter keyboard for an infinite amount of time will almost surely type a particular finite text, such as the Bible.*

# Bieberbach Conjecture

**Ludwig Georg Elias Moses Bieberbach** (1886–1982), **Louis de Branges de Bourcia** (b. 1932)

The Bieberbach conjecture is associated with two colorful personalities: the vicious Nazi mathematician Ludwig Bieberbach, who made the conjecture in 1916, and French-American Louis de Branges, a loner mathematician who proved the conjecture in 1984, although some mathematicians were initially skeptical of de Branges' work because he had earlier announced false results. Author Karl Sabbagh writes of de Branges, "He may not be a crank, but he is cranky. 'My relationships with my colleagues are disastrous,' he told me. And he does seem to have left a trail of disgruntled, irritated, and even contemptuous colleagues behind him if only because he makes no concessions to students and colleagues who are not familiar with the field in which he works."

Bieberbach was an active Nazi and involved in the repression of Jewish colleagues, including German mathematicians Edmund Landau and Issai Schur. Bieberbach said that "representatives of overly different races do not mix as students and teachers.... I find it surprising that Jews are still members of academic commissions."

The Bieberbach conjecture states that if a function provides a one-to-one association between points in the unit circle and points in a simply connected region of the plane, the coefficients of the power series that represents the function are never larger than the corresponding power. In other words, we are given $f(z) = a_0 + a_1z + a_2z^2 + a_3z^3 + \ldots$ If $a_0 = 0$ and $a_1 = 1$, then $|a_n| \leq n$ for each $n \geq 2$. A "simply connected region" may be quite complicated, but it may not contain any holes.

De Branges says about his mathematical approach, "My mind is not very flexible. I concentrate on one thing, and I am incapable of keeping an overall picture. [If I omit something] then I have to be very careful with myself that I don't fall into some sort of a depression...." The Bieberbach conjecture is significant partly because it challenged mathematicians for 68 years, and during this time, it inspired significant research.

**SEE ALSO** Riemann Hypothesis (1859) and Poincaré Conjecture (1904).

*Bieberbach began working as a Privatdozent (private lecturer) at the University of Königsberg in 1910. Shown here is one of the old buildings of the university, later destroyed in World War II. Königsberg Cathedral is in the background.*

1916

# Johnson's Theorem

**Roger Arthur Johnson** (1890–1954)

Johnson's theorem states that if three identical circles pass through a common point, then their other three intersections must lie on another circle that is of the same size as the original three circles. The theorem is notable not only for its simplicity but also because it was apparently not "discovered" until 1916 by the American geometer Roger Johnson. David Wells writes that this relatively recent finding in the history of mathematics "suggests a wealth of geometrical properties still lie hidden, waiting to be discovered."

Johnson is the author of *Johnson's Modern Geometry: An Elementary Treatise on the Geometry of the Triangle and the Circle*. He received his Ph.D. from Harvard in 1913, and from 1947 to 1952 he served as chairman of the Mathematics Department of the Brooklyn branch of Hunter College, which later became Brooklyn College.

The idea that very simple yet profound mathematics can still be discovered even today is not as far-fetched as it sounds. For example, mathematician Stanislaw Ulam, in the mid to late 1900s, seemed to overflow with simple but novel ideas that quickly led to new branches of mathematics such as those that include **cellular automata** theory and the Monte Carlo method. Another example of simplicity and profundity are **Penrose Tilings**, the pattern of tiles discovered around 1973 by Roger Penrose. These tiles can completely cover an infinite surface in a pattern that is always non-repeating (aperiodic). Aperiodic tiling was first considered merely a mathematical curiosity, but physical materials were later found in which the atoms were arranged in the same pattern as a Penrose tiling, and now the field plays an important role in chemistry and physics. We should also consider the intricate and strikingly beautiful behavior of the **Mandelbrot Set**, a complicated fractal object described by a simple formula, $z = z^2 + c$, and unearthed toward the end of the twentieth century.

SEE ALSO Borromean Rings (834), Buffon's Needle (1777), Sangaku Geometry (c. 1789), Cellular Automata (1952), Penrose Tiles (1973), Fractals (1975), and Mandelbrot Set (1980).

*According to Johnson's theorem, if three identical circles pass through a common point, then their other three intersections must lie on another circle that is of the same size as the original three circles.*

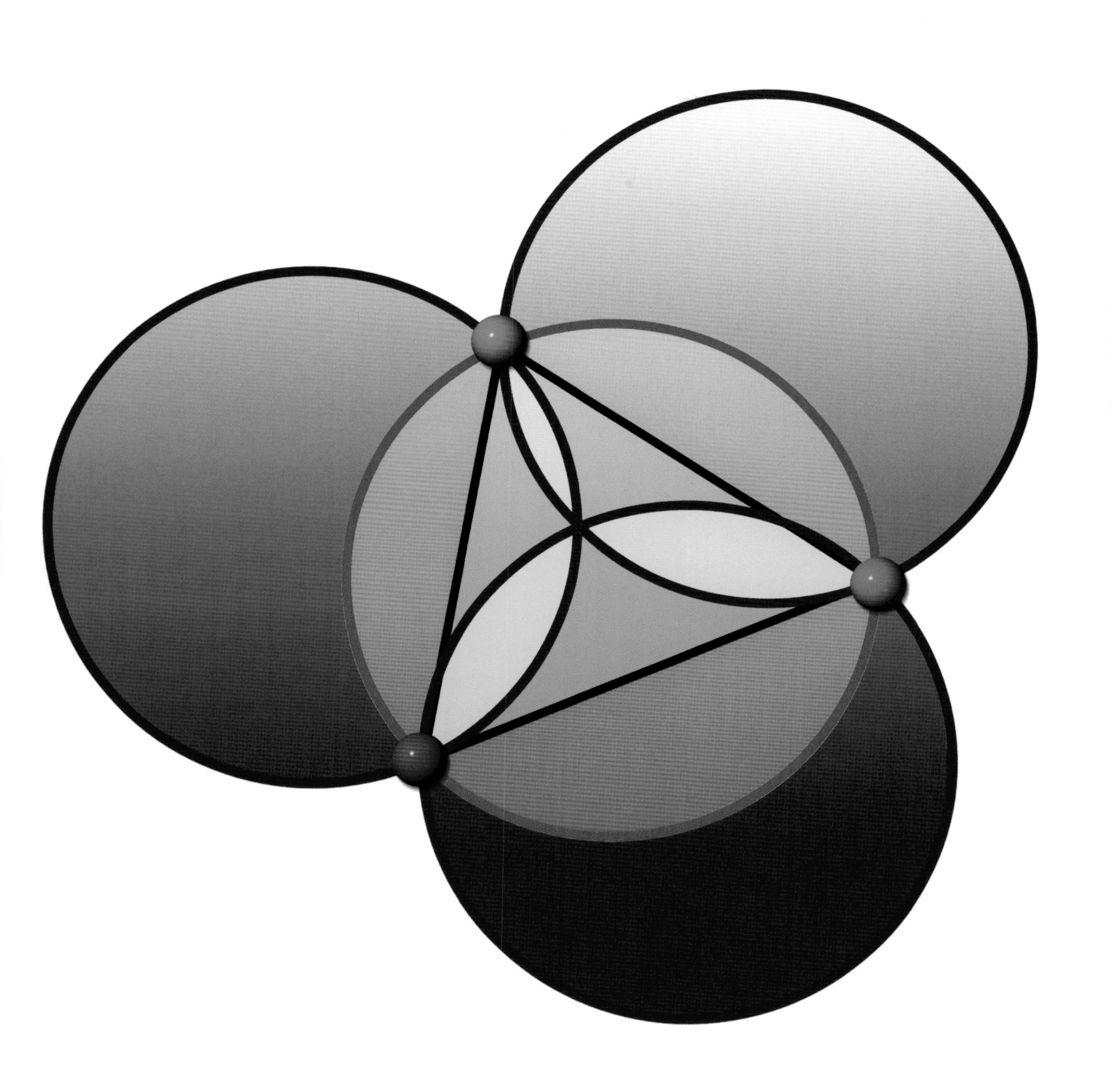

1918

# Hausdorff Dimension

**Felix Hausdorff** (1868–1942)

The Hausdorff dimension was introduced in 1918 by the mathematician Felix Hausdorff and can be used to measure the fractional dimensions of fractal sets. In everyday life, we usually think about the whole-number topological dimensions of smooth objects. For example, a plane is two-dimensional because a point on a plane can be described by two independent parameters, for example, locations along an $x$- and $y$-axis. A straight line is one-dimensional.

For certain more complicated sets and curves, the Hausdorff dimension provides another way to define dimension. For example, imagine a line that zigzags and twists in such an intricate way that it partially fills the plane. Its Hausdorff dimension increases beyond 1 and takes on values that get closer and closer to 2 the more the line fills the plane.

Space-filling curves like the infinitely convoluted **Peano Curves** have a Hausdorff dimension of 2. The Hausdorff dimensions of coastlines vary from about 1.02 for the coastline of South Africa to 1.25 for the west coast of Great Britain. In fact, one definition of a fractal is a set for which the Hausdorff dimension exceeds the topological dimension. The use of fractional dimensions to quantify roughness, scaling behavior, and intricacy has been demonstrated in such diverse areas as art, biology, and geology.

Hausdorff, a Jew, was a professor of mathematics at the University of Bonn, and was one of the founders of modern topology and famous for his work in functional analysis and set theory. In 1942, when he was about to be sent to a concentration camp by the Nazis, he committed suicide together with his wife and sister-in-law. The day before, Hausdorff wrote a friend, "Forgive us. We wish you and all our friends will experience better times." Many of the approaches used to compute the Hausdorff dimension for intricate sets were formulated by another Jew, the Russian mathematician Abram Samoilovitch Besicovitch (1891–1970), and hence the term Hausdorff-Besicovitch dimension is sometimes used.

**SEE ALSO** Peano Curve (1890), Koch Snowflake (1904), Coastline Paradox (c. 1950), and Fractals (1975).

*The Hausdorff dimension can be used to measure the fractional dimensions of fractal sets such as represented by this intricate fractal pattern rendered by Paul Nylander.*

1919

# Brun's Constant

**Viggo Brun** (1885–1978)

Martin Gardner writes, "No branch of number theory is more saturated with mystery than the study of prime numbers: those exasperating, unruly integers that refuse to be divided evenly by any integers except themselves and 1. Some problems concerning primes are so simple that a child can understand them and yet so deep and far from solved that many mathematicians now suspect they have no solution....Perhaps number theory, like quantum mechanics, has its own uncertainty principle that makes it necessary, in certain areas, to abandon exactness for probabilistic formulations."

Primes often occur as pairs of consecutive odd integers such as 3 and 5. In 2008, the largest known *twin primes* contained more than 58,000 digits each. Although infinitely many twin primes may exist, this conjecture remains unproven. Perhaps because the *twin prime conjecture* is a major unsolved problem, the movie *The Mirror Has Two Faces* features a math professor played by Jeff Bridges who explains the conjecture to Barbra Streisand.

In 1919, Norwegian mathematician Viggo Brun proved that if we add together the reciprocals of successive twin primes, the sum converges to a specific numerical value, now called Brun's constant: $B = (1/3 + 1/5) + (1/5 + 1/7) + \ldots \approx 1.902160\ldots$Given that the sum of the reciprocals of all prime numbers diverges to infinity, it is fascinating that the twin prime sum converges—that is, approaches a definite finite value. This in turn suggests the relative "scarcity" of twin primes, even though an infinite set of twin primes may exist. Today, the quest for twin primes, as well as for increasingly accurate values for $B$, continues in several universities. Other than the first pair, all pairs of twin primes have the form $(6n - 1, 6n + 1)$.

Andrew Granville remarks, "Prime numbers are the most basic objects in mathematics. They also are among the most mysterious, for after centuries of study, the structure of the set of prime numbers is still not well understood...."

**SEE ALSO** Cicada-Generated Prime Numbers (c. 1 Million B.C.), Sieve of Eratosthenes (240 B.C.), Harmonic Series Diverges (c. 1350), Goldbach Conjecture (1742), Constructing a Regular Heptadecagon (1796), Gauss's *Disquisitiones Arithmeticae* (1801), Proof of the Prime Number Theorem (1896), Polygon Circumscribing (c. 1940), Gilbreath's Conjecture (1958), Ulam Spiral (1963), and Andrica's Conjecture (1985).

*A graph of the number of twin primes less than* x. *The range of the* x*-axis is from 0 to 800, and the rightmost plateau, at the top of the graph, occurs at a value of 30.*

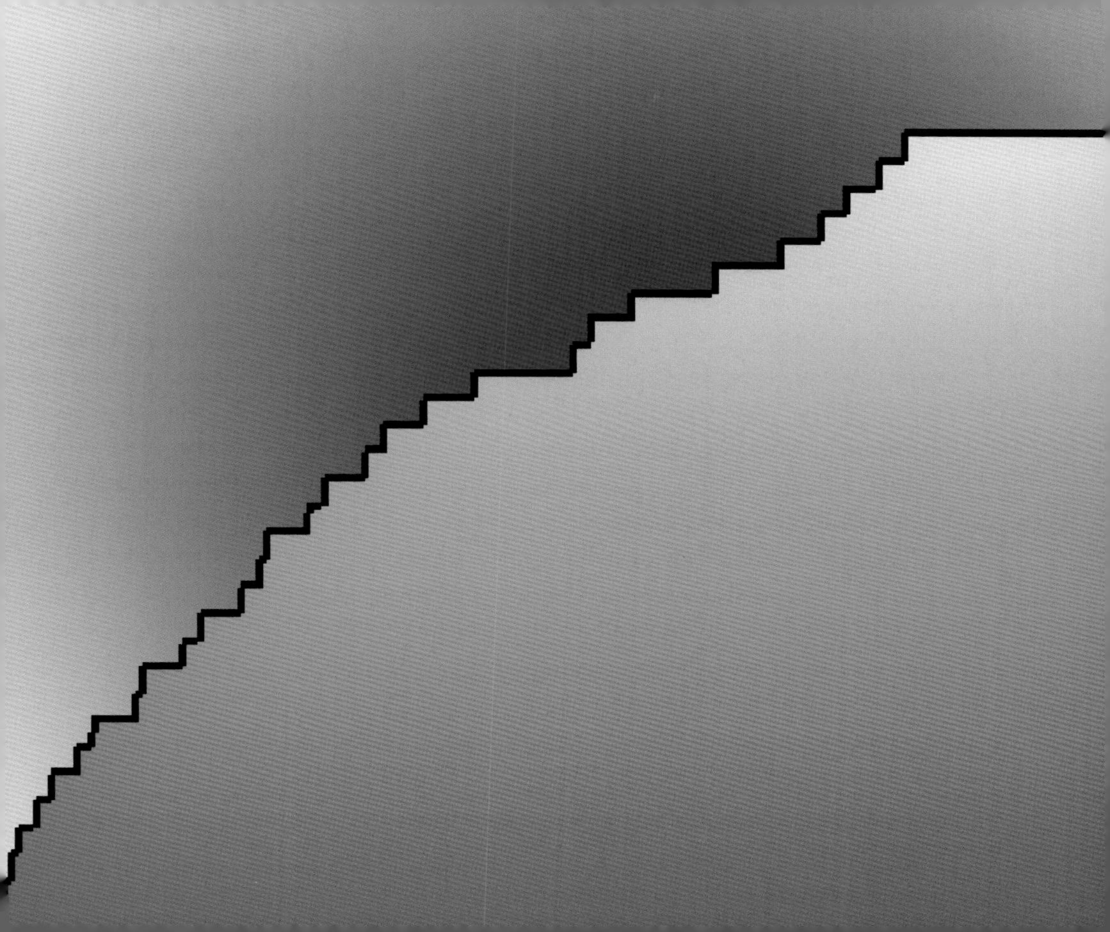

c.1920

# Googol

**Milton Sirotta** (1911–1981), **Edward Kasner** (1878–1955)

The term *googol*, which stands for the number 1 followed by 100 zeros, was coined by nine-year-old Milton Sirotta. Milton and his brother Edwin worked for most of their lives in their father's factory in Brooklyn, New York, pulverizing apricot pits to form an abrasive used for industrial purposes. Sirotta was the nephew of American mathematician Edward Kasner, who popularized the term after he asked Milton to make up a word for a very large number. The word *googol* first appeared in print publications in 1938.

Kasner is famous for being the first Jew appointed to a faculty position in the sciences at Columbia University and for his coauthoring of the book *Mathematics and the Imagination*, in which he introduced *googol* to a wide nontechnical audience. Although googol is of no special significance in mathematics, it has proven to be very useful for comparing large quantities, and for stimulating awe in the public mind as to the wonders of mathematics and the vast universe in which we live. It has also changed the world in other ways. Larry Page, one of the founders of the Internet search engine Google, was intrigued by mathematics and named his company after googol, after accidentally misspelling the word.

A little more than a googol different ways exist to arrange 70 items in a sequence, such as 70 people waiting in line to enter a doorway. Most scientists agree that if we could count all the atoms in all the stars we can see, we would have far less than a googol atoms. A googol years are required for all the black holes in the universe to evaporate. However, the number of possible chess games is *more* than a googol. The term *googolplex* is 1 followed by a googol number of zeros. It has more *digits* than there are atoms in stars in the visible universe.

**SEE ALSO** Archimedes: Sand, Cattle & Stomachion (c. 250 B.C.), Cantor's Transfinite Numbers (1874), and Hilbert's Grand Hotel (1925).

*A little more than a googol different ways exist for arranging the 70 beads in sequence, assuming that each bead is different and that the necklace remains open.*

1920

# Antoine's Necklace

**Louis Antoine** (1888–1971)

Antoine's necklace is a gorgeous mathematical object that may be represented as chains within chains within chains.... The necklace can be constructed by first considering a solid torus, or doughnut shape. Within the torus, we construct a chain C of $n$ components (links). Next, we modify each link of chain C so that it is actually another chain $C_1$ of $n$ solid tori. In each link of $C_1$, we construct a smaller chain of solid tori embedded in each link. Continue the process forever to create the delicate necklace of tori whose diameters decrease to zero.

Mathematicians refer to Antoine's necklace as homeomorphic with a Cantor set. Two geometrical objects are called homeomorphic if the first can be deformed into the second by stretching and bending. For example, we can smoothly deform a malleable, clay doughnut into the shape of a coffee cup without any tearing through the clay and pasting parts together again. The hole in the doughnut becomes the space in the handle for the coffee cup. The Cantor set, introduced by German mathematician Georg Cantor in 1883, is a special set of points with infinitely many gaps between them.

French mathematician Louis Antoine lost his sight at the age of 29 in World War I. Mathematician Henri Lebesgue advised Antoine to study two- and three-dimensional topology, because "in such a study, the eyes of the spirit and the habit of concentration will replace the lost vision." Antoine's necklace is notable because it is the first "wild embedding" of a set in three-dimensional space. Using Antoine's ideas, James Alexander invented his famous **Horned Sphere**.

Beverly Brechner and John Mayer write, "The tori are used to construct Antoine's Necklace, but *no torus is actually contained in* Antoine's Necklace. Only the 'beads,' the intersections of (infinitely many) solid tori, are left. Antoine's Necklace is totally disconnected...because for any two different points, there is some stage of construction such that the two points will lie in different tori...."

**SEE ALSO** Königsberg Bridges (1736), Alexander's Horned Sphere (1924), Menger Sponge (1926), and Fractals (1975).

*Rendering of Antoine's necklace, by computer scientist and mathematician Robert Scharein. In the next stage of construction, each component ring would be replaced with a linked chain of rings. Given an infinite number of stages, what remains is Antoine's necklace.*

1921

# Noether's *Idealtheorie*

**Amalie Emmy Noether** (1882–1935)

Despite the horrible prejudice they faced, several women have fought against the establishment and persevered in mathematics. German mathematician Emmy Noether was described by Albert Einstein as "the most significant creative mathematical genius thus far produced since the higher education of women began."

In 1915, while at the University of Göttingen, Germany, Noether's first significant mathematical breakthrough was in theoretical physics. In particular, Noether's theorem dealt with symmetry relationships in physics and their relationship to conservation laws. This and related work was an aid to Einstein when he developed his general theory of relativity, which focused on the nature of gravity, space, and time.

After Noether had received her Ph.D., she attempted to teach at Göttingen, but her opponents said that men could not expect to learn "at the feet of a woman." Her colleague David Hilbert replied to her detractors, "I do not see that the sex of the candidate is against her admission as a privatdozent [licensed lecturer]. After all, the university senate is not a bathhouse."

Noether is also known for her contributions to noncommutative algebras, where the order in which terms are multiplied affects the results. She is most famous for her study of "chain conditions on ideals of rings," and, in 1921, Noether published *Idealtheorie in Ringbereichen*, which is of major importance in the development of modern abstract algebra. This area of mathematics examines the general properties of operations and often unifies logic and number theory with applied mathematics. Alas, in 1933, her mathematical achievements were utterly dismissed when the Nazis terminated her from the University of Göttingen because she was Jewish.

She fled Germany and joined the faculty at Bryn Mawr College in Pennsylvania. According to journalist Siobhan Roberts, Noether "made weekly trips to lecture at Princeton's institute, and to visit her friends Einstein and Herman Weyl." Her influence was far and wide, and many of her ideas appeared in papers written by students and colleagues.

**SEE ALSO** The Death of Hypatia (415) and The Doctorate of Kovalevskaya (1874).

*Amalie Emmy Noether, author of* Idealtheorie in Ringbereichen (Theory of Ideals in Ring Domains), *which was of major importance in the development of modern abstract algebra. Noether also developed some of the mathematics of general relativity but often toiled without pay.*

1921

# Lost in Hyperspace

## **George Pólya** (1887–1985)

Imagine a robotic beetle placed in a twisting tube. The creature executes an infinite random walk by walking forever as it moves randomly one step forward or one step back in the tube. Assume that the tube is infinitely long. What is the probability that the random walk will eventually take the beetle back to its starting point?

In 1921, Hungarian mathematician George Pólya proved that the answer is one—infinite likelihood of return for a one-dimensional random walk. If the beetle were placed at the origin of a two-space universe (a plane), and then the beetle executed an infinite random walk by taking a random step north, south, east, or west, the probability that the random walk would eventually take the beetle back to the origin is also one.

Pólya also showed that our three-dimensional world is special: Three-dimensional space is the first Euclidean space in which it is possible for the beetle to get hopelessly lost. The beetle, executing an infinite random walk in a three-space universe, will eventually come back to the origin with a 0.34 or 34 percent probability. In higher dimensions, the chances of returning are even slimmer, about $1/(2n)$ for large dimensions $n$. This $1/(2n)$ probability is the same as the probability that the beetle would return to its starting point on its second step. If the beetle does not make it home in early attempts, it is probably lost in space forever.

Pólya's parents were Jewish but had converted to Roman Catholicism the year before his birth. He was born in Budapest, Hungary, and in the 1940s became a professor of mathematics at Stanford University. His book *How to Solve It* sold more than one million copies, and he is considered by many to be among the most influential mathematicians of the twentieth century.

**SEE ALSO** Dice (3000 B.C.), Law of Large Numbers (1713), Buffon's Needle (1777), Laplace's *Théorie Analytique des Probabilités* (1812), and Murphy's Law and Knots (1988).

*An insect randomly walks one step forward or one step back in an infinite tube. What is the probability that the random walk will eventually take the insect back to its starting point?*

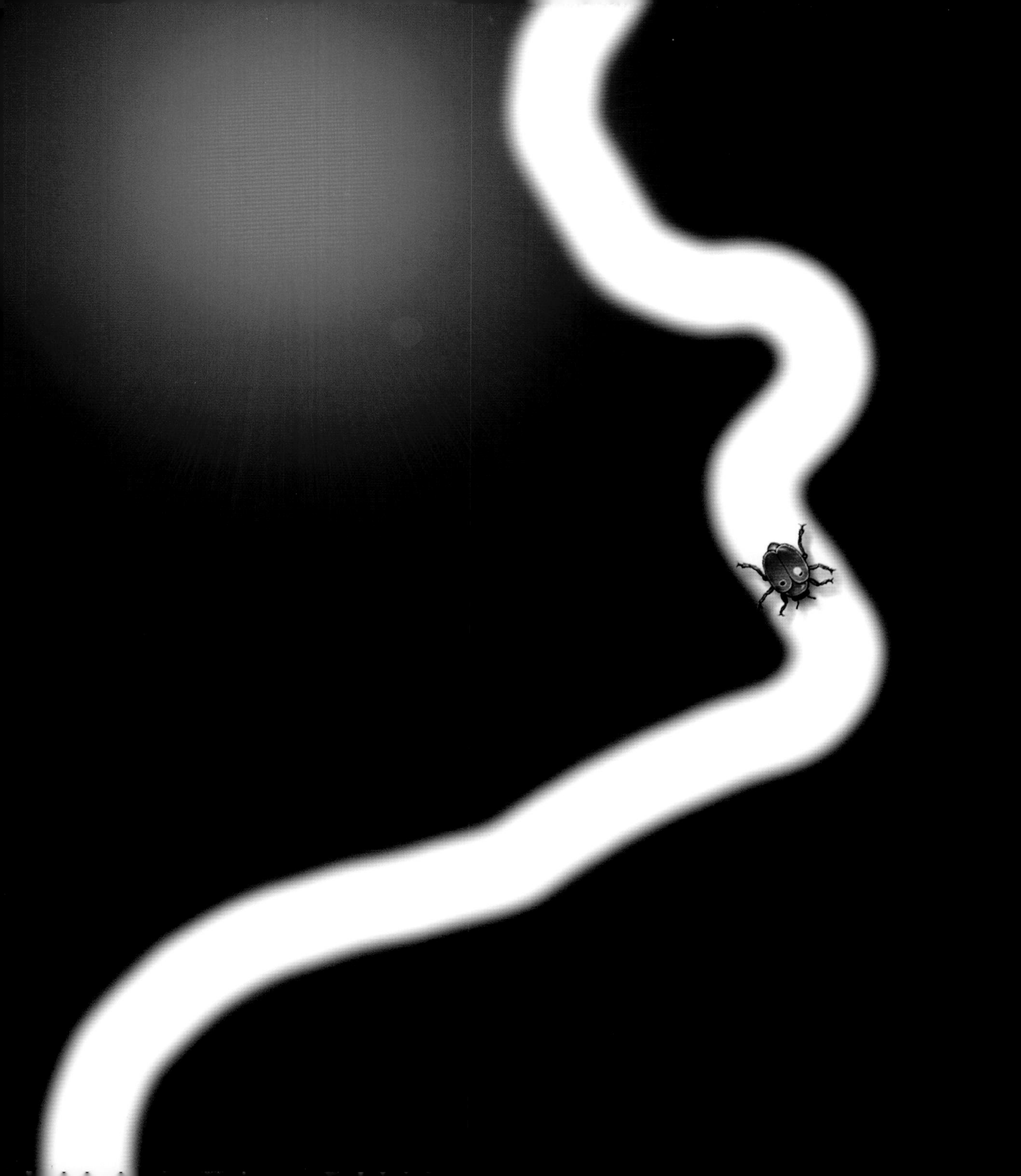

1922

# Geodesic Dome

**Walther Bauersfeld** (1879–1959), **Richard Buckminster "Bucky" Fuller** (1895–1983)

A geodesic dome can be created by triangulating a **Platonic Solid** or other polyhedron so that it has flat triangular faces and so that it can more closely approximate a sphere or hemisphere. Several designs for such domes exist. As an example, consider a regular dodecahedron with its twelve pentagonal faces. Place a point in the middle of each pentagon, and connect it with five lines to the vertices of the pentagon. Raise the point so that it touches an imaginary sphere around the decahedron. You've now created a new polyhedron with 60 triangular faces and a simple example of a geodesic sphere. Closer approximations to spheres may be created by dividing the faces into more triangles.

The triangular faces distribute stress across the entire structure, and in theory the domes can grow to extremely large sizes due to their rigidity and strength. The first true geodesic dome was designed by German engineer Walther Bauersfeld for a planetarium in Jena, Germany, which opened to the public in 1922. In the late 1940s, American architect R. Buckminster Fuller independently invented the geodesic dome, and he received a U.S. patent for his design. The U.S. Army was so impressed with his structures that they had him oversee the design of domes for military use. Aside from strength, domes were desirable because they enclosed a great volume for little surface area, which made them efficient in terms of building materials and decreased heat loss. Fuller himself lived in a geodesic dome for part of his life, and he noted that its low air resistance would help it withstand hurricanes. Always the dreamer, Fuller formulated the ambitious plan to place a geodesic dome, 2 miles (3.2 kilometers) in diameter and 1 mile (1.6 kilometers) high at its center, over New York City so that the weather could be regulated and inhabitants protected from rain and snow!

**SEE ALSO** Platonic Solids (350 B.C.), Archimedean Semi-Regular Polyhedra (c. 240 B.C.), Euler's Formula for Polyhedra (1751), Icosian Game (1857), Pick's Theorem (1899), Császár Polyhedron (1949), Szilassi Polyhedron (1977), Spidrons (1979), and Solving of the Holyhedron (1999).

*The United States Pavilion with a geodesic dome, featured at the 1967 World Exhibition ("Expo 67") in Montreal, Canada. The sphere had a diameter of 250 feet (76 meters).*

1924

# Alexander's Horned Sphere

**James Waddell Alexander** (1888–1971)

Alexander's horned sphere is an example of a convoluted, intertwined surface for which it is visually difficult to define an inside and outside. Introduced by mathematician James Waddell Alexander in 1924, Alexander's horned sphere is formed by successively growing pairs of horns that are almost interlocked and whose end points approach each other. The initial steps of the construction can be visualized with your fingers. Move the thumb and forefinger of each of your hands close to one another, and then grow a smaller thumb and forefinger on each of these, and continue this budding without limit! The object is a fractal, composed of interlocking pairs of "fingers" that trace orthogonal (perpendicular) circles of decreasing radii.

Although difficult to visualize, Alexander's horned sphere (together with its inside) is homeomorphic to a ball. (Two geometrical objects are called homeomorphic if the first can be deformed into the second by stretching and bending.) Thus, Alexander's horned sphere can be stretched into a ball without puncturing or breaking it. Martin Gardner writes, "The infinitely regressing, interlocking horn forms, at the limit, what topologists call a 'wild structure'.... Although it is equivalent to the simply connected surface of a ball, it bounds a region that is not simply connected. A loop of elastic cord circling the base of a horn cannot be removed from the structure even in infinity of steps."

Alexander's horned sphere is more than a mind-boggling curiosity—it is a concrete and important demonstration that the Jordan-Schönflies theorem does not extend to higher dimensions. This theorem states that simple closed curves separate a plane into an inside bounded region and an outside unbounded region and that these regions are homeomorphic to the inside and outside of a circle. The theorem is invalid in three dimensions.

**SEE ALSO** Jordan Curve Theorem (1905), Antoine's Necklace (1920), and Fractals (1975).

*A portion of Alexander's horned sphere, rendered by Cameron Browne. Introduced by mathematician James Waddell Alexander in 1924, Alexander's horned sphere is a fractal, composed of an infinite number of interlocking pairs of "fingers."*

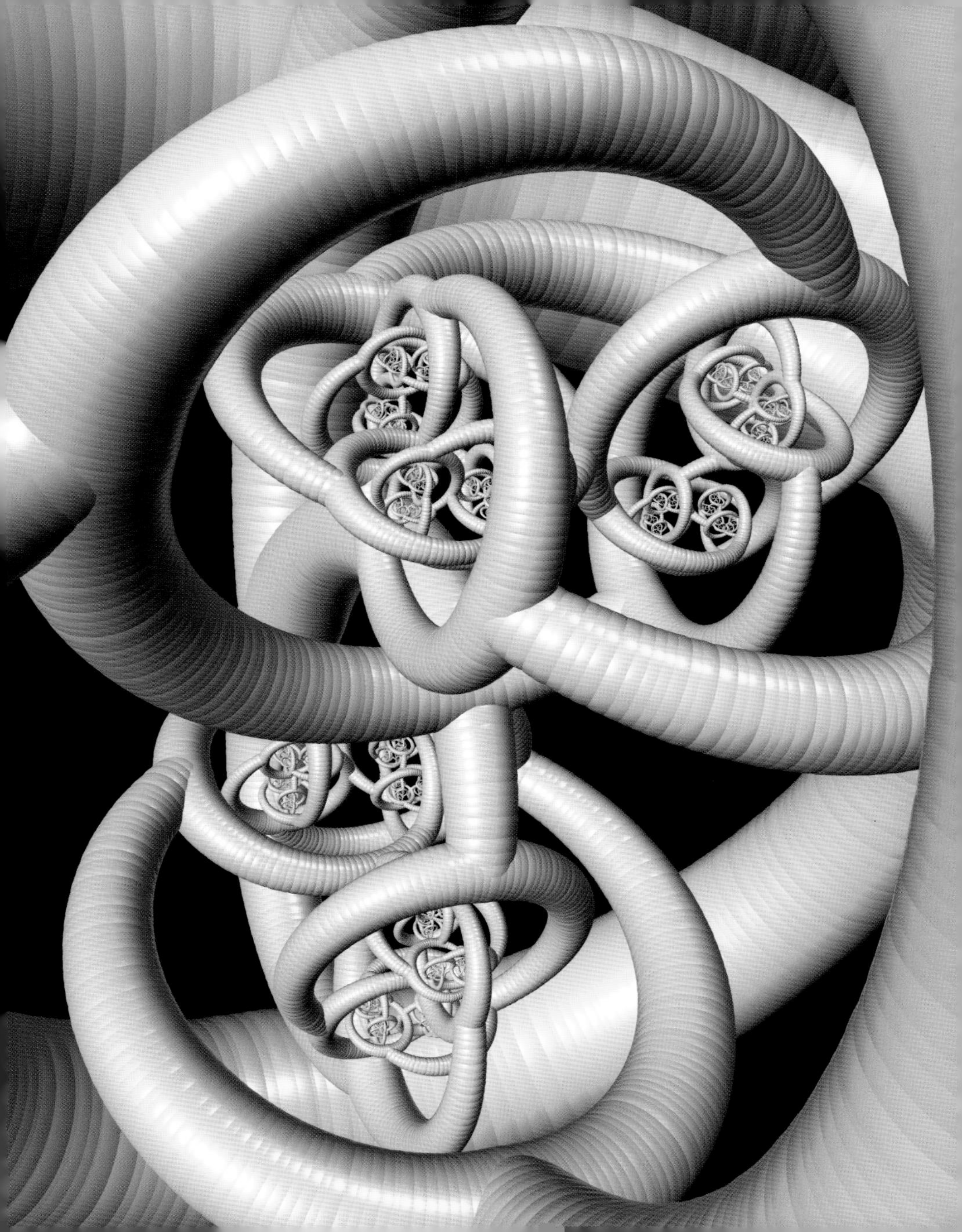

1924

# Banach-Tarski Paradox

**Stefan Banach** (1892–1945), **Alfred Tarski** (1902–1983)

The famous and seemingly bizarre Banach-Tarski (BT) paradox was first stated by Polish mathematicians Stefan Banach and Alfred Tarski in 1924. The paradox (which is actually a proof) shows how it is possible to take a mathematical representation of a ball, break it into several pieces, and then reassemble those pieces to make two identical copies of the ball. Moreover, it shows how one can decompose a pea-sized ball and then reassemble the pieces to make another ball the size of the moon! (In 1947, Robinson showed that five is the minimal number of pieces required.)

This paradox, built on the early work of Felix Hausdorff, shows that the kinds of quantities that can be measured in our physical universe are not necessarily preserved when a ball, as defined by mathematicians, with an infinite set of points is chopped into pieces and reassembled in a different way using just translations and rotations. In the BT paradox, the unmeasurable subsets (pieces) involved are very complicated and convoluted, lacking straightforward counterparts to boundaries and volume in the physical world. The paradox does not hold in two dimensions but does hold in all dimensions higher than two.

The BT paradox depends on the **Axiom of Choice** (AC). Because the paradox result seems so strange, some mathematicians have suggested that AC must be wrong. On the other hand, acceptance of AC is so useful in numerous branches of mathematics that mathematicians often quietly use it and proceed with their proofs and theorems.

In 1939, the brilliant Banach was elected president of the Polish Mathematical Society, but a few years later, during the Nazi occupation, Banach was compelled to feed his own blood to lice for a German study of infectious diseases. Tarski had converted to Roman Catholicism because it would be difficult for a Jew to obtain a serious position in the Polish universities. During World War II, the Nazis murdered nearly all his extended family.

**SEE ALSO** Zeno's Paradoxes (c. 445 B.C.), Aristotle's Wheel Paradox (c. 320 B.C.), St. Petersburg Paradox (1738), Barber Paradox (1901), Zermelo's Axiom of Choice (1904), Hausdorff Dimension (1918), Hilbert's Grand Hotel (1925), Birthday Paradox (1939), Coastline Paradox (c. 1950), Newcomb's Paradox (1960), and Parrondo's Paradox (1999).

*The Banach-Tarski paradox shows how it is possible to take a mathematical representation of a ball, break it into several pieces, and then reassemble those pieces to make two identical copies of the ball.*

1925

# Squaring a Rectangle

**Zbigniew Moroń** (1904–1971)

A difficult puzzle that has captivated mathematicians for at least a hundred years involves the operation of "squaring" a rectangle and a square, the latter of which is also known as a "perfect square dissection." The general problem is to tile a rectangle or square using square tiles all of *different* sizes expressed as integers. This may sound easy, and you can even experiment with a pencil, paper, and graph paper, but it turns out that very few tile arrangements work.

The first squared *rectangle* was discovered in 1925 by Polish mathematician Zbigniew Moroń. In particular, Moroń found a 33 × 32 rectangle that can be tiled with nine different squares with lengths 1, 4, 7, 8, 9, 10, 14, 15, and 18. He also discovered a 65 × 47 rectangle tiled with 10 square tiles with lengths 3, 5, 6 11, 17, 19, 22, 23, 24, and 25. For years, mathematicians claimed that perfect square dissections of squares were impossible to construct.

In 1936, four students at Trinity College—R. L Brooks, C. A. B. Smith, A. H. Stone, and W. T. Tutte—became fascinated by the topic, and finally, in 1940, these mathematicians discovered the first squared square consisting of 69 tiles! With further effort, Brooks reduced the number of tiles to 39. In 1962, A. W. J. Duivestijn proved that any squared square must contain at least 21 tiles, and in 1978, he had found such a square and proved that it was the only one.

In 1993, S. J. Chapman found a tiling of the Möbius band using just 5 square tiles. A cylinder can also be tiled with squares of different sizes, but this requires at least 9 tiles.

**SEE ALSO** Wallpaper Groups (1891), Voderberg Tilings (1936), and Penrose Tiles (1973).

*Polish mathematician Zbigniew Moroń discovered this 65 × 47 rectangle that is tiled with 10 square tiles with side lengths 3, 5, 6 11, 17, 19, 22, 23, 24, and 25.*

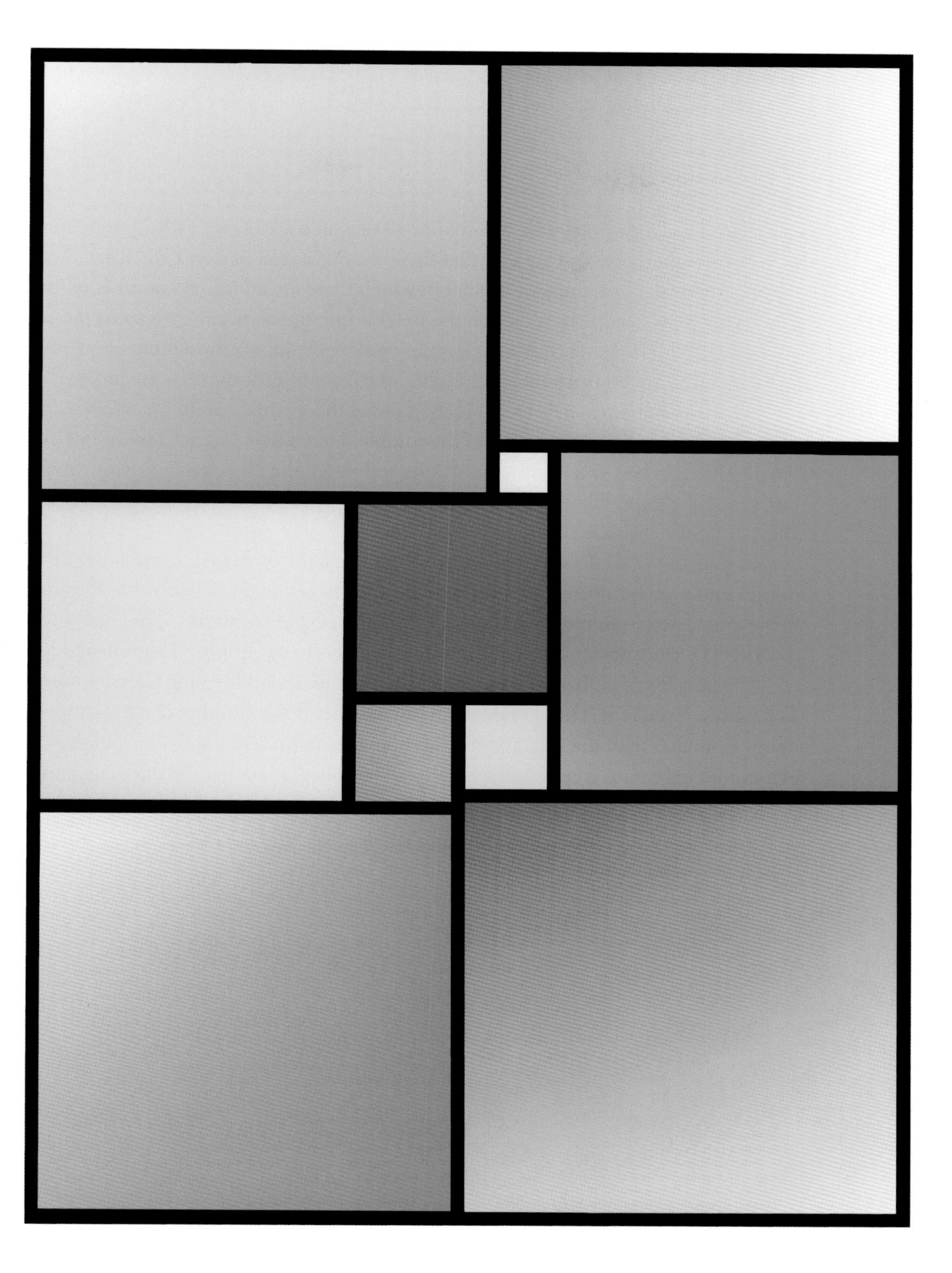

1925

# Hilbert's Grand Hotel

**David Hilbert** (1862–1943)

Imagine an ordinary hotel with 500 rooms, all of which are occupied by guests. You arrive in the afternoon and are told that there are no vacancies. You sadly leave. There is no paradox here. Next, imagine a hotel in which there are an infinite number of rooms, each of which is occupied. Although the hotel is full, the clerk can give you a room. How can this be? Later on the same day, an endless stream of conventioneers arrives, and the clerk is able to give them all rooms, making a huge fortune in the process!

German mathematician David Hilbert posed these paradoxes in the 1920s to illustrate the mysterious properties of the infinite. Here's how you get a room in Hilbert's Grand Hotel. When you arrive alone at the full hotel, the clerk can give you a room by moving the guest who is in Room 1 into Room 2, and then moving the original guest in Room 2 into Room 3, and so on. Room 1 is now vacant for you. In order to accommodate the endless stream of conventioneers, the current occupants are all moved into even-numbered rooms by moving the original guest in Room 1 to Room 2, the original guest in Room 2 into Room 4, the original guest in Room 3 into Room 6, and so forth. The clerk now may assign the conventioneers to the empty odd-numbered rooms.

The paradox of Hilbert's Grand Hotel can be understood by using Cantor's theory of **Transfinite Numbers**. Thus, while in an ordinary hotel, the number of odd-numbered rooms is smaller than the total number of rooms, in an infinite hotel, the "number" of odd-numbered rooms is no smaller than total "number" of rooms. (Mathematicians use the term *cardinality* when referring to the size of these sets of rooms.)

**SEE ALSO** Zeno's Paradoxes (c. 445 B.C.), Cantor's Transfinite Numbers (1874), Peano Axioms (1889), and Hilbert's 23 Problems (1900).

*In Hilbert's Grand Hotel, the hotel is fully occupied, yet the clerk can give you a room. How can this be?*

HOTEL

1926

# Menger Sponge

**Karl Menger** (1902–1985)

The Menger sponge is a fractal object with an infinite number of cavities—a nightmarish object for any dentist to contemplate. The object was first described by Austrian mathematician Karl Menger in 1926. To construct the sponge, we begin with a "mother cube" and subdivide it into 27 identical smaller cubes. Next, we remove the cube in the center and the six cubes that share faces with it. This leaves behind 20 cubes. We continue to repeat the process forever. The number of cubes increases by $20^n$, where $n$ is the number of iterations performed on the mother cube. The second iteration gives us 400 cubes, and by the time we get to the sixth iteration, we have 64,000,000 cubes.

Each face of the Menger sponge is called a Sierpiński carpet. Fractal antennae based on the Sierpiński carpet are sometimes used as efficient receivers of electromagnetic signals. Both the carpets and the entire cube have fascinating geometrical properties. For example, the sponge has an infinite surface area while enclosing zero volume.

According to the Institute for Figuring, with each iteration, the Sierpiński carpet face "dissolves into a foam whose final structure has no area whatever yet possesses a perimeter that is infinitely long. Like the skeleton of a beast whose flesh has vanished, the concluding form is without substance—it occupies a planar surface, but no longer fills it." This porous remnant hovers between a line and a plane. Whereas a line is one-dimensional and a plane two-dimensional, the Sierpiński carpet has a "fractional" dimension of 1.89. The Menger sponge has a fractional dimension (technically referred to as the **Hausdorff Dimension**) between a plane and a solid, approximately 2.73, and it has been used to visualize certain models of a foam-like space-time. Dr. Jeannine Mosely has constructed a Menger sponge model from more than 65,000 business cards that weighs about 150 pounds (70 kilograms).

**SEE ALSO** Pascal's Triangle (1654), Prince Rupert's Problem (1816), Hausdorff Dimension (1918), Antoine's Necklace (1920), Ford Circles (1938), and Fractals (1975).

*A child exploring inside a Menger sponge with its infinite number of cavities. This collaborative artwork by fractal enthusiasts Gayla Chandler and Paul Bourke makes use of Bourke's computer-generated sponge that he merged with an image of a human child.*

# Differential Analyzer

**Vannevar Bush** (1890–1974)

Differential equations play a crucial role in physics, engineering, chemistry, economics, and numerous other disciplines. These equations are relevant whenever a function expresses continuously changing quantities along with some rate of change, expressed as derivatives. Only the simplest differential equations yield solutions that are expressed by compact and explicit formulas with a finite number of basic functions like sines and **Bessel Functions**.

In 1927, American engineer Vannevar Bush and his colleagues developed a differential analyzer (DA), an analog computer with wheel-and-disk components that could solve, via integration methods, differential equations with several independent variables. The DA was among the first advanced computing devices to be used for practical applications.

Earlier versions of these kinds of devices had their roots in the work of Lord Kelvin and his **harmonic analyzer** (1876). In the United States, researchers working at the Wright-Patterson Air Force Base and Moore School of Electrical Engineering at the University of Pennsylvania built the DA devices, in part for creating artillery firing tables, prior to the invention of **ENIAC** (electronic numerical integrator and computer).

Over the years, the DA has had many applications, ranging from soil erosion studies and the building of blueprints for dams to the design of bombs used to destroy German dams during World War II. These devices have even been featured in science fiction films such as the 1956 classic *Earth vs. the Flying Saucers*!

In his 1945 essay "As We May Think," Bush described his vision of the *memex*, a futuristic machine that would enhance human memory by allowing humans to store and retrieve information linked by associations, in a manner similar to hypertext on the Web today. He wrote, "It is a far cry from the **abacus** to the modern keyboard accounting machine. It will be an equal step to the arithmetical machine of the future.... Relief must be secured from laborious detailed manipulation of higher mathematics.... Man's spirit should be elevated...."

**SEE ALSO** Abacus (c. 1200), Bessel Functions (1817), Harmonograph (1857), Harmonic Analyzer (1876), ENIAC (1946), Curta Calculator (1948), and Ikeda Attractor (1979).

*A differential analyzer at the Lewis Flight Propulsion Laboratory, in 1951. The analyzer was among the first advanced computing devices to be used for practical applications such as the design of bombs used to destroy German dams during World War II.*

1928

# Ramsey Theory

**Frank Plumpton Ramsey** (1903–1930)

Ramsey theory is concerned with finding order and patterns in systems. Author Paul Hoffman writes, "The idea behind Ramsey theory is that complete disorder is impossible.... Any mathematical 'object' can be found if sought in a large enough universe. The Ramsey theorist wants to know the smallest universe that is guaranteed to contain a certain object."

Ramsey theory is named after the English mathematician Frank Ramsey. He started this branch of mathematics in 1928 while exploring a problem in logic. As Hoffman suggested, Ramsey theorists often seek the number of elements in a system that is necessary for a particular property to hold. Except for some interesting work by Paul Erdös, it was not until the late 1950s that research in Ramsey theory began to make rapid progress.

One example of the simplest application deals with the **Pigeonhole Principle** that states that if we have $m$ pigeon homes and $n$ pigeons, we can be sure that at least one home houses more than one pigeon if $n > m$. For a more complicated example, consider a scattering of $n$ points on a paper. Each point is connected to every other point with a straight line that is either red or blue. Ramsey's theorem—which is just one foundational result in combinatorics and Ramsey theory—shows that $n$ must be 6 in order to *ensure* that either a blue triangle or red triangle appears on the paper.

Another way to think about Ramsey theory involves the so-called party problem. For example, what is the smallest party that is guaranteed to have either at least 3 attendees who are (pairwise) mutual strangers or at least 3 of them who are (pairwise) mutual acquaintances? The answer is 6. Determining the party size necessary to ensure the presence of at least 4 mutual friends or at least 4 mutual strangers is much more difficult, and solutions for higher party sizes may never be known.

**SEE ALSO** Archimedes: Sand, Cattle & Stomachion (c. 250 B.C.), Euler's Polygon Division Problem (1751), Thirty-Six Officers Problem (1779), Pigeonhole Principle (1834), Birthday Paradox (1939), and Császár Polyhedron (1949).

*Five points connected to each other with straight lines that are either red or blue. In this depiction, no all-red or all-blue triangle exists between points. Six points are required to ensure that either a blue or red triangle is formed.*

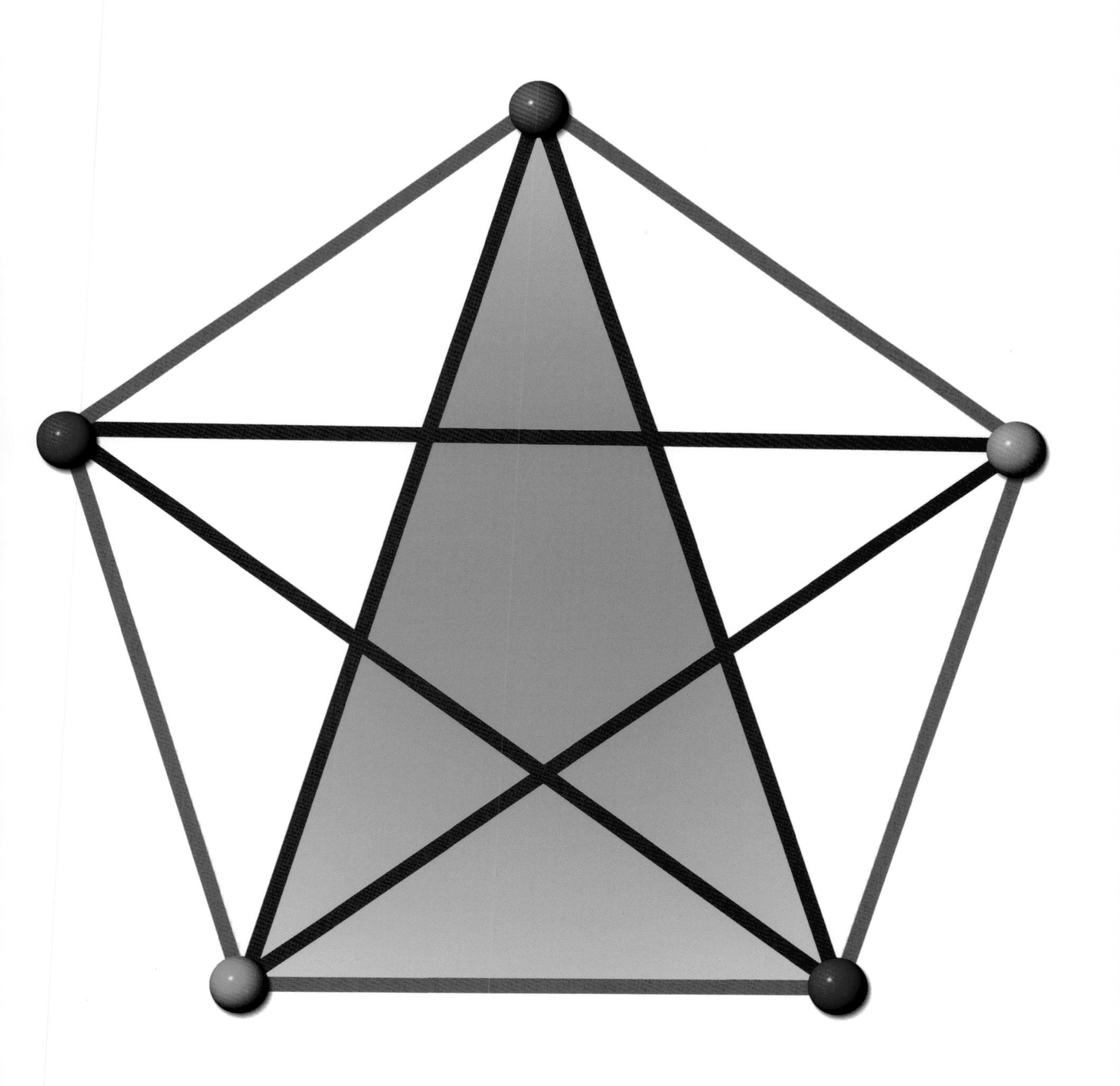

1931

# Gödel's Theorem

**Kurt Gödel** (1906–1978)

Austrian mathematician Kurt Gödel was an eminent mathematician and among the most brilliant logicians of the twentieth century. The implications of his incompleteness theorem are vast, applying not only to mathematics but also touching on areas such as computer science, economics, and physics. When Gödel was at Princeton University, one of his closest friends was Albert Einstein.

Gödel's theorem, published in 1931, had quite a sobering effect upon logicians and philosophers because it implies that within any rigidly logical mathematical system, propositions or questions exist that cannot be proved or disproved on the basis of axioms within that system, and therefore it is possible for basic axioms of arithmetic to give rise to contradictions. This makes mathematics essentially "incomplete." The repercussions of this fact continue to be felt and debated. Moreover, Gödel's theorem put an end to centuries of attempting to establish axioms that would provide a rigorous basis for all of mathematics.

Author Hao Wang writes on this very subject in his book *Reflections on Kurt Gödel*: "The impact of Gödel's scientific ideas and philosophical speculations has been increasing, and the value of their potential implications may continue to increase. It may take *hundreds of years* for the appearance of more definite confirmations or refutations of some of his larger conjectures." Douglas Hofstadter notes that a second theorem of Gödel's also suggests the inherent limitation of mathematical systems and "implies that the only versions of formal number theory which assert their own consistency are inconsistent."

In 1970, Gödel's mathematical proof of the existence of God began to circulate among his colleagues. The proof was less than a page long and caused quite a stir. Toward the end of his life, Gödel was paranoid and felt that people were trying to poison him. He stopped eating and died in 1978. During his life, he had also suffered from nervous breakdowns and hypochondria.

**SEE ALSO** Aristotle's *Organon* (c. 350 B.C.), Boolean Algebra (1854), Venn Diagrams (1880), *Principia Mathematica* (1910–1913), and Fuzzy Logic (1965).

*Albert Einstein and Kurt Gödel. Photo by Oskar Morgenstern, Institute of Advanced Study Archives, Princeton, 1950s.*

# Champernowne's Number

**David Gawen Champernowne** (1912–2000)

If you were to concatenate, or link together, positive integers, 1, 2, 3, 4,..., and lead with a decimal point, we would obtain Champernowne's number, 0.1234567891011121314.... Like $\pi$, and $e$, Champernowne's number is **transcendental**—that is, it is not the root of any polynomial with integer coefficients. We also know this number to be "**normal**" in base 10, which means that any finite pattern of numbers occurs with the frequency expected for a completely random sequence. David Champernowne demonstrated that this number is normal by showing that not only will the digits 0 through 9 occur exactly with a 10 percent frequency in the limit, but each possible block of two digits will occur with 1 percent frequency in the limit, each block of three digits will occur with 0.1 percent frequency, and so forth.

Cryptographers have noted that Champernowne's number does not trigger some of the simplest, traditional statistical indicators of nonrandomness. In other words, simple computer programs, which attempt to find regularity in sequences, may not "see" the regularity in Champernowne's number. This deficit reinforces the notion that statisticians must be very cautious when declaring a sequence to be random or patternless.

Champernowne's number is the first constructed example of a normal number. It was produced in 1933 by David Champernowne while he was still an undergraduate student at the University of Cambridge. In 1937, German mathematician Kurt Mahler proved that Champernowne's constant is transcendental. Today, we know that the binary Champernowne constant, obtained by concatenating the binary (0 and 1) representations of the integers, is normal in base 2.

Hans Von Baeyer suggests that by translating the 0s and 1s to Morse code, "every conceivable finite sequence of words is buried somewhere in the string's tedious gobbledygook, . . . every love letter and every novel ever written. . . . You may have to travel out along the string for billions of light years before you find them, but they are all in there somewhere. . . ."

**SEE ALSO** Transcendental Numbers (1844) and Normal Number (1909).

*The first 100,000 binary digits of Champernowne's number in binary, adapted from the work of Adrian Belshaw and Peter Borwein. The 0s in the sequence are converted to −1s, and then digit pairs (±1, ±1) used to walk (±1, ±1) in the plane. The graph's* x-axis *range is (0, 8,400).*

1935

# Bourbaki: Secret Society

**Henri Cartan** (1904–2008), **Claude Chevalley** (1909–1984), **Szolem Mandelbrojt** (1899–1983), **André Weil** (1906–1998), **and others**

Historian of science Amir Aczel once wrote that Nicolas Bourbaki was "the greatest mathematician of the twentieth century" who "changed the way we think about mathematics.... He was responsible for the emergence of the 'New Math' that swept through American education in the middle of the century...." His treatises "form a towering foundation for much of the modern mathematics.... No working mathematician... today is free of the influence of the seminal work of Nicolas Bourbaki."

However, Bourbaki, the genius mathematician and author of dozens of acclaimed works, never existed! Bourbaki was not an individual, but rather a secret society of mathematicians, almost all French, formed in 1935. The group attempted to produce a completely self-contained, extremely logical and rigorous treatment of all essential modern mathematics—from beginning to end—by publishing books on set theory, algebra, topology, functions, integration, and more. The founding members of the secret group included the brilliant mathematicians Henri Cartan, Jean Coulomb, Jean Delsarte, Claude Chevalley, Jean Dieudonné, Charles Ehresmann, René de Possel, Szolem Mandelbrojt, and André Weil. Members felt that older mathematicians were needlessly clinging to old practices; thus, members of Bourbaki had to resign by age 50.

While writing their collaborative books, any member had the right to veto any aspect that he deemed inappropriate. Shouting matches ensued. At each meeting, their works would be read aloud and scrutinized, line by line. In 1983, Bourbaki published its last volume, titled *Spectral Theory*. Today, *L'Association des Collaborateurs de Nicolas Bourbaki* still organizes Bourbaki seminars every year.

Author Maurice Mashaal wrote that "Bourbaki never invented revolutionary techniques nor proved grandiose theorems—and neither did it try to do so. What the group did bring... was a new vision of mathematics, a profound reorganization and reclarification of its components, lucid terminology and notation, and distinctive style."

**SEE ALSO** *Principia Mathematica* (1910–1913).

*World War I cemetery in France near Verdun. The aftermath of the war presented a difficult challenge for aspiring French mathematicians. Vast numbers of students and young teachers were killed, which was one motivation for several young Parisian math students to create the Bourbaki group.*

1936

# Fields Medal

## **John Charles Fields** (1863–1932)

The Fields Medal is the most famous and influential award in mathematics. Like the Nobel Prize for other realms of achievement, the Fields Medal grew from a desire to elevate mathematics above national hostilities. The medal is awarded every four years and rewards past achievements and stimulates future research.

The award is sometimes referred to as the "Nobel Prize of mathematicians" because there is no actual Nobel Prize awarded for mathematics; however, the Fields Medal is awarded only to mathematicians 40 years of age and younger. The monetary amount is relatively small, only about $13,500 in 2006 compared to the Nobel Prize, which is more than $1 million. The award was established by Canadian mathematician John Charles Fields and first awarded in 1936. When Fields died, his will specified that $47,000 be added to the funds for the gold medal.

The front of the medal depicts the Greek geometer Archimedes. The Latin phrase on the back translates to "The mathematicians having congregated from the whole world awarded [this medal] because of outstanding writings."

Mathematician Alexander Grothendieck boycotted his own Fields Medal ceremony in 1966, because it was held in Moscow and he wished to protest the Soviet military presence in Eastern Europe. In 2006, Russian mathematician Grigori Perelman rejected the prize when he was awarded the medal for "his contributions to geometry and his revolutionary insights into the analytical and geometric structure of the Ricci flow," which led to the proof of the **Poincaré Conjecture**. He refused, saying that the prize was irrelevant.

Interestingly, roughly 25 percent of medalists have been Jewish, and almost half have held appointments at the Institute for Advanced Study, in Princeton, New Jersey. Alfred Nobel (1833–1896), Swedish chemist and inventor of dynamite, created the Nobel Prize; however, because he was an inventor and industrialist, he did not establish a prize in mathematics, because he personally had little interest in mathematics or theoretical sciences.

**SEE ALSO** Archimedes: Sand, Cattle & Stomachion (c. 250 B.C.), Poincaré Conjecture (1904), Langlands Program (1967), Catastrophe Theory (1968), and Monster Group (1981).

*The Fields Medal is sometimes referred to as the "Nobel Prize of mathematicians"; however, the Fields Medal is awarded only to mathematicians 40 years of age and younger.*

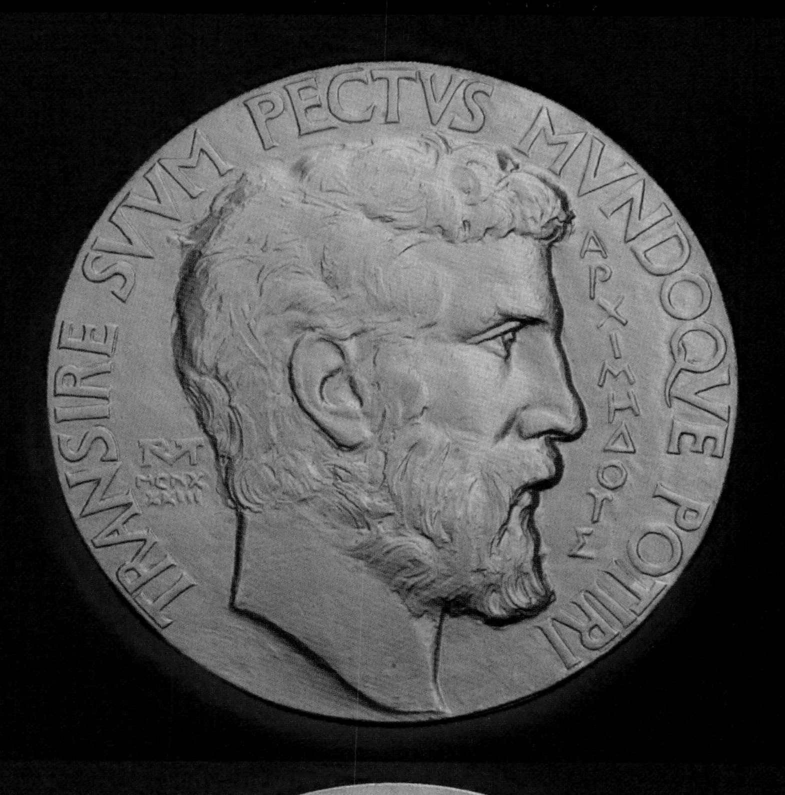
TRANSIRE SVVM PECTVS MVNDOQVE POTIRI
ΑΡΧΙΜΗΔΟΥΣ
RM
MCMXXXIII

CONGREGATI
EX TOTO ORBE
MATHEMATICI
OB SCRIPTA INSIGNIA
TRIBVERE

1936

# Turing Machines

## **Alan Turing** (1912–1954)

Alan Turing was a brilliant mathematician and computer theorist who was forced to become a human guinea pig and subjected to drug experiments to "reverse" his homosexuality. This persecution occurred despite the fact that his code-breaking work helped shorten World War II and led to his award of the Order of the British Empire.

When Turing had called the police to investigate a burglary at his home in England, a homophobic police officer suspected that Turing was homosexual. Turing was forced to either go to jail for a year or take experimental drug therapy. To avoid imprisonment, he agreed to be injected with estrogen hormone for a year. His death at age 42, two years after his arrest, was a shock to his friends and family. Turing was found in bed. The autopsy indicated cyanide poisoning. Perhaps he had committed suicide, but to this day we are not certain.

Many historians consider Turing to be the "father of modern computer science." In his landmark paper, "On Computable Numbers, with an Application to the Entscheidungs Problem" (written in 1936), he proved that Turing machines (abstract symbol-manipulating devices) would be capable of performing any conceivable mathematical problem that is represented as an algorithm. Turing machines help scientists better understand the limits of computation.

Turing is also the originator of the Turing test, which caused scientists to think more clearly about what it means to call a machine "intelligent" and whether machines may one day "think." Turing believed that machines would eventually be able to pass his test by demonstrating they could converse with people in such a natural way that people could not tell if they were talking to a machine or a human.

In 1939, Turing invented an electromechanical machine that could help break the Nazi codes produced by their Enigma code machine. Turing's machine, called the "Bombe," was enhanced by mathematician Gordon Welchman, and it became the main tool for deciphering Enigma communications.

**SEE ALSO** ENIAC (1946), Information Theory (1948), and Public-Key Cryptography (1977).

*A replica of a Bombe machine. Alan Turing invented this electromechanical device to help break the Nazi codes produced by their Enigma code machine.*

BOMBE REBUILD PROJECT

1936

# Voderberg Tilings

## **Heinz Voderberg** (1911–1942)

A tessellation, or tiling, of a plane consists of a collection of smaller shapes, called *tiles*, that fills a surface with no overlaps and no gaps between the tiles. Perhaps the most obvious tessellations are those seen on tiled floors in which the tiles are shaped like squares or hexagons. Hexagonal tiling is the basic structure of a honeycomb, perhaps "useful" to the bees because of the efficiency of this tiling in terms of material required to create a lattice of cells within a given area. Eight different kinds of tessellations of the plane exist that employ two or more convex regular polygons such that the same polygons, in the same order, surround each polygon vertex.

Tessellations are common in the art of Dutch artist M. C. Escher as well as in ancient Islamic art. In fact, tessellations are thousands of years old and can be traced to the Sumerian civilization (about 4000 B.C.), in which building walls were decorated by tiling designs constructed from clay.

The Voderberg tiling, discovered by Heinz Voderberg in 1936, is special because it is the earliest-known spiral tessellation of the plane. The attractive pattern is made from a single repeating tile in the form of an irregularly shaped nonagon—that is, a nine-sided polygon. As the nonagon is repeated, it forms an infinite spiral strip, which, when joined with another strip, covers the plane with no gaps. The Voderberg tiling is referred to as *monohedral* because it is a tessellation in which all tiles are the same.

In the 1970s, a wonderful new set of spiral tilings were discussed by mathematicians Branko Grünbaum and Geoffrey C. Shephard. Their tiles can be used to produce one-, two-, three-, and six- armed spirals that tile the plane. In 1980, Marjorie Rice and Doris Schattschneider described additional ways to create spiral tilings, containing multiple arms, from pentagonal tiles.

**SEE ALSO** Wallpaper Groups (1891), Squaring a Rectangle (1925), Penrose Tiles (1973), and Spidrons (1979).

*A spiral Voderberg tiling, rendered by Teja Krašek. This kind of tiling is referred to as* monohedral *because it is a tessellation in which all tiles are the same.*

1937

# Collatz Conjecture

**Lothar Collatz** (1910–1990)

Imagine walking in a blinding hailstorm in which the hailstones drift up and down in the wisps and eddies of wind. Sometimes the stones shoot up for as far as your eye can see and then come plummeting back to Earth, smashing into the ground like little meteorites.

Hailstone number problems have fascinated mathematicians for several decades and are studied because they are so simple to calculate yet apparently intractably hard to solve. To compute a sequence of hailstone numbers—also referred to as $3n + 1$ numbers—start by choosing any positive integer. If your number is even, divide it by 2. If it is odd, multiply by 3 and add 1. Next, take your answer and repeat the rule. For example, the hailstone sequence for 3 is 3, 10, 5, 16, 8, 4, 2, 1, 4,... (The "..." indicates that the sequence continues forever as 4, 2, 1, 4, 2, 1, 4, and so forth.)

Like hailstones falling from the sky through storm clouds, this sequence drifts down and up, sometimes in seemingly haphazard patterns. Also, like hailstones, hailstone numbers always seem eventually to fall back down to the "ground" (the integer "1"). The Collatz conjecture, named after German mathematician Lothar Collatz who proposed it in 1937, states that this process will eventually fall to 1 for *any* starting positive integer. So far, mathematicians have not found a way to prove this conjecture, although the conjecture has been checked by computer for all start values up to $19 \times 2^{58} \approx 5.48 \times 10^{18}$.

A variety of awards have been offered to anyone who can prove or disprove the conjecture. Mathematician Paul Erdös commented on the complexity of $3n + 1$ numbers, "Mathematics is not yet ready for such problems." The amiable and modest Collatz received many honors for his contributions to mathematics, and he died in 1990 in Bulgaria, while attending a mathematics conference concerning computer arithmetic.

**SEE ALSO** Erdös and Extreme Collaboration (1971), Ikeda Attractor (1979), and The On-Line Encyclopedia of Integer Sequences (1996).

*Fractal Collatz pattern. Although the behavior of 3n+1 numbers is usually studied for integers, it is possible to extend the mathematical mappings to complex numbers and represent the intricate fractal behavior through coloration in the complex plane.*

1938

# Ford Circles

**Lester Randolph Ford, Sr.** (1886–1975)

Imagine a frothy milkshake with an infinite number of bubbles of all sizes, touching one another but not interpenetrating. The bubbles become smaller and smaller, always filling in the cracks and spaces between larger ones. One form of such mysterious froth was discussed by mathematician Lester Ford in 1938, and it turns out that they characterize the very fabric of our "rational" number system. (*Rational numbers* are numbers like ½ that can be expressed as fractions.)

To create the Ford froth, begin by choosing any two integers, $h$ and $k$. Draw a circle with radius $1/(2k^2)$ and centered at $(h/k\ ,\ 1/(2k^2))$. For example, if you select $h = 1$ and $k = 2$, you draw a circle centered at (0.5, 0.125) and with radius 0.125. Continue to place circles for different values of $h$ and $k$. As your picture becomes denser, you'll notice that none of your circles intersect, although some will be tangent to one another (that is, just kiss one another). Any circle has an infinitude of circles that kisses it.

Consider a godlike archer positioned above the Ford froth with an appropriately large $y$ value. To simulate the shooting of the arrow, draw a vertical line from the location of your archer (for example, at $x = a$) down to the $x$-axis. (This line is perpendicular to the $x$-axis.) If $a$ is a rational number, the line must pierce some Ford circle and hit the horizontal $x$-axis exactly at the circle's point of tangency. However, when the archer's position is at an *irrational number* (a non-repeating, endless decimal value like $\pi$ = 3.1415…), it must *leave* every circle that it enters and then must enter another circle. Thus, the archer's arrow must pass through an infinity of circles! A deeper mathematical study of Ford circles shows that they provide excellent visualizations of different levels of infinity and of **Cantor's Transfinite Numbers**.

**SEE ALSO** Cantor's Transfinite Numbers (1874), Menger Sponge (1926), and Fractals (1975).

*Ford circles, rendered by Jos Leys. The image is rotated 45° so that the x-axis extends from bottom left to upper right. The circles become smaller and smaller, always filling in the cracks and spaces between larger ones.*

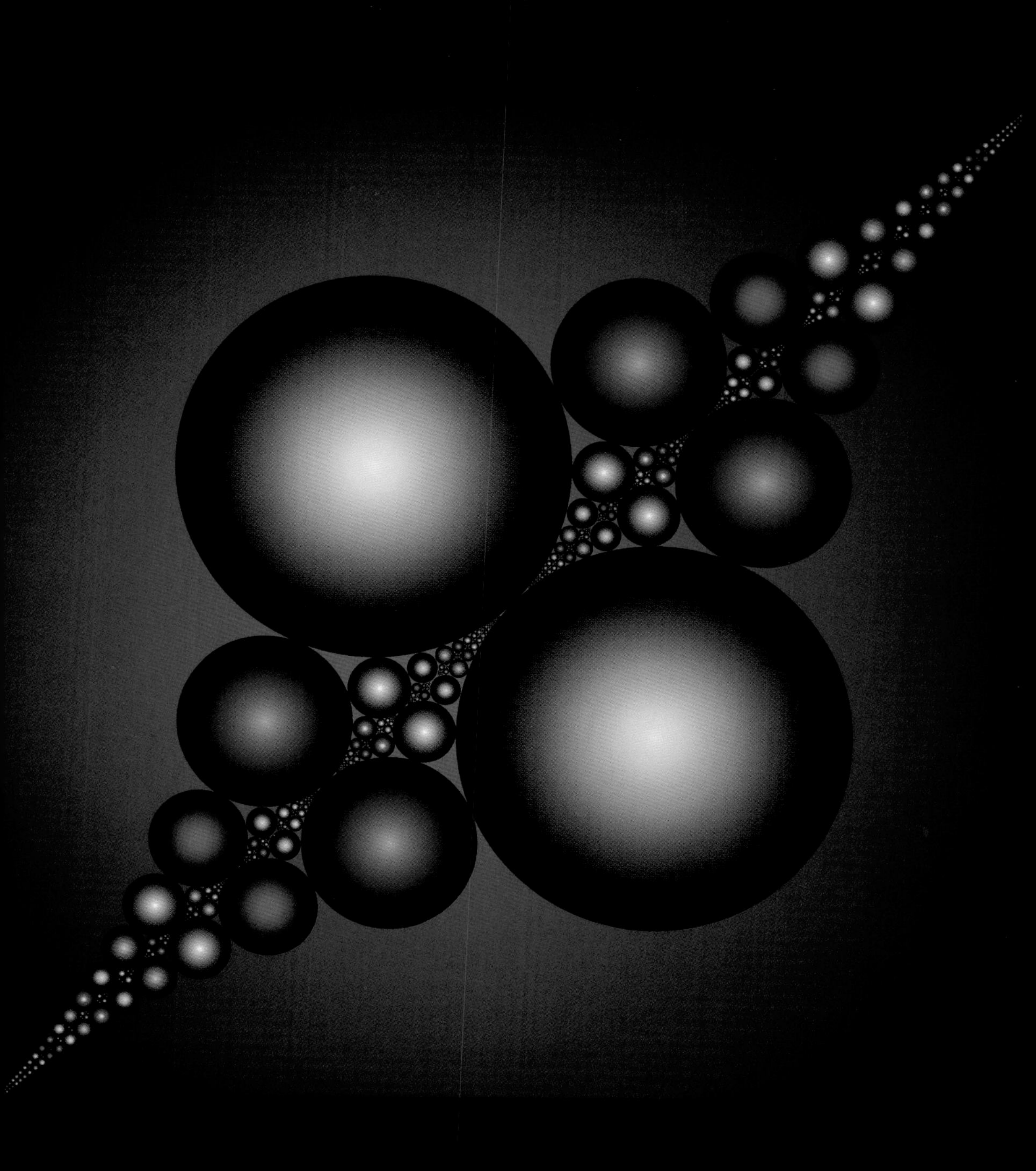

1938

# The Rise of Randomizing Machines

**William Thomson, Baron Kelvin of Largs** (1824–1907), **Sir Maurice George Kendall** (1907–1983), **Bernard Babington Smith** (1905–1993), **Leonard Henry Caleb Tippett** (1902–1985), **Frank Yates** (1902–1995), **Sir Ronald Aylmer Fisher, FRS** (1890–1962)

In modern science, random number generators are useful in simulating natural phenomena and in sampling data. Before the rise of modern electronic computers, researchers had to be creative in their approaches for obtaining random numbers. For example, in 1901, Lord Kelvin used numbers written on small papers drawn from a bowl to generate random numbers. However, he found this approach "unsatisfactory," writing, "The best mixing we could make in the bowl seemed to be quite insufficient to secure equal chances for all the billets."

In 1927, British statistician Leonard Tippett provided researchers with a table of 41,600 random digits that he had constructed by taking the middle digits of numbers representing the area of parishes in England. In 1938, British statisticians Ronald Fischer and Frank Yates published 15,000 additional random numbers using two decks of playing cards to select digits in logarithms.

In 1938 and 1939, British statistician Maurice Kendall started his research with British psychologist Bernard Babington Smith to produce random numbers by machine. Their randomizing machine was the first such mechanical device used to produce a table of 100,000 random digits. They also formulated a series of rigorous tests to determine if the digits were indeed statistically random. Kendall and Smith's numbers were commonly used until the RAND Corporation published *A Million Random Digits with 100,000 Normal Deviates* in 1955. RAND used a roulette-wheel-like machine similar to the Kendall and Smith machine, and verified the digits as statistically random using similar mathematical tests.

Kendall and Smith used a motor connected to a circular piece of cardboard, about 10 inches (25 centimeters) in diameter. The disk was divided into 10 segments "as equal in size as we could make them," numbered consecutively from 0 to 9. The disk was illuminated by a neon lamp. A capacitor charged and the lamp eventually produced a flash. The operator of the randomizing machine would see a number and record it.

**SEE ALSO** Dice (c. 3000 B.C.), Buffon's Needle (1777), and Von Neumann's Middle-Square Randomizer (1946).

*The complex and unpredictable motions of wax blobs within lava lamps has been used as a source of random numbers. Such a system for generating random numbers is mentioned in U.S. Patent 5,732,138, issued in 1998.*

1939

# Birthday Paradox

**Richard von Mises** (1883–1953)

Martin Gardner writes, "Since the beginning of history, unusual coincidences have strengthened belief in the influence on life of occult forces. Events that seemed to miraculously violate the laws of probability were attributed to the will of gods or devils, God or Satan, or at the very least to mysterious laws unknown to science and mathematics." One problem that has intrigued coincidence researchers is the *birthday paradox*.

Imagine you are in a large living room that people gradually enter. How many people must be in the room before the probability that some share a birthday becomes at least 50 percent? This problem, posed in 1939 by Austrian-born American mathematician Richard von Mises, is significant because its solution is counterintuitive to most people, because it is one of the most explored probability problems in classrooms today, and because variations of the birthday problem serve as useful models for analyzing amazing coincidences in everyday life.

Assuming 365 days each year, the answer to the problem is a mere 23 people. In other words, if a room is filled with 23 or more randomly chosen people, there is more than a 50 percent probability that some pair of people will have the same birthday. For 57 or more people, the probability is more than 99 percent. The probability becomes 100 percent if there are at least 366 people in the room, due to the **Pigeonhole Principle**. We assume that the 365 possible birthdays are equally likely, and leap days are ignored. The formula for calculating the probability of at least two of the $n$ persons having the same birthday is $1-[365!/[365^n(365-n)!]$, which can be approximated by $1-e^{-n^2/(2\cdot365)}$.

Just 23 people may have been fewer than you expected because we were not seeking two particular people or a specific birth date. A match on *any* date for *any* two people is sufficient. In fact, 253 different pairings are possible among 23 people, any of which could lead to a match.

**SEE ALSO** Zeno's Paradoxes (c. 445 B.C.), Aristotle's Wheel Paradox (c. 320 B.C.), Law of Large Numbers (1713), St. Petersburg Paradox (1738), Pigeonhole Principle (1834), Barber Paradox (1901), Banach-Tarski Paradox (1924), Hilbert's Grand Hotel (1925), Ramsey Theory (1928), Coastline Paradox (c. 1950), Newcomb's Paradox (1960), and Parrondo's Paradox (1999).

*How many people must be in the room before the probability that some share a birthday becomes at least 50 percent? Assuming 365 days each year, the counterintuitive answer to the problem is a mere 23 people.*

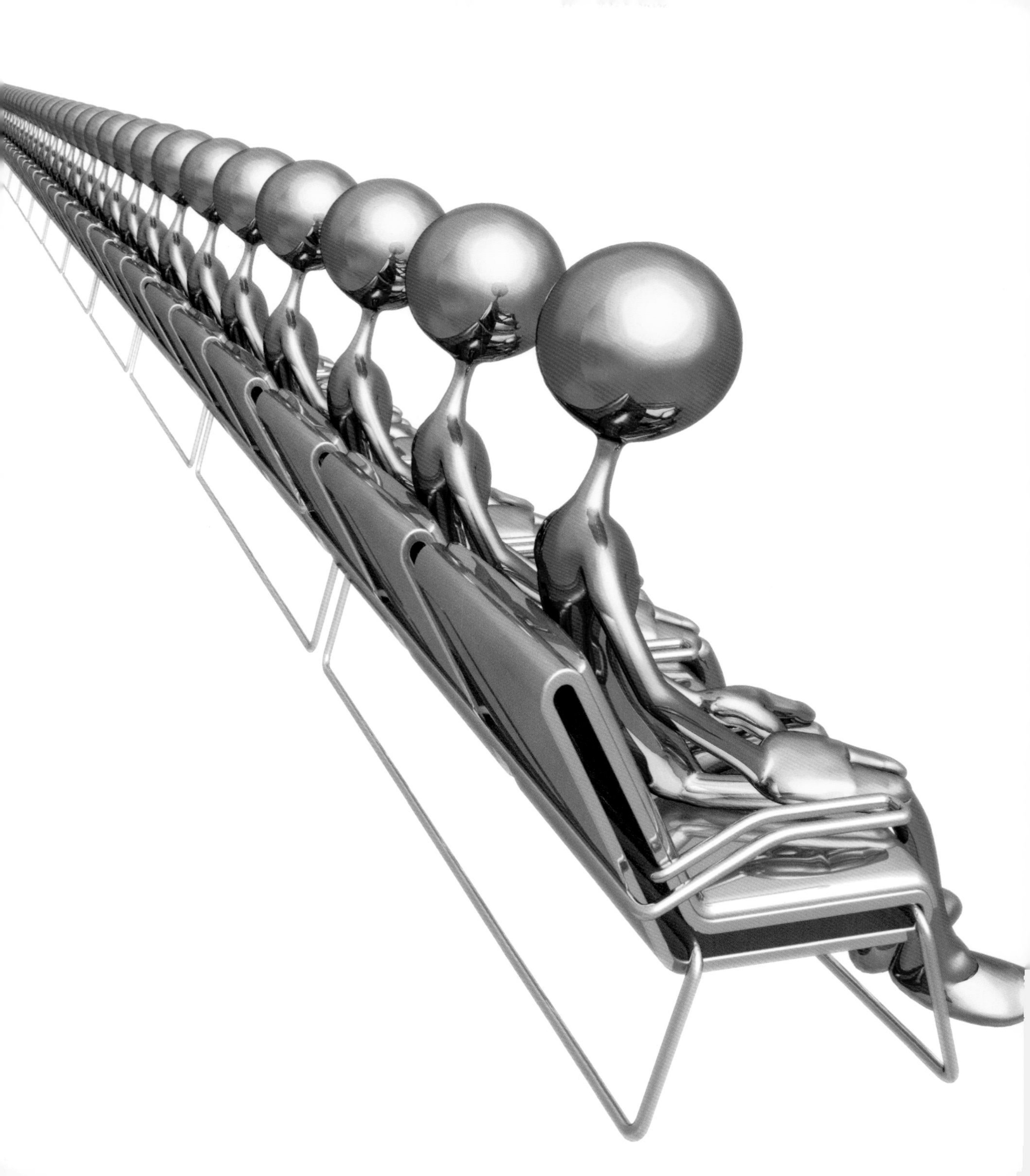

c.1940

# Polygon Circumscribing

**Edward Kasner** (1878–1955), **James Roy Newman** (1907–1966)

Draw a circle, with a radius equal to 1 inch (about 2.5 centimeters). Next, circumscribe (surround) the circle with an equilateral triangle. Next, circumscribe the triangle with another circle. Then circumscribe this second circle with a square. Continue with a third circle, circumscribing the square. Circumscribe this circle with a regular pentagon. Continue this procedure indefinitely, each time increasing the number of sides of the regular polygon by one. Every other shape used is a circle that grows continually in size as it encloses the assembly of predecessors. If you were to repeat this process, always adding larger circles at the rate of a circle a minute, how long would it take for the largest circle to have a radius equal to the radius of our solar system?

By continually surrounding the shapes with circles, it would seem that the radii should grow larger and larger, becoming infinite as we continue the process. However, the assembly of nested polygons and circles will never grow as large as the solar system, never grow as large as the Earth, never grow as large as a typical adult bicycle tire. Although the circles initially grow very quickly in size, the rate of growth gradually slows down, and the radii of the resulting circles approach a limiting value given by the infinite product: $R = 1/[\cos(\pi/3) \times \cos(\pi/4) \times \cos(\pi/5)\ldots]$.

Perhaps most intriguing was the controversy over the limiting value of $R$. It seems simple enough to compute. According to mathematicians Edward Kasner and James Newman, who first reported a value in the 1940s, $R$ is approximately equal to 12. A value of 12 is also mentioned in a German article published in 1964.

Christoffel J. Bouwkamp published a paper in 1965 that reports the true value of $R = 8.7000$. I find it fascinating that, until 1965, mathematicians still assumed that the correct value of $R$ was 12. The correct value of $R$ with 17 digits is 8.7000366252081945....

**SEE ALSO** Zeno's Paradoxes (c. 445 B.C.), Wheat on a Chessboard (1256), Harmonic Series Diverges (c. 1350), Discovery of Series Formula for π (c. 1500), and Brun's Constant (1919).

*A central circle is surrounded by alternating polygons and circles, as described in the text (the red lines are thickened in the illustration for artistic effect). Is it possible to make the pattern grow as large as a typical adult bicycle tire?*

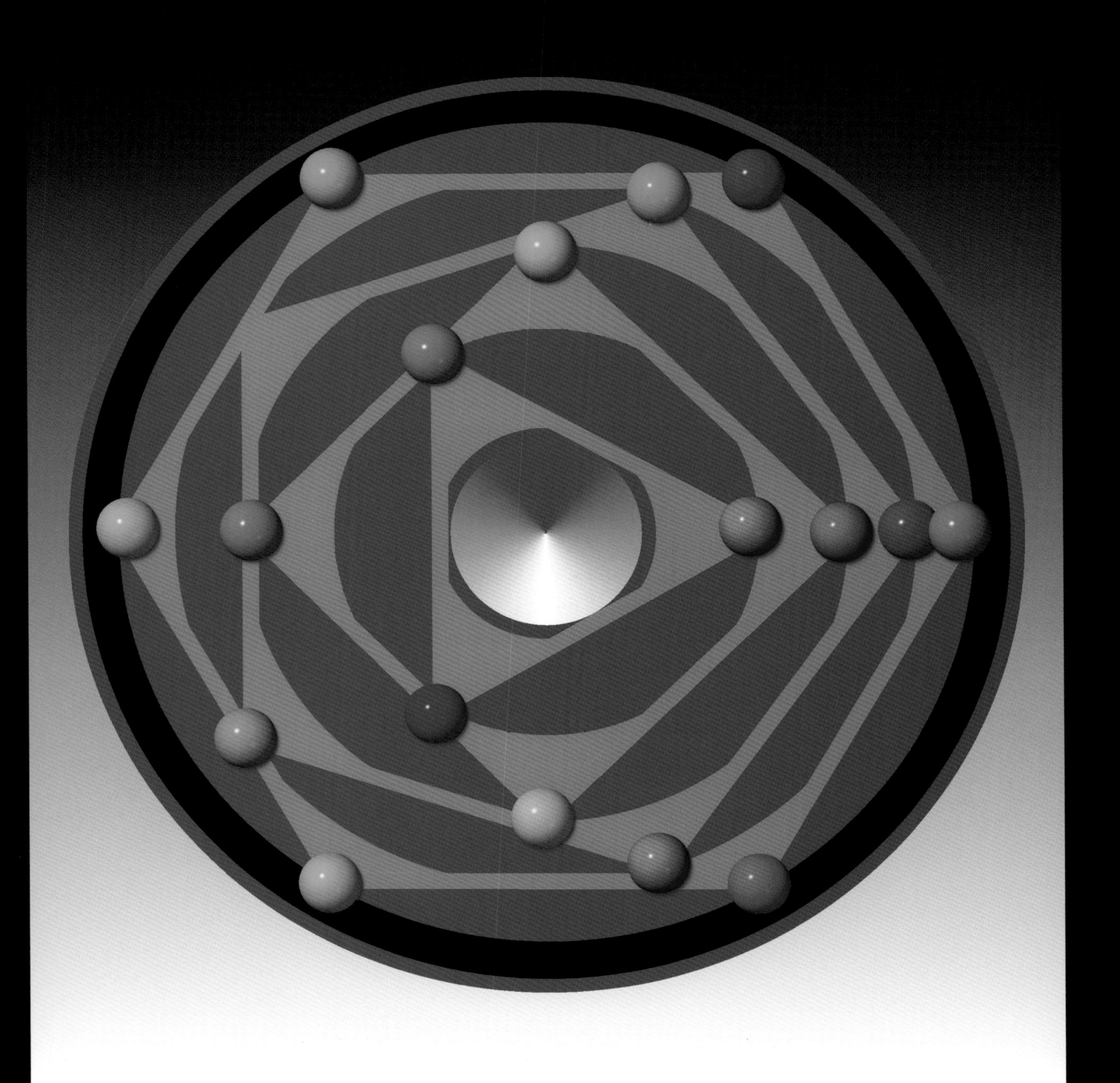

1942

# Hex

**Piet Hein** (1905–1996), **John Forbes Nash, Jr.** (b. 1928)

Hex is a two-player board game played on a hexagonal grid, usually in the shape of an 11 × 11 diamond. It was invented by Danish mathematician and poet Piet Hein in 1942 and independently by American mathematician John Nash in 1947. Nash, a Nobel Prize winner, is perhaps best known to the public as the subject of the Hollywood movie *A Beautiful Mind*, which highlights his mathematical prowess and battles with schizophrenia. According to the book *A Beautiful Mind*, Nash promoted a 14 × 14 board as the optimal size.

Players use differently colored pieces (for example, red and blue) and alternatively place them in the hexagonal cells. Red's goal is to form a red path connecting two opposite sides of the board. Blue's goal is to form a path connecting the other opposite sides. The four corner hexagons belong to both sides. Nash discovered that the game can never end in a tie, and the game favors the first player, who can have a winning strategy. One way of making the game fairer is to allow the second player to choose his or her color after the first player makes the first move or after the first three moves.

In 1952, Parker Brothers marketed a version of the game to the general public that used hexagonal playing pieces. A winning strategy for the first player has been demonstrated for several sizes of playing boards. Although the game seems simple, mathematicians have used it for more profound applications, such as for proving the **Brouwer Fixed-Point Theorem**.

Hein became internationally famous for his designs, poems, and mathematical games. When the Germans invaded Denmark in 1940, he was forced to go underground because he was the head of an anti-Nazi group. In 1944, he explained his creative approach: "Art is the solution to problems which cannot be formulated clearly before they have been solved."

**SEE ALSO** Brouwer Fixed-Point Theorem (1909), Pig Game Strategy (1945), Nash Equilibrium (1950), and Instant Insanity (1966).

*Hex board game played on a hexagonal grid. Red's goal is to form a red path connecting two opposite sides of the board. Blue's goal is to form a path connecting the other opposite sides. In this example, Red wins.*

1945

# Pig Game Strategy

**John Scarne** (born Orlando Carmelo Scarnecchia) (1903–1985)

Pig is a game with simple rules but with surprisingly complex strategies and analyses. It is significant as a metaphor for many apparently simple problems that have led to rich mathematical research years later, and as a teaching tool used by numerous educators when discussing game strategy.

Pig was first described in print in 1945 by John Scarne—American magician, game expert, card manipulator, and inventor—but the game has its roots in older "folk games" with several variations. To play Pig, a player rolls a die until either a 1 is rolled or the player "holds" and tallies the sum of the rolls during her turn. If the player rolls a 1, nothing is added to the player's score during her turn, and the opponent now has his turn. The first player to attain a score of 100 or more wins. Example: You roll a 3. You decide to roll again and roll a 1. Thus, you add nothing to your score and hand the die to your opponent. He rolls the sequence 3-4-6 and decides to hold. Thus, he adds 13 to his score and hands the die back to you.

Pig is considered a "jeopardy" dice game because players must decide if they should jeopardize previous gains by rolling for possible additional gains. In 2004, computer scientists Todd W. Neller and Clifton Presser of Gettysburg College in Pennsylvania analyzed Pig in detail to elucidate a strategy for optimal play. Using mathematics and computer graphics, they revealed an intricate, nonintuitive strategy for winning and showed why playing to maximize points for a single turn clearly is different from playing to win. About their findings and visualizations of optimal policies, they poetically write, "Seeing the 'landscape' of this policy is like seeing the surface of a distant planet sharply for the first time, having previously seen only fuzzy images."

**SEE ALSO** Dice (c. 3000 B.C.), Nash Equilibrium (1950), Prisoner's Dilemma (1950), Newcomb's Paradox (1960), and Instant Insanity (1966).

*The simple game of Pig has surprisingly complicated strategies and analyses. Pig was first described in print in 1945 by American magician and inventor John Scarne.*

1946

# ENIAC

## **John Mauchly** (1907–1980) **and J. Presper Eckert** (1919–1995)

ENIAC, short for Electronic Numerical Integrator and Computer, was built at the University of Pennsylvania by American scientists John Mauchly and J. Presper Eckert. This device was the first electronic, reprogrammable, digital computer that could be used to solve a large range of computing problems. The original purpose of ENIAC was to calculate artillery firing tables for the U.S. Army; however, its first important application involved the design of the hydrogen bomb.

ENIAC was unveiled in 1946, having cost nearly $500,000, and it was in nearly continuous use until it was turned off on October 2, 1955. The machine contained more than 17,000 vacuum tubes and around 5 million hand-soldered joints. An IBM card reader and card punch machine were used for input and output. In 1997, a team of engineering students led by Professor Jan Van der Spiegel created a "replica" of the 30-ton ENIAC on a single integrated circuit!

Other important electrical computing machines of the 1930s and 1940s include the American Atanasoff-Berry Computer (demonstrated in December, 1939), the German Z3 (demonstrated in May, 1941), and the British Colossus computer (demonstrated in 1943); however, these machines were either not fully electronic or not general purpose.

The authors of the ENIAC patent (No. 3,120,606; filed in 1947) write, "With the advent of everyday use of elaborate calculations, speed has become paramount to such a high degree that there is no machine on the market today capable of satisfying the full demand of modern computational methods....The present invention is intended to reduce to seconds such lengthy computations...."

Today, computer use has invaded most areas of mathematics, including numerical analysis, number theory, and probability theory. Mathematicians, of course, increasingly use computers in their research and in their teaching, sometimes using computer graphics to gain insight. Famous mathematical proofs have been done with the aid of the computer.

**SEE ALSO** Abacus (1200), Slide Rule (1621), Babbage Mechanical Computer (1822), Differential Analyzer (1927), Turing Machines (1936), Curta Calculator (1948), and HP-35: First Scientific Pocket Calculator (1972).

*U.S. Army photo of ENIAC, the first electronic, reprogrammable, digital computer that could be used to solve a large range of computing problems. Its first important application involved the design of the hydrogen bomb.*

1946

# Von Neumann's Middle-Square Randomizer

**John von Neumann** (1903–1957)

Scientists use random number generators for tackling a wide variety of problems, such as developing secret codes, modeling the movement of atoms, and conducting accurate surveys. A pseudorandom number generator (PRNG) is an algorithm that produces a sequence of numbers that emulate the statistical properties of random numbers.

The *middle-square method*, developed by mathematician John von Neumann in 1946, is one of the most famous and earliest computer-based PRNGs. He started with a number such as 1946 and squared it to produce 3786916, which can be written as 03786916. He removed the middle four digits, 7869, and continued the process of squaring and removal. In actual practice, von Neumann used 10-digit numbers and followed the same rules.

Von Neumann, famous for his collaborative research in thermonuclear reactions that led to the hydrogen bomb, understood that his simple randomizing approach had flaws and that the sequences would eventually repeat, but he was satisfied with the method for many applications. In 1951, von Neumann cautioned users of these schemes, "Anyone who considers arithmetical methods of producing random digits is, of course, in a state of sin." Nevertheless, he preferred this approach to better hardware-based random number generators that did not record their values, thus making it difficult to repeat procedures to identify problems. In any case, von Neumann did not have access to sufficient computer memory to store many "random" values. Indeed, his wonderfully simple approach produced numbers on the ENIAC computer hundreds of times faster than reading numbers from punch cards.

More recent and useful PRNGs make use of the linear congruential method of the form $X_{n+1} = (aX_n + c) \bmod m$. Here, $n \geq 0$, $a$ is the multiplier, $m$ the modulus, $c$ the increment, and $X_0$ the starting value. The Mersenne twister PRNG algorithm, developed in 1997 by Makoto Matsumoto and Takuji Nishimura, is also desirable for many of today's applications.

**SEE ALSO** Dice (c. 3000 B.C.), Buffon's Needle (1777), The Rise of Randomizing Machines (1938), and ENIAC (1946).

*John von Neumann in the 1940s. Von Neumann developed the middle-square method, a famous, early computer-based pseudorandom number generator.*

1947

# Gray Code

**Frank Gray** (1887–1969), **Émile Baudot** (1845–1903)

A Gray code represents numbers in a positional notation so that when the numbers are in *counting order*, any adjacent number pair will differ in some single digit by 1, and at one position only. For example, 182 and 172 could be adjacent counting numbers in a decimal Gray code (the middle digits differ by 1), but not 182 and 162 (no digits differ by 1), nor 182 and 173 (more than one digit pair differs by 1).

One simple, famous, and useful Gray code is called the reflected binary Gray code, which consists of only 0s and 1s. Martin Gardner explains that to convert a standard binary number to its reflected Gray equivalent, we first examine the rightmost digit and then consider each digit in turn. If the next digit to the left is 0, let the original digit stand. If the next digit to the left is 1, change the original digit. (The digit at the extreme left is assumed to have a 0 on its left and therefore remains unchanged.) For example, applying this conversion to the number 110111 gives the Gray number 101100. We can then convert all the standard binary numbers to create the Gray sequence that starts 0, 1, 11, 10, 110, 111, 101, 100, 1100, 1101, 1111, . . .

The reflected binary code was originally designed to make it easier to prevent erroneous output from electromechanical switches. In this application, a slight change in position only affects one bit. Today, Gray codes are used to facilitate error correction in digital communications, such as in TV signal transmission, and to make transmission systems less susceptible to noise. The French engineer Émile Baudot used Gray codes in telegraphy in 1878. The code is named after Bell Labs research physicist Frank Gray, who made extensive use of these codes in his engineering patents. Gray had invented a method to convert analog signals to binary Gray code using vacuum tubes. Today, Gray codes also have important applications in graph theory and number theory.

**SEE ALSO** Boolean Algebra (1854), Gros's *Théorie du Baguenodier* (1872), Tower of Hanoi (1883), and Information Theory (1948).

*Diagram from Frank Gray's U.S. patent 2,632,058, filed in 1947 and issued in 1953. In this patent, Gray introduced his famous code, referring to it as a "reflected binary code." The code was later named after Gray by other researchers.*

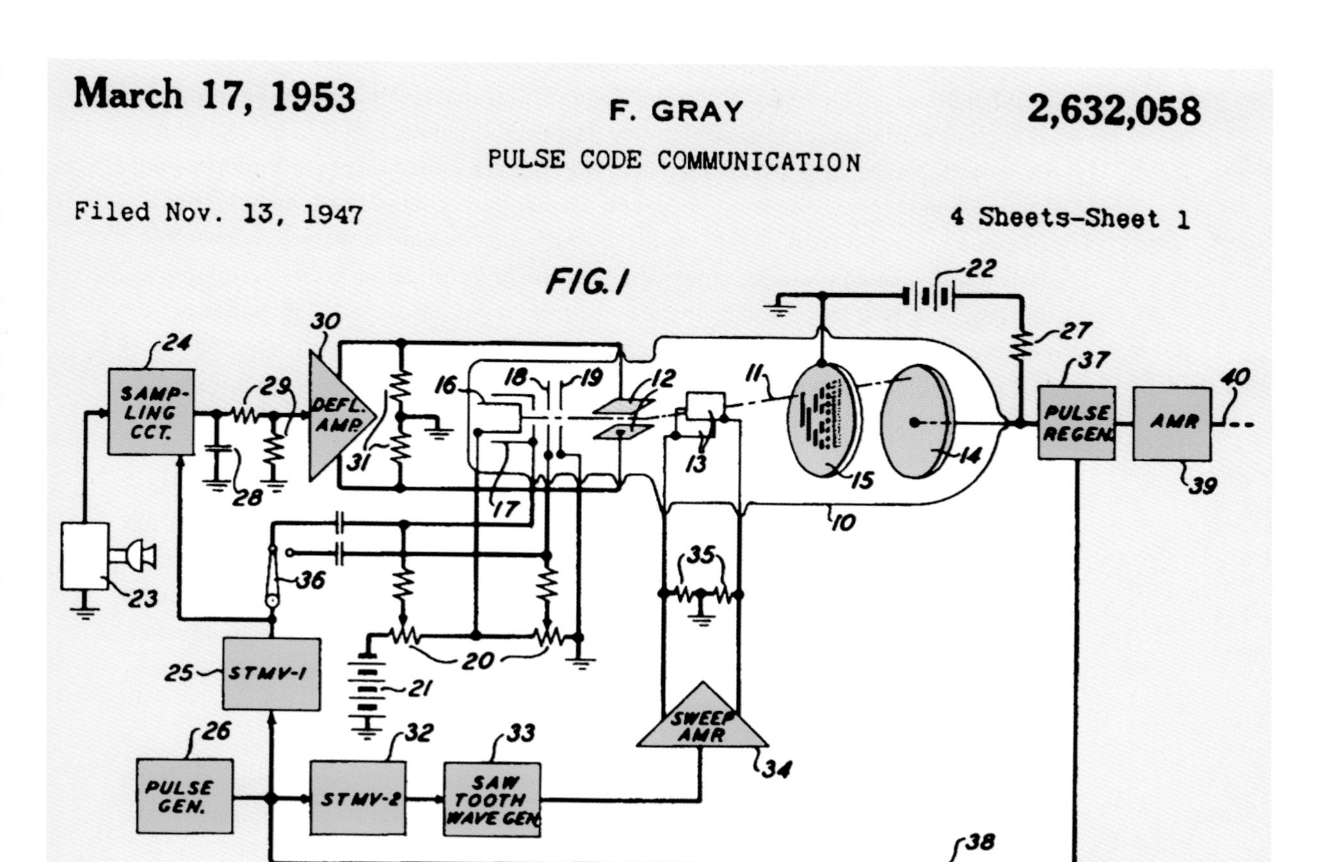
March 17, 1953
F. GRAY
2,632,058
PULSE CODE COMMUNICATION
Filed Nov. 13, 1947
4 Sheets-Sheet 1
FIG. 1
SAMPLING CCT.
DEFL. AMP.
PULSE REGEN.
AMP.
STMV-1
PULSE GEN.
STMV-2
SAW TOOTH WAVE GEN.
SWEEP AMP.

1948

# Information Theory

## **Claude Elwood Shannon** (1916–2001)

Teenagers watch TV, cruise the Internet, spin their DVDs, and chat endlessly on the phone usually without ever realizing that the foundations for this Information Age were laid by American mathematician Claude Shannon, who in 1948 published "A Mathematical Theory of Communication." Information theory is a discipline of applied mathematics involving the quantification of data, and it helps scientists understand the capacity of various systems to store, transmit, and process information. Information theory is also concerned with data compression and with methods for reducing noise and error rates to enable as much data as possible to be reliably stored and communicated over a channel. The measure of information, known as information entropy, is usually expressed by the average number of bits needed for storage or communication. Much of the mathematics behind information theory was established by Ludwig Boltzmann and J. Willard Gibbs for the field of thermodynamics. Alan Turing also used similar ideas when breaking of the German Enigma ciphers during World War II.

Information theory affects a diverse array of fields, ranging from mathematics and computer science to neurobiology, linguistics, and black holes. Information theory has practical applications such as breaking codes and recovering from errors due to scratches in movie DVDs. According to a 1953 issue of *Fortune*: "It may be no exaggeration to say that man's progress in peace, and security in war, depend more on fruitful applications of Information Theory than on physical demonstrations, either in bombs or in power plants, that Einstein's famous equation works."

Claude Shannon died in 2001, at the age of 84, after a long struggle with Alzheimer's disease. At one point in his life, he had been an excellent juggler, unicyclist, and chess player. Sadly, due to his affliction, he was unable to observe the Information Age that he helped create.

**SEE ALSO** Boolean Algebra (1854), Turing Machines (1936), and Gray Code (1947).

*Information theory helps technologists understand the capacity of various systems to store, transmit, and process information. Information theory has applications in fields ranging from computer science to neurobiology.*

30
31
32
33
34
35
36
5
6
7
8
9
10
105
105
107
107
31
32
33
34
35
36

1948

# Curta Calculator

**Curt Herzstark** (1902–1988)

The Curta is considered by many historians of science to be the first commercially successful portable mechanical calculator. Developed by Austrian Jew Curt Herzstark while a prisoner in the Buchenwald concentration camp, the handheld Curta could perform multiplication, addition, subtraction, and division. The cylindrical body of the Curta was usually held in the left hand and contained eight sliders for number entry.

In 1943, Herzstark was accused of "helping Jews" and "indecent contacts with Aryan women." He eventually ended up at Buchenwald, where news of his technical expertise and ideas on calculating machines led the Nazis to demand that he make a drawing of his calculator designs; they had hoped to give the device to Hitler as a gift at the end of the war.

After the war, in 1946, Herzstark was invited by the prince of Liechtenstein to found a manufacturing plant for the devices, which became widely available to the public in 1948. For a time, the Curtas were among the best portable calculators available, and were in frequent use until the rise of electronic calculators in the 1970s.

The Type I Curta had an 11-digit result counter. The larger Type II Curta, introduced in 1954, had a 15-digit result counter. Over a period of about 20 years, approximately 80,000 of the Curta I and 60,000 of the Curta II devices were built.

Astronomer and author Cliff Stoll writes, "Johannes Kepler, Isaac Newton, and Lord Kelvin all complained about the time they had to waste doing simple arithmetic.... Oh, for a pocket calculator that could add, subtract, multiply and divide! One with digital readouts and memory. A simple, finger-friendly interface. But none were available until 1947. Then, for a quarter of a century, the finest pocket calculators came from Liechtenstein. In this diminutive land of Alpine scenery and tax shelters, Curt Herzstark built the most ingenious calculating machine ever to grace an engineer's hand: the Curta calculator."

**SEE ALSO** Abacus (1200), Slide Rule (1621), Babbage Mechanical Computer (1822), Ritty Model I Cash Register (1879), Differential Analyzer (1927), and HP-35: First Scientific Pocket Calculator (1972).

*The Curta calculator may be the first commercially successful portable mechanical calculator. The handheld device was developed by Curt Herzstark while a prisoner in the Buchenwald concentration camp. The Nazis hoped to give the device to Adolph Hitler as a gift.*

CURTA

1949

# Császár Polyhedron

**Ákos Császár** (b. 1924)

Polyhedra are solids built from a collection of polygons joined at their edges. How many polyhedra exist with every pair of vertices joined by an edge? Apart from the tetrahedron (triangular pyramid), the Császár polyhedron is the only known polyhedron that is considered to have no diagonals, where a diagonal is defined as a line joining any two vertices not connected by an edge. Note that the tetrahedron has four vertices, six edges, four faces, and no diagonals. An edge joins every pair of corners.

The Császár polyhedron was first described in 1949 by the Hungarian mathematician Ákos Császár. Using the theory of combinatorics (the study of the ways of choosing and arranging objects from collections), mathematicians now know that other than the tetrahedron, any other no-diagonal polyhedron must have at least one hole (tunnel). The Császár polyhedron has one hole (difficult to visualize without a model to hold) and is topologically equivalent to a torus (doughnut). This polyhedron has 7 vertices, 14 faces, and 21 edges, and is the dual of the **Szilassi polyhedron**. For dual polyhedra, the vertices of one polyhedron correspond to the faces of the other polyhedron.

David Darling writes, "It isn't known if there are any other polyhedra in which every pair of vertices is joined by an edge. The next possible figure would have 12 faces, 66 edges, 44 vertices, and 6 holes, but this seems an unlikely configuration—as, indeed, to an even greater extent, does any more complex member of this curious family."

Martin Gardner remarks on the wide-ranging applications of the Császár polyhedron, "In studying the skeletal structure of a bizarre solid... [we find] some remarkable isomorphisms that involve the seven-color map on a torus, the smallest 'finite projective plane,' the solution of an old puzzle about triplets of seven girls, the solution of a bridge-tournament problem about eight teams, and the construction of a new kind of magic square known as a Room square."

**SEE ALSO** Platonic Solids (350 B.C.), Archimedean Semi-Regular Polyhedra (c. 240 B.C.), Euler's Formula for Polyhedra (1751), Icosian Game (1857), Pick's Theorem (1899), Geodesic Dome (1922), Ramsey Theory (1928), Szilassi Polyhedron (1977), Spidrons (1979), and Solving of the Holyhedron (1999).

*Császár polyhedron. Aside from the tetrahedron, the Császár polyhedron is the only known polyhedron that is considered to have no diagonals, where a diagonal is defined as a line joining any two vertices not connected by an edge.*

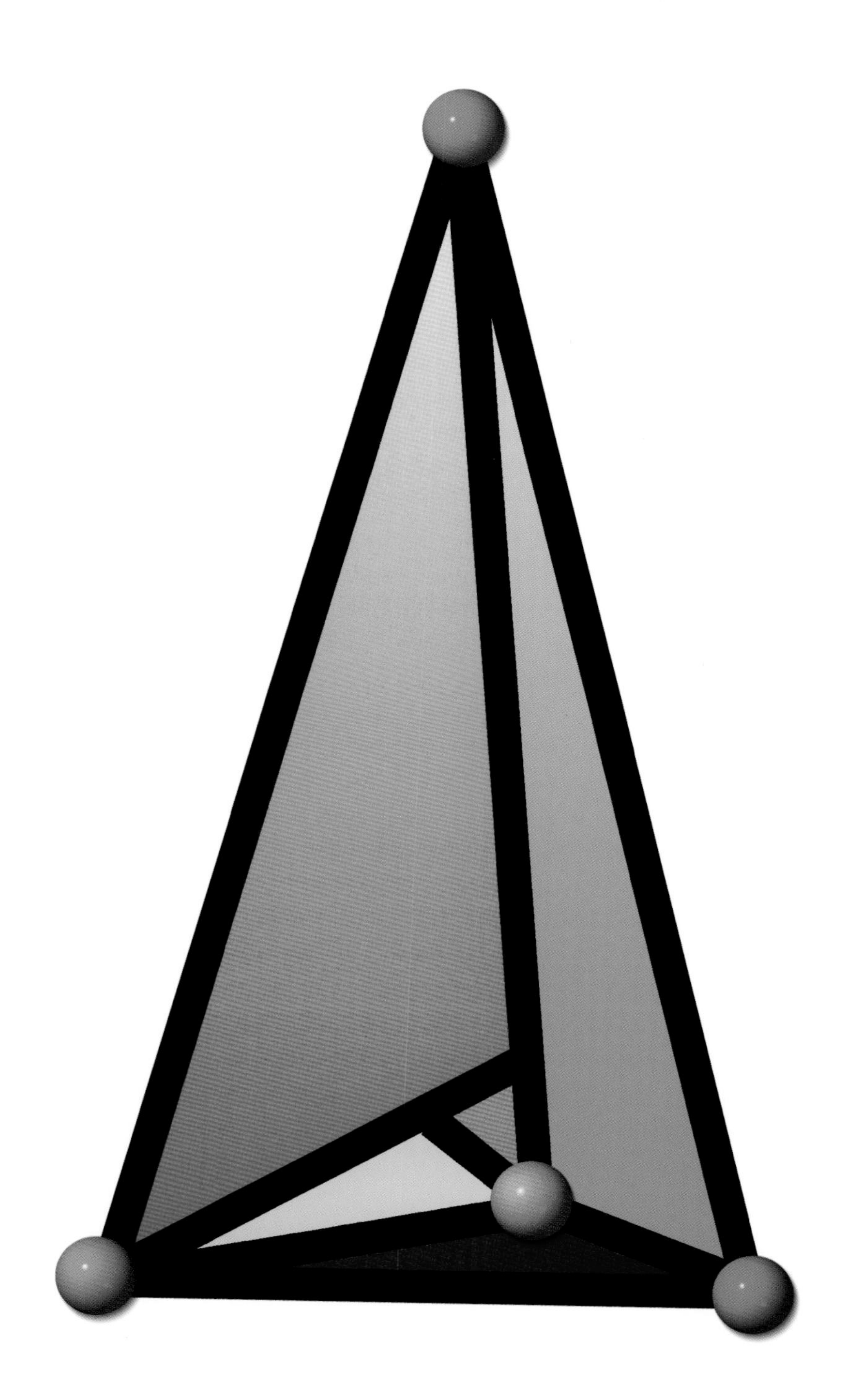

1950

# Nash Equilibrium

**John Forbes Nash, Jr.** (b. 1928)

American mathematician John Nash received the 1994 Nobel Prize in Economics. His prize-winning work appeared almost half a century earlier in his slender 27-page doctoral thesis written at the age of 21.

In game theory, the Nash equilibrium concerns games involving two or more players, where no player has anything to gain by changing his strategy on his own. If each player has chosen a strategy and no player can benefit by changing his strategy while the other players' strategies remain unchanged, then the current set of strategy choices is part of a Nash equilibrium. In 1950, Nash was the first to show in his dissertation, "Non-cooperative Games," that Nash equilibria for mixed strategies must exist for all finite games with an arbitrary number of players.

Game theory had made great strides in the 1920s with the work of John von Neumann, which culminated in his book *Theory of Games and Economic Behavior*, coauthored with Oskar Morgenstern. They focused on "zero-sum" games for which the interests of two players were strictly opposed. Today, game theory has relevance in studying human conflict and bargaining and to the behavior of animal populations.

As for Nash, in 1958, *Fortune* singled him out for his achievements in game theory, algebraic geometry, and nonlinear theory, calling him the most brilliant of the younger generation of mathematicians. He seemed destined for continued achievements, but in 1959, he was institutionalized and diagnosed as schizophrenic. He believed that aliens had made him emperor of Antarctica and that an ordinary thing, such as a sentence in the newspaper, could have a hidden and extra-important significance. Nash once remarked: "I would not dare to say that there is a direct relation between mathematics and madness, but there is no doubt that great mathematicians suffer from maniacal characteristics, delirium, and symptoms of schizophrenia."

**SEE ALSO** Hex (1942), Pig Game Strategy (1945), Prisoner's Dilemma (1950), Newcomb's Paradox (1960), and Checkers Is Solved (2007).

LEFT: *Nobel Prize–winner John Nash. This photo was taken in 2006 at a game theory symposium at the University of Cologne in Germany.* RIGHT: *The mathematics of game theory may be used to model real-world scenarios in fields that range from social sciences to international relations and biology. Recent studies have applied the Nash equilibrium to the modeling of honeybee hives that compete for habitat resources.*

c.1950

# Coastline Paradox

**Lewis Fry Richardson** (1881–1953), **Benoit Mandelbrot** (1924–2010)

If one were to attempt to measure a coastline or the boundary of two nations, the value of the measurement would depend on the length of the measuring stick used. As the measuring stick decreased in length, the measurement would become sensitive to smaller and smaller wiggles in the boundary, and, in principle, the length of the coastline would approach infinity as the length of the stick approached zero. British mathematician Lewis Richardson considered this phenomenon during his attempt to correlate the occurrence of wars with the nature of the boundary separating two or more nations. (He found that the number of a country's wars was proportional to the number of countries it bordered.) Franco-American mathematician Benoit Mandelbrot built upon Richardson's work and suggested that the relationship between the measuring-stick length ($\varepsilon$) and the apparent total length ($L$) of a coastline could be expressed by the parameter $D$, the fractal dimension.

One can appreciate $D$ by studying the relationship between the number $N$ of measuring sticks and the length $\varepsilon$. For a smooth curve such as a circle, we have $N(\varepsilon) = c/\varepsilon$, where $c$ is a constant. However, for a fractal curve like a coastline, this relationship becomes $N(\varepsilon) = c/\varepsilon^D$. If we multiply both sides of the formula by $\varepsilon$, the relation can be expressed in terms of the length of the measuring stick: $L(\varepsilon) = \varepsilon/\varepsilon^D$. $D$ corresponds somewhat to the traditional notion of dimension (a line is one dimensional, a plane two dimensional), except that $D$ can be a fraction. Because a coastline is convoluted at different size scales, it slightly "fills" a surface, and its dimension lies between a line and a plane. The fractal structure implies that repeated magnification of its graph reveals ever-finer levels of detail. Mandelbrot gives $D = 1.26$ for the coastline of Britain. Of course, for real-world objects, we never can actually use infinitesimally small measuring sticks, but this "paradox" shows how natural features exhibit fractional dimensions over a range of measurement scales.

**SEE ALSO** Weierstrass Function (1872), Koch Snowflake (1904), Hausdorff Dimension (1918), and Fractals (1975).

*As one uses increasingly small measuring sticks to measure the length of the coastline of England, the length of the coastline appears to approach infinity. This "paradox" shows how natural features exhibit fractional dimensions over a range of measurement scales.*

1950

# Prisoner's Dilemma

**Melvin Dresher** (1911–1992), **Merrill Meeks Flood** (1908–1991), **Albert W. Tucker** (1905–1995)

Imagine an angel dealing with two prisoners. *Both* Cain and Abel are suspected of having illegally snuck back into the Garden of Eden. Insufficient evidence exists against either of them. If *neither* human confesses, the angel has to lower the "charges" to trespassing, and the two brothers are condemned to wander the desert for a mere six months. If just one brother confesses, then the confessor goes free, and the other is doomed to crawl and eat dust for thirty years. On the other hand, if *both* Cain and Abel confess, each will get reduced sentences of five years of wandering. Cain and Abel are separated so that they cannot communicate. What should Cain and Abel do?

At first, the solution to their dilemma seems straightforward: Neither Cain nor Abel should confess so that they both end up with the minimum punishment—wandering in the desert for six months. However, it's quite possible that if Cain wishes to cooperate, then Abel will be tempted to double-cross Cain at the last minute, thereby achieving the best possible outcome, which is freedom. One important game-theoretical approach shows that the scenario leads each suspect to confess even though it will bring a harsher punishment than the strategy of cooperation and no confession. Cain and Abel's dilemma explores the conflict between the good of the individual and the good of the group.

The Prisoner's Dilemma was first formally identified in 1950 by Melvin Dresher and Merrill M. Flood. Albert W. Tucker researched the dilemma to understand and illustrate the difficulty of analyzing *non-zero-sum games*—dilemmas in which one person's victory is not necessarily the other person's defeat. Tucker's work has since given rise to an enormous related literature in disciplines ranging from philosophy and biology to sociology, political science, and economics.

**SEE ALSO** Zeno's Paradoxes (c. 445 B.C.), Aristotle's Wheel Paradox (c. 320 B.C.), St. Petersburg Paradox (1738), Barber Paradox (1901), Banach-Tarski Paradox (1924), Hilbert's Grand Hotel (1925), Birthday Paradox (1939), Pig Game Strategy (1945), Nash Equilibrium (1950), Newcomb's Paradox (1960), and Parrondo's Paradox (1999).

*The Prisoners' Dilemma was first formally identified in 1950 by Melvin Dresher and Merrill M. Flood. The dilemma helps researchers illustrate the difficulty of analyzing non-zero-sum games in which one person's victory is not necessarily the other person's defeat.*

1952

# Cellular Automata

**John von Neumann** (1903–1957), **Stanisław Marcin Ulam** (1909–1984), **John Horton Conway** (b. 1937)

Cellular automata are a class of simple mathematical systems that can model a variety of physical processes with complex behaviors. Applications include the modeling of the spread of plant species, the propagation of animals such as barnacles, the oscillations of chemical reactions, and the spread of forest fires.

Some of the classic cellular automata consist of a grid of cells that can exist in two states, occupied or unoccupied. The occupancy of one cell is determined from a simple mathematical analysis of the occupancy of neighbor cells. Mathematicians define the rules, set up the game board, and let the game play itself out on a checkerboard world. Though the rules governing the creation of cellular automata are simple, the patterns they produce are very complicated and sometimes seem almost random, like a turbulent fluid flow or the output of a cryptographic system.

Early work in this area began with Stanislaw Ulam in the 1940s, when he modeled the growth of crystals using a simple lattice. Ulam suggested that mathematician John von Neumann use a similar approach to modeling self-replicating systems, such as robots that could build other robots, and around 1952, von Neumann created the first 2-D cellular automata, with 29 states per cell. Von Neumann proved mathematically that a particular pattern could make endless copies of itself within the given cellular universe.

The most famous two-state, two-dimensional cellular automaton is the Game of Life invented by John Conway, and popularized by Martin Gardner in *Scientific American*. Despite its simple rules, an amazing diversity of behaviors and forms are generated including gliders—that is, arrangements of cells that move themselves across their universe and can even interact to perform computations. In 2002, Stephen Wolfram published *A New Kind of Science*, which reinforced the idea that cellular automata can have significance in virtually all disciplines of science.

**SEE ALSO** Turing Machines (1936) and Mathematical Universe Hypothesis (2007).

*Cone snail with cellular-automata patterns on its shell, resulting from that activation and inhibition of neighboring pigment cells. The pattern resembles the output of a one-dimensional cellular automaton, referred to as a* Rule 30 *automaton.*

# Martin Gardner's Mathematical Recreations

**Martin Gardner** (1914–2010)

> *"Perhaps an angel of the Lord surveyed an endless sea of chaos, then troubled it gently with his finger. In this tiny and temporary swirl of equations, our cosmos took shape."*
>
> —Martin Gardner, *Order and Surprise*

The authors of *Winning Ways for Your Mathematical Plays* wrote that Martin Gardner "brought more mathematics to more millions than anyone else." Allyn Jackson, deputy editor of the American Mathematics Society, wrote that Gardner "opened the eyes of the general public to the beauty and fascination of mathematics and inspired many to go on to make the subject their life's work." Indeed, several famous concepts in math were first brought to world attention through Gardner's works before they appeared in other publications.

Martin Gardner is an American author who wrote the "Mathematical Games" column in *Scientific American* from 1957 to 1981. He has also published more than 65 books. Gardner attended the University of Chicago, where he earned a bachelor's degree in philosophy. The bulk of his vast education came through his wide reading and correspondence.

According to many modern mathematicians, Gardner is the most important person to have nurtured interest in mathematics in the United States for a substantial part of the twentieth century. Douglas Hofstadter once called Gardner "one of the great intellects produced in this country in this century." Gardner's "Mathematical Games" covered such subjects as flexagons, Conway's Game of Life, polyominoes, the soma cube, **Hex**, tangrams, **Penrose Tiles**, **public-key cryptography**, the works of M.C. Escher, and **fractals.**

Gardner's first article in *Scientific American*, on the topic of hexaflexagons (flexible folding objects), ran in December, 1956. Gerry Piel, the publisher, called Gardner into his office and asked him if enough similar material existed to make a regular magazine feature. Gardner replied that he thought so. The next issue—January, 1957—ran the first of the columns.

**SEE ALSO** Hex (1942), Cellular Automata (1952), Penrose Tiles (1973), Fractals (1975), Public-Key Cryptography (1977), and *NUMB3RS* (2005).

LEFT: *One logo used for the 2008 Gathering for Gardner conference. This biannual conference is held in honor of Martin Gardner to promote the exposition of new ideas in recreational mathematics, magic, puzzles, art, and philosophy. (The logo is by Teja Krašek.)* RIGHT: *Martin Gardner stands by all his words: Six shelves contain his publications, dating back to 1931. (The photo was taken in his Oklahoma home in March 2006.)*

The Magic Numbers of Dr. Matrix
MARTIN GARDNER
Order and Surprise
Martin Gardner
GARDNER · ORDER & SURPRISE
The New AMBIDEXTROUS UNIVERSE
MARTIN GARDNER
GREAT ESSAYS IN SCIENCE
Great Essays in Science
CLASSIC BRAINTEASERS
The Flight of Peter Fromm

1958

# Gilbreath's Conjecture

**Norman L. Gilbreath** (b. 1936)

In 1958, after scribbling on a napkin, American mathematician and magician Norman L. Gilbreath presented a mystifying hypothesis concerning prime numbers. Gilbreath wrote the first few prime numbers—that is, numbers larger than 1, such as 5 or 13, that are divisible only by themselves or 1. Next, he continued to subtract successive terms and record the unsigned differences

2, 3, 5, 7, 11, 13, 17, 19, 23, 29, 31,...

1, 2, 2, 4, 2, 4, 2, 4, 6, 2,...

1, 0, 2, 2, 2, 2, 2, 2, 4,...

1, 2, 0, 0, 0, 0, 0, 2,...

1, 2, 0, 0, 0, 0, 2,...

1, 2, 0, 0, 0, 2,...

1, 2, 0, 0, 2,...

1, 2, 0, 2,...

1, 2, 2,...

1, 0,...

1,...

Gilbreath's conjecture is that, after the initial row, the first number in each row is always one. No one has ever found an exception, despite searches out to several hundred billion rows. Mathematician Richard Guy once wrote, "It does not seem likely that we shall see a proof of Gilbreath's conjecture in the near future, although the conjecture is probably true." Mathematicians are unsure if the conjecture is particularly relevant to prime numbers or whether it applies to any sequence that begins with 2 and is followed by odd numbers that increase at a sufficient rate with sufficient gaps between them.

Although Gilbreath's conjecture is historically not as significant as many of the other entries in this book, it is a marvelous example of the kinds of simple-to-state problems that even amateur mathematicians can offer, but that may require mathematicians centuries to solve. A proof may someday be within our grasp when humanity better understands the distribution of the gaps between prime numbers.

**SEE ALSO** Cicada-Generated Prime Numbers (c. 1 Million B.C.), Sieve of Eratosthenes (240 B.C.), Goldbach Conjecture (1742), Constructing a Regular Heptadecagon (1796), Gauss's *Disquisitiones Arithmeticae* (1801), Riemann Hypothesis (1859), Proof of the Prime Number Theorem (1896), Brun's Constant (1919), Ulam Spiral (1963), and Andrica's Conjecture (1985).

*Norman Gilbreath, 2007, while at Cambridge University. The great number theorist Paul Erdös said that he thought the Gilbreath conjecture was true, but that it would probably be 200 years before it was proved.*

1958

# Turning a Sphere Inside Out

**Stephen Smale** (b. 1930), **Bernard Morin** (b. 1931)

For many years, topologists knew that it was theoretically possible to turn a sphere inside out (or "evert" it), yet they didn't have the slightest idea how to do it. When computer graphics became available to researchers, mathematician and graphics expert Nelson Max produced an animated film finally illustrating the transformation of the sphere. Max's 1977 movie *Turning a Sphere Inside Out* was based on the 1967 sphere eversion work of Bernard Morin, a blind French topologist. The animation focuses on how the eversion can be performed by passing the surface through itself without making any holes or creases. Mathematicians had believed that the problem was insoluble until around 1958, when American mathematician Stephen Smale proved otherwise. However, no one could clearly visualize the motion without the graphics.

When we discuss the eversion of a sphere, we're not talking about turning a beach ball inside out by pulling the deflated ball through its opening and then inflating it again. Instead, we are referring to a sphere with no orifice. Mathematicians try to visualize a sphere made out of a thin membrane that can stretch and even pass through itself without ripping or developing a sharp kink or crease. The task of avoiding such sharp creases makes the mathematical sphere eversion so difficult.

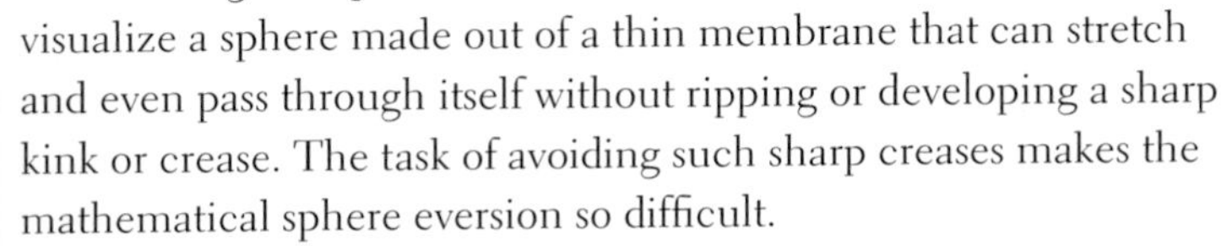

In the late 1990s, mathematicians went a step further and discovered a geometrically *optimal* path—one that minimizes the energy needed to contort the sphere through its transformation. This optimal sphere eversion, or *optiverse*, is now the star of a colorful computer-graphics movie titled *The Optiverse*. However, we can't use the principles in the movie to turn a real sealed balloon inside out. Because real balls and balloons are not made of a material that can pass through itself, it is not possible to turn such objects inside out without poking a hole through them.

**SEE ALSO** The Möbius Strip (1858), Klein Bottle (1882), and Boy's Surface (1901).

LEFT: *Today, mathematicians know precisely how to turn a sphere inside out. However, for many years, topologists were unable to show how to accomplish this formidable geometrical task.* RIGHT: *Carlo H. Sequin's physical model of one mathematical stage of the sphere eversion process. (The sphere had started out as green on the outside and red on the inside.)*

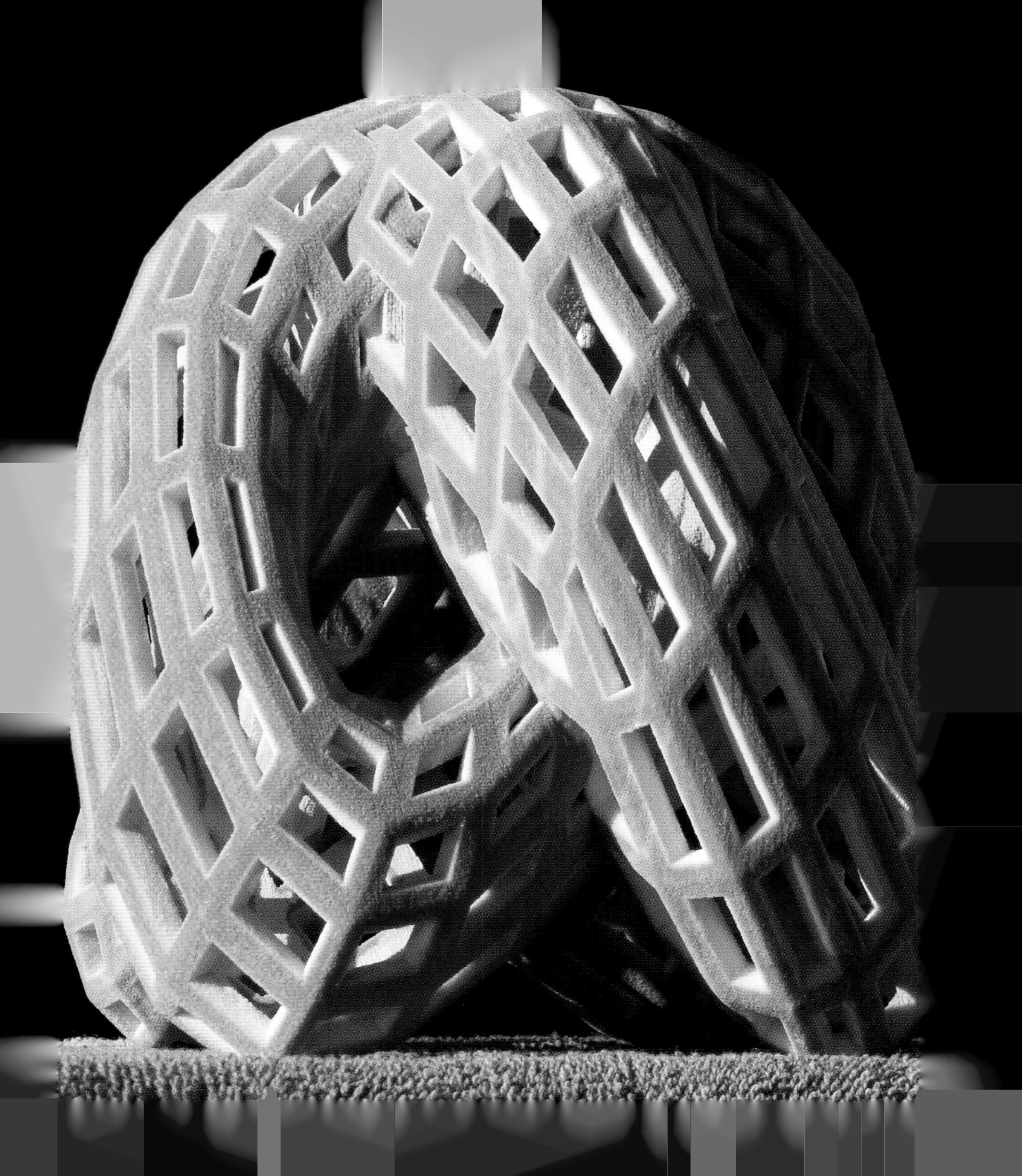

1958

# Platonic Billiards

**Lewis Carroll** (1832–1898), **Hugo Steinhaus** (1887–1972), **Matthew Hudelson** (b. 1962)

The Platonic billiards question has intrigued mathematicians for more than a century, and a complete solution had to wait for nearly fifty years after it was solved for the case of a cube. Imagine a billiard ball bouncing around inside a cube. Friction and gravity are neglected for this theoretical discussion. Can we find a path such that the ball returns to its starting point after hitting each wall once? The problem was initially posed by English author and mathematician Lewis Carroll (1832–1898).

In 1958, Polish mathematician Hugo Steinhaus widely published a solution that showed that such paths existed for cubes, and in 1962, mathematicians John Conway and Roger Hayward discovered similar paths inside a regular tetrahedron. Each leg of the path between walls has the same length for the cube and the tetrahedron. In theory, the ball bounces along the path forever. However, no one was sure if these kinds of paths existed for other **Platonic Solids**.

Finally, in 1997, American mathematician Matthew Hudelson demonstrated intriguing paths for a billiard ball bouncing around inside Platonic solids—the eight-sided octahedron, 12-sided dodecahedron, and the 20-sided icosahedron. These Hudelson paths make contact with each side of the inner walls and finally return to their starting points and starting directions of travel. Hudelson used a computer to assist him in his research. His challenge was particularly difficult, considering the large number of possibilities that had to be investigated for the dodecahedron and icosahedron. In order to obtain a better intuition about the problem for these shapes, Hudelson wrote a program that generated more than 100,000 random initial trajectories, and he studied those that hit all 12 sides in the dodecahedron and that hit all 20 sides in the icosahedron.

**SEE ALSO** Platonic Solids (350 B.C.) and Outer Billiards (1959).

*Mathematicians have discovered billiard-ball return shots within five Platonic solids. For example, a closed "bouncing ball" path exists that makes contact with each inner wall of the 20-sided icosahedron, rendered here by Teja Krašek.*

1959

# Outer Billiards

**Bernhard Hermann Neumann** (1909–2002), **Jürgen Moser** (1928–1999), **Richard Evan Schwartz** (b. 1966)

The concept of *outer billiards* (OB) was developed in the 1950s by German-born British mathematician Bernhard Neumann. German American mathematician Jürgen Moser popularized OB in the 1970s as a simplified model for planetary motions. To experiment with OB, draw a polygon. Place a point $x_0$ outside the polygon. Think of this as the starting point of a billiard ball. The ball moves along a straight line just touching a polygon vertex and continues to travel to a new point $x_1$ such that the vertex is at the midpoint of the line between $x_0$ and $x_1$. Continue the procedure with the next vertex in a clockwise fashion.

Neumann asked whether such a trajectory, or orbit, around the convex polygon could be unbounded so that the ball eventually runs away toward infinity. For *regular* polygons, all trajectories are bounded and do not meander further and further from the polygon. If the vertices of the polygons have rational coordinates (for example, they can be expressed in terms of fractions), the trajectories are bounded and periodic, eventually returning to their starting points.

In 2007, Richard Schwartz of Brown University finally showed that Neumann's OB could lead to an unbounded trajectory in the Euclidean plane, demonstrating this for a quadrilateral called the Penrose kite, which is used in Penrose tiling. Schwartz also discovered three large, octagonal regions within which trajectories bounce periodically from one region to the other. Other regions led to behavior that converged to a set of points from which the trajectories are unbounded. As with other modern proofs in mathematics, Schwartz's initial proof relied on a computer.

As for Neumann, he was awarded his doctorate by the University of Berlin in 1932. When Hitler came to power in 1933, Neumann understood the dangers of being a Jew and fled to Amsterdam and then to Cambridge.

**SEE ALSO** Platonic Billiards (1958) and Penrose Tiles (1973).

*Richard Schwartz demonstrates that the dynamics of outer billiards around a Penrose kite (the orange central polygon) may be visualized by an intricate tiling pattern. The colors of various polygonal regions provide an indication of the behavior of trajectories with endpoints in these regions.*

1960

# Newcomb's Paradox

**William A. Newcomb** (1927–1999), **Robert Nozick** (1938–2002)

Before you are two closed arks, or boxes, labeled "Ark 1" and "Ark 2." An angel explains that Ark 1 contains a golden goblet worth $1,000. Ark 2 contains either a spider worth absolutely nothing or the *Mona Lisa* painting worth millions of dollars. You have two choices: Take what is in *both* arks, or take *only* what is in Ark 2.

Now the angel makes your choice perplexing. "We have made a prediction about what you will decide. We are almost certainly correct. When we *expect* you to choose both arks, we put only the worthless spider in Ark 2. When we expect you to take only Ark 2, we have placed the *Mona Lisa* inside it. Ark 1 always contains $1,000, no matter what we think you will do."

At first, you think that you should select only Ark 2. The angels are excellent predictors, and therefore you will get the *Mona Lisa*. If you take both arks, the angel will very likely have anticipated your choice and have put a spider in Ark 2. You will only get the $1,000 cup and a spider.

But now the angel confuses you. "Forty days ago, we made a prediction about which you would choose. We *already* have either put the *Mona Lisa* or the spider in Ark 2, and we're not going to tell you."

Now you think you should take *both* arks and get everything possible. It seems foolish for you to select *only* Ark 2, because if you do so, you can't get more than the *Mona Lisa*. Why give up the $1,000?

This is the essence of Newcomb's paradox, formulated in 1960 by physicist William A. Newcomb. The puzzle was further elucidated by philosopher Robert Nozick in 1969. Experts still tear their hair out over this dilemma and disagree as to your best strategy.

**SEE ALSO** Zeno's Paradoxes (c. 445 B.C.), Aristotle's Wheel Paradox (c. 320 B.C.), St. Petersburg Paradox (1738), Barber Paradox (1901), Banach-Tarski Paradox (1924), Hilbert's Grand Hotel (1925), Prisoner's Dilemma (1950), and Parrondo's Paradox (1999).

*Newcomb's paradox, formulated in 1960 by physicist William A. Newcomb. Would you take both boxes, knowing that the angels are super-intelligent predictors and almost certainly correct?*

1960

# Sierpiński Numbers

**Wacław Franciszek Sierpiński** (1882–1969)

Mathematician Don Zagier writes that "there is no apparent reason why one number is prime and another not. To the contrary, upon looking at these numbers one has the feeling of being in the presence of one of the inexplicable secrets of creation." In 1960, Polish mathematician Wacław Sierpiński proved that there are infinitely many odd integers $k$, called Sierpiński numbers, such that $k \times 2^n + 1$ is *never* prime for every positive integer $n$. Ivars Peterson writes, "That's a strange result. There appears to be no obvious reason why these particular expressions never yield a prime." Given this background, the Sierpiński problem may be stated as "What is the smallest Sierpiński number?"

In 1962, American mathematician John Selfridge discovered the smallest-known Sierpiński number, $k = 78{,}557$. In particular, he proved that when $k = 78{,}557$, all numbers of the form $k \times 2^n + 1$ are divisible by one of the following: 3, 5, 7, 13, 19, 37, or 73.

In 1967, Sierpiński and Selfridge conjectured that 78,557 is the smallest Sierpiński number and, thus, is the answer to the Sierpiński problem. Today, mathematicians wonder if a smaller Sierpiński number will ever be discovered. If we were able to scan all values of $k < 78{,}557$ and find a prime number for each, then we would know for sure. As of February, 2008, there were a mere six candidate numbers that had not been eliminated as possible smaller Sierpiński numbers. "Seventeen Or Bust," a distributed computing project, is testing these remaining numbers. For example, in October of 2007, "Seventeen Or Bust" proved that $33{,}661 \times 2^{7{,}031{,}232} + 1$, a 2,116,617-digit number, is prime, thus eliminating $k = 33{,}661$ as a possible Sierpiński number. If mathematicians are able to find a prime of the proper form for all the remaining $k$, the Sierpiński problem will be solved and the nearly 50-year quest ended.

**SEE ALSO** Cicada-Generated Prime Numbers (c. 1 Million B.C.), Sieve of Eratosthenes (240 B.C.), Goldbach Conjecture (1742), Constructing a Regular Heptadecagon (1796), Gauss's *Disquisitiones Arithmeticae* (1801), Proof of the Prime Number Theorem (1896), Brun's Constant (1919), Gilbreath's Conjecture (1958), Ulam Spiral (1963), Erdös and Extreme Collaboration (1971), and Andrica's Conjecture (1985).

*Logo of "Seventeen Or Bust," a distributed-computing project devoted to determining if 78,557 is the smallest Sierpiński number. For years, their system has harnessed the computational power of hundreds of computers around the world, working together on the problem.*

# Seventeen Or Bust

A Distributed Attack on the Sierpinski Problem

1963

# Chaos and the Butterfly Effect

**Jacques Salomon Hadamard** (1865–1963), **Jules Henri Poincaré** (1854–1912), **Edward Norton Lorenz** (1917–2008)

To ancient humans, chaos represented the unknown, the spirit world—menacing, nightmarish visions that reflected man's fear of the uncontrollable and the need to give shape and structure to his apprehensions. Today, chaos theory is an exciting, growing field that involves the study of wide-ranging phenomena exhibiting a sensitive dependence on initial conditions. Although chaotic behavior often seems "random" and unpredictable, it often obeys strict mathematical rules derived from equations that can be formulated and studied. One important research tool to aid in the study of chaos is computer graphics. From chaotic toys with randomly blinking lights to wisps and eddies of cigarette smoke, chaotic behavior is generally irregular and disorderly; other examples include weather patterns, some neurological and cardiac activity, the stock market, and certain electrical networks of computers. Chaos theory has also often been applied to a wide range of visual art.

In science, certain famous and clear examples of chaotic physical systems exist, such as thermal convection in fluids, panel flutter in supersonic aircraft, oscillating chemical reactions, fluid dynamics, population growth, particles impacting on a periodically vibrating wall, various pendula and rotor motions, nonlinear electrical circuits, and buckled beams.

The early roots of chaos theory started around 1900 when mathematicians such as Jacques Hadamard and Henri Poincaré studied the complicated trajectories of moving bodies. In the early 1960s, Edward Lorenz, a research meteorologist at the Massachusetts Institute of Technology, used a system of equations to model convection in the atmosphere. Despite the simplicity of his formulas, he quickly found one of the hallmarks of chaos—that is, extremely minute changes of the initial conditions led to unpredictable and different outcomes. In his 1963 paper, Lorenz explained that a butterfly flapping its wings in one part of the world could later affect the weather thousands of miles away. Today, we call this sensitivity the *butterfly effect*.

**SEE ALSO** Catastrophe Theory (1968), Feigenbaum Constant (1975), Fractals (1975), and Ikeda Attractor (1979).

*Chaotic mathematical pattern, created by Roger A. Johnston. Although chaotic behavior may seem "random" and unpredictable, it often obeys mathematical rules derived from equations that can be studied. Very small changes of the initial conditions can lead to very different outcomes.*

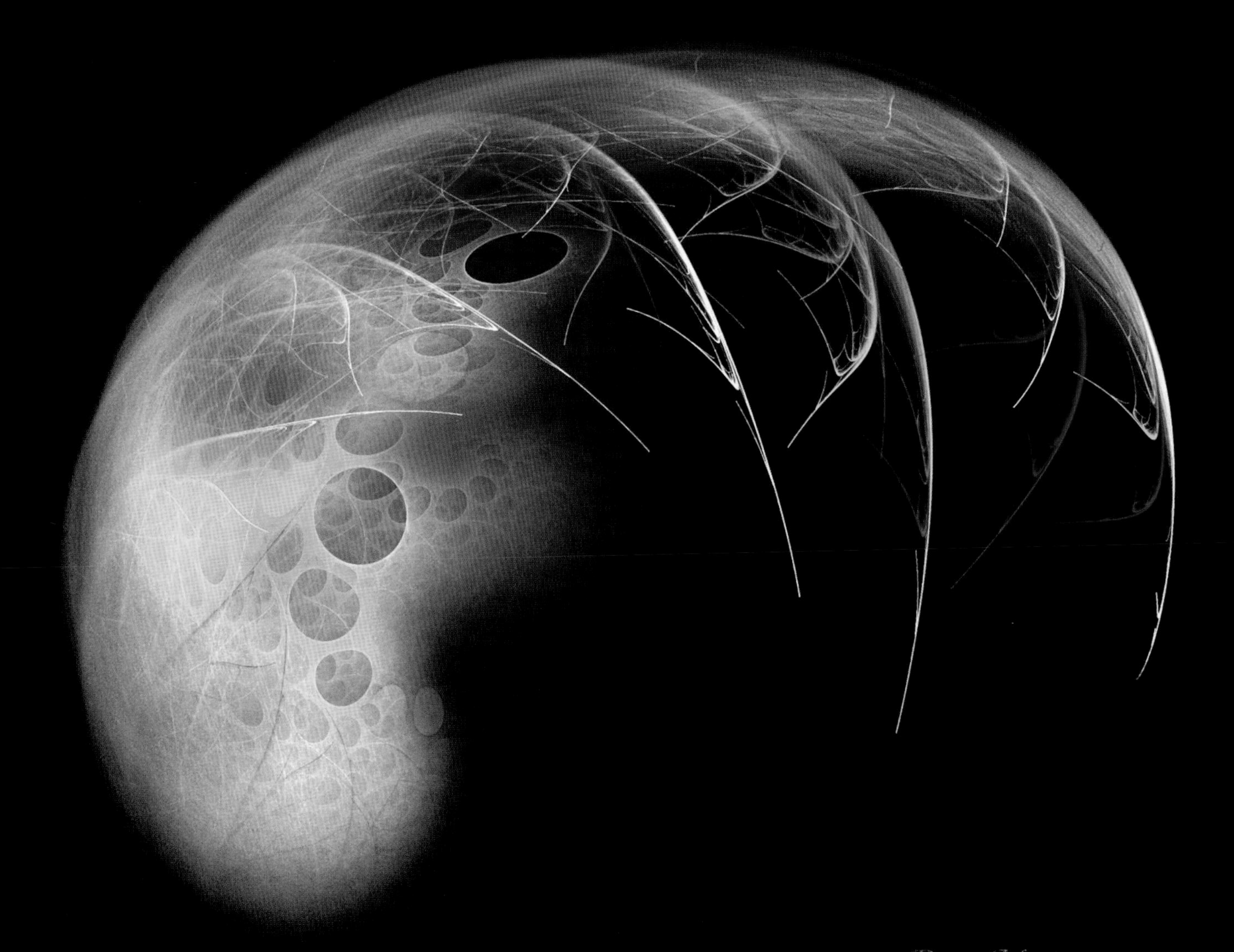

Roger Johnston

1963

# Ulam Spiral

**Stanisław Marcin Ulam** (1909–1984)

In 1963, while doodling on paper during a boring meeting, Polish-born American mathematician Stanisław Ulam discovered a remarkable spiral that reveals patterns in prime numbers. (A prime number is a number larger than 1, such as 5 or 13, that is divisible only by itself or 1.) Starting with 1 at the center of the counterclockwise spiral, Ulam wrote consecutive natural numbers. He then circled all the prime numbers. As the spiral grew larger, he noticed that the prime numbers tended to form diagonal patterns.

As later computer graphics made quite clear, although some kind of diagonal structures may simply arise from diagonals alternatively containing odd and even numbers, it is intriguing that the prime numbers tend to lie on some diagonal lines more than others. Perhaps more important than the discovery of patterns, Ulam's simple demonstration highlights the use of the computer as a kind of microscope that allows mathematicians to visualize structures that may lead to new theorems. This kind of investigation in the early 1960s gradually gave rise to the explosion in experimental mathematics toward the end of the twentieth century.

Martin Gardner writes, "Ulam's spiral grids have added a touch of fantasy to speculations about the enigmatic blend of order and haphazardry in the distribution of primes. . . . Ulam's doodlings in the twilight zone of mathematics are not to be taken lightly. It was he who made the suggestion that led him and Edward Teller to think of the 'idea' that made possible the first thermonuclear bomb."

In addition to his mathematical contributions and his work on the Manhattan Project to develop the first nuclear weapon during World War II, Ulam is also famous for his work on spacecraft propulsion systems. He had escaped with his brother from Poland on the eve of the Second World War, but the rest of his family died in the Holocaust.

**SEE ALSO** Cicada-Generated Prime Numbers (c. 1 Million B.C.), Sieve of Eratosthenes (240 B.C.), Goldbach Conjecture (1742), Gauss's *Disquisitiones Arithmeticae* (1801), Riemann Hypothesis (1859), Proof of the Prime Number Theorem (1896), Johnson's Theorem (1916), Brun's Constant (1919), Gilbreath's Conjecture (1958), Sierpiński Numbers (1960), Erdös and Extreme Collaboration (1971), Public-Key Cryptography (1977), and Andrica's Conjecture (1985).

*A 200 × 200 Ulam spiral plot. Several diagonal patterns are highlighted in yellow. Ulam's simple plot demonstrates the use of the computer as a kind of microscope that allows mathematicians to visualize structures that may lead to new theorems.*

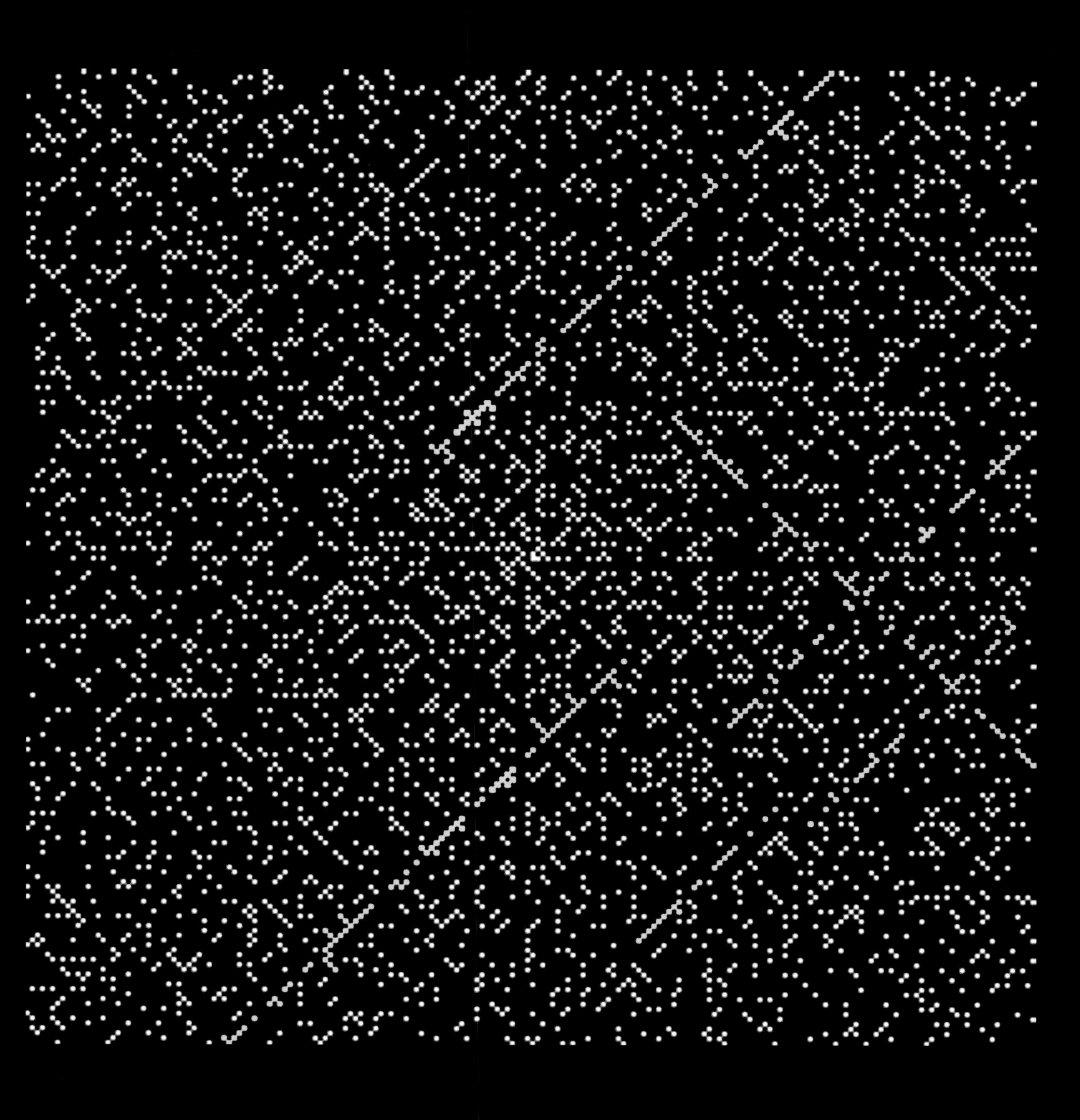

# Continuum Hypothesis Undecidability

**Georg Cantor** (1845–1918), **Paul Joseph Cohen** (1934–2007)

In the entry on **Cantor's Transfinite Numbers**, we discussed the smallest transfinite number called *aleph-nought*, written as $\aleph_0$, which "counts" the number of integers. Even though there is an infinite number of integers, *rational numbers* (numbers that can be expressed as fractions), and *irrational numbers* (like the square root of 2), the infinite number of irrationals is in some sense greater than the infinite number of rationals and integers. Similarly, there are more *real* numbers (which include rational and irrational numbers) than there are integers.

To denote this difference, mathematicians refer to the infinity of rationals or integers as $\aleph_0$ and the infinite number of irrationals or real numbers as $C$. There is a simple relationship between $C$ and $\aleph_0$, namely $C = 2^{\aleph_0}$. Here, $C$ is the cardinality of the set of real numbers, which are sometimes called *the continuum*.

Mathematicians also contemplate greater infinities, symbolized by $\aleph_1$, $\aleph_2$, etc. Here, the set theory symbol $\aleph_1$ stands for the smallest infinite set larger than $\aleph_0$, and so forth. Cantor's continuum hypothesis states that $C = \aleph_1 = 2^{\aleph_0}$ ; however, the question of whether or not $C$ truly equals $\aleph_1$ is considered undecidable in our present set theory. In other words, great mathematicians such as Kurt Gödel proved that the hypothesis was a consistent assumption with the standard axioms of set theory. However, in 1963, American mathematician Paul Cohen proved that it was also consistent to assume the continuum hypothesis is false! Cohen was born in Long Branch, New Jersey, into a Jewish family, and he graduated in 1950 from Stuyvesant High School in New York City.

Interestingly, the number of rational numbers is the same as the number of integers, and the number of irrationals is the same as the number of real numbers. (Mathematicians usually use the term *cardinality* when discussing the "number" of infinite numbers.)

**SEE ALSO** Aristotle's Wheel Paradox (c. 320 B.C.), Cantor's Transfinite Numbers (1874), and Gödel's Theorem (1931).

*Various infinitudes, while difficult to contemplate, may be explored using computer graphics, as in this rendition of Gaussian rational numbers. Here, sphere positions represent the complex fraction* p/q. *The spheres touch the complex plane at location* p/q *and have radii equal to* $1/(2q\bar{q})$.

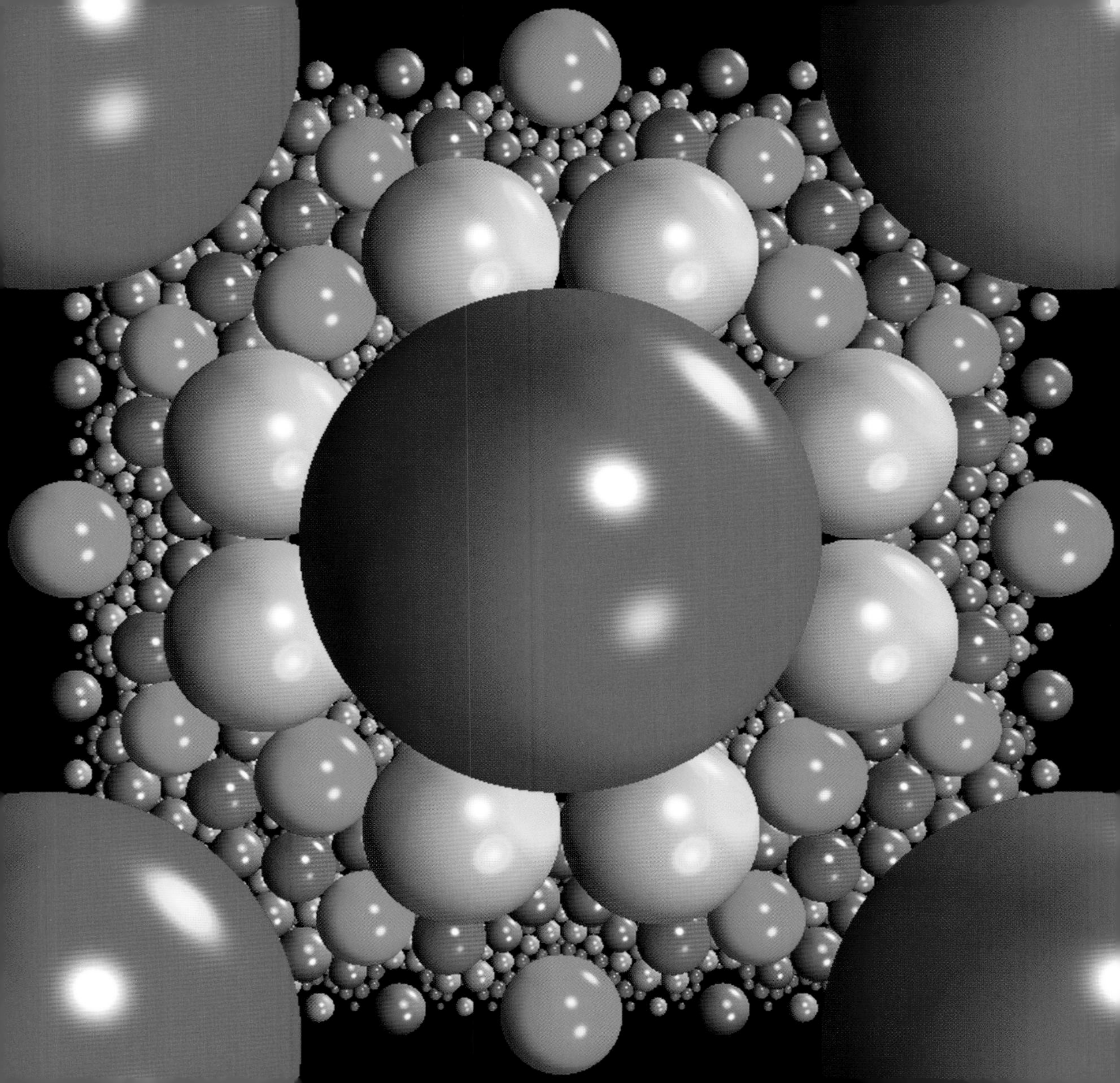

# Superegg

**Piet Hein** (1905–1996)

Around 1965, Danish scientist, designer, and inventor Piet Hein promoted the superegg, also known as the super-ellipsoid, as an object of beauty and of fascination because it stood on either end with a spooky stability. The 3-D shape is produced by using a superellipse, defined by the formula $|x/a|^{2.5} + |y/b|^{2.5} = 1$ for $a/b = 4/3$, and revolving this shape around the $x$-axis. More generally, we can give the equation of the super-ellipsoid as $(|x|^{2/a} + |y|^{2/a})^{a/b} + |z|^{2/b} = 1$ where $a$ and $b$ are greater than zero.

Hein's supereggs, made of various materials, were popular as toys and novelty items in the 1960s. Today, the design is ubiquitous. Supereggs are used as candlestick holders, in furniture designs, and as liquid-filled stainless-steel drink coolers that are tossed into beverage glasses. Hein's superegg was "laid" for the first time in 1965, when a handheld version was manufactured and sold by Skjøde in Skjern, Denmark. In 1971, the world's largest superegg, made of metal and weighing almost a ton, was placed outside Kelvin Hall in Glasgow.

French mathematician Gabriel Lamé (1795–1870) worked with the more general form of the superellipse before Hein, but Hein was the first to create the superegg and is famous for popularizing his own versions in architecture, furniture, and even urban planning.

The superellipse was also used as the shape for a roundabout road in Stockholm, Sweden. The ellipse was unsuitable because its pointed ends would interfere with smooth traffic flow in the roughly rectangular space. In 1959, Hein was asked for his opinion. Martin Gardner writes about the Stockholm road: "Hein's curves proved to be strangely satisfying, neither too rounded nor too orthogonal, a happy blend of elliptical and rectangular beauty. Stockholm immediately accepted the 2.5-exponent superellipse [with $a/b = 6/5$] as the basic motif of its new center. . . ."

**SEE ALSO** Astroid (1674).

*Piet Hein's superegg, sitting across the moat from Egeskov Castle in Kvaerndrup, Fyn Island, Denmark. The castle, built in the mid-1550s, is one of the best-preserved Renaissance "water castles." Originally, the castle could only be accessed via a drawbridge.*

1965

# Fuzzy Logic

**Lotfi Zadeh** (b. 1921)

Classical two-valued logic is concerned with conditions that are either true or false. Fuzzy logic (FL) allows a continuous range of truth values and was introduced by mathematician and computer scientist Lotfi Zadeh, who grew up in Iran and moved to the United States in 1944. FL has a wide range of practical applications and is derived from fuzzy set theory, which focuses on members of a set that have *degrees* of membership. Zadeh published his groundbreaking mathematical paper on fuzzy sets in 1965, and in 1973 provided the details of FL.

As an example, consider a temperature-monitoring system for a device. A membership function may exist for the concepts cold, warm, and hot. A single measurement may consist of three values such as "not cold," "slightly warm," and "slightly hot"—which may be used to control the device. Zadeh believed that if feedback controllers could be programmed to make use of imprecise, noisy input, they could be more effective and easier to implement. In some sense, this approach is similar to how people often make decisions.

FL methodology had a difficult start, and Zadeh could not easily find a technical journal to publish his 1965 paper, perhaps due to a reluctance to let "vagueness" creep into the engineering field. Author Kazuo Tanaka writes, "A turning point for fuzzy logic came in 1974 [when] Ebraham Mamdani of the University of London applied fuzzy logic to...the control of a simple steam engine...." In 1980, FL was used to control a cement kiln. Various Japanese companies have used FL to control water purification processes and train systems. FL has also since been used to control steel mills, self-focusing cameras, washing machines, fermentation processes, automobile engine controls, anti-lock braking systems, color-film developing systems, glass processing, computer programs used in financial trading, and systems used for recognizing subtle differences in written and spoken languages.

**SEE ALSO** Aristotle's *Organon* (c. 350 B.C.), Boolean Algebra (1854), Venn Diagrams (1880), *Principia Mathematica* (1910–1913), and Gödel's Theorem (1931).

*Fuzzy logic has been used in the design of efficient washing machines. For example, U.S. patent 5,897,672, issued in 1999, describes the use of fuzzy logic for detecting the relative proportion of various fabric types present in a clothes load in a clothes washer.*

SELECT
P
1
2
3
4
5
6
7
8
9
2

1966

# Instant Insanity

**Frank Armbruster** (b. 1929)

As a child, I could never solve the colorful cube game called Instant Insanity. I shouldn't have felt too bad, because 41,472 different ways exist for arranging the four cubes in a row, only 2 of which are solutions. A trial-and-error approach could never have worked.

The puzzle looks deceptively simple, consisting of four cubes with one of four colors on each of their six faces. The goal is to arrange the four cubes in a row so that only one color appears along each side of the row of cubes. Because each cube has 24 orientations, a maximum of $4! \times 24^4 = 7{,}962{,}624$ positions exist. However, the number may be reduced to 41,472 solutions, partly because the cubes can be stacked in any order without making a difference for the solution.

Mathematicians have represented the colored faces of the cubes in terms of a graph in order to understand efficient ways of solving the puzzle. Using this approach, each cube is represented by a graph of the colors that appear on opposite pairs of faces. According to mathematics journalist Ivars Peterson, "Those familiar with graph theory can typically work out the solution in minutes. Indeed, the puzzle serves as a neat lesson in logical thinking."

The Instant Insanity craze skyrocketed after educational consultant Frank Armbruster licensed his version of the puzzle to Parker Brothers, and it sold more than 12 million copies in the late 1960s. A similar colored cube puzzle was also popular around 1900, when it was called the Great Tantalizer. Armbruster wrote to me, "When I was given a sample of the Great Tantalizer in 1965, I saw the potential for using it to teach combinations and permutations. My first sample was made of wood with painted sides. I sold my subsequent plastic version, packaged in the solved condition, and a customer suggested its name, which I trademarked. Parker Brothers then made me an offer I couldn't refuse."

**SEE ALSO** Gros's *Théorie du Baguenodier* (1872), Fifteen Puzzle (1874), Tower of Hanoi (1883), Hex (1942), and Rubik's Cube (1974).

*Frank Armbruster, holding his famous puzzle Instant Insanity. There are 41,472 different ways for arranging the four cubes in a row, only two of which are solutions. More than 12 million puzzles sold in the late 1960s.*

1967

# Langlands Program

**Robert Phelan Langlands** (b.1936)

In 1967, Robert Langlands, a 30-year-old Princeton mathematics professor, wrote a letter to the famous number theorist André Weil (1906–1998), asking Weil's opinion on some new mathematical ideas. "If you are willing to read [my letter] as pure speculation, I would appreciate that. If not—I'm sure you have a waste basket." According to *Science* writer Dana Mackenzie, Weil never wrote back, but Langlands' letter turned out to be a "Rosetta stone" linking two different branches of mathematics. In particular, Langlands posited that an equivalence existed between Galois representations (which describe relationships among solutions to equations studied in number theory) and automorphic forms (highly symmetric functions like the cosine function).

The Langlands program is so fertile a territory that it led to two Fields Medals for other mathematicians. Langlands' conjectures sprang, in part, from an effort to find general versions of patterns that govern how whole numbers can be broken down into sums of products of other whole numbers.

According to *The Fermat Diary*, the Langlands program may be considered a grand unified theory of mathematics that suggests that "the mathematics of algebra, which involves equations, and the mathematics of analytics, which involves the study of smooth curves and continuous variations, are intimately related." The conjectures in the Langlands program "are like a cathedral, the way they fit together so beautifully." However, the conjectures are very difficult to prove, and some mathematicians feel that it may take centuries to complete the Langlands program.

Mathematician Stephen Gelbart writes, "[The] Langlands program is a synthesis of several important themes in classical number theory. It is also—and more significantly—a program for future research. This program emerged around 1967 in the form of a series of conjectures, and it has subsequently influenced research in number theory in much the same way the conjectures of A. Weil shaped the course of algebraic geometry since 1948."

**SEE ALSO** Group Theory (1832) and Fields Medal (1936).

LEFT: *Robert Langlands.* RIGHT: *The Langlands program links two different branches of mathematics and involves conjectures that are said to be "like a cathedral" because they exhibit such an elegant fit. The Langlands program may be considered a grand unified theory of mathematics which may take centuries to completely elucidate.*

1967

# Sprouts

**John Horton Conway** (b. 1937) **and Michael S. Paterson** (b. 1942)

The game of Sprouts was invented in 1967 by mathematicians John H. Conway and Michael S. Paterson, when both were at the University of Cambridge. The addictive game has fascinating mathematical properties. Conway wrote to Martin Gardner, "The day after Sprouts sprouted, it seemed that everyone was playing it.... peering over ridiculous to fantastic Sprout positions. Some people were already attacking Sprouts on toruses, Klein bottles, and... thinking of higher-dimensional versions."

To play Sprouts against an opponent, start by placing several dots on a page. To make a move, draw a curve between two spots or a loop from a spot to itself. Your curve may not cross another curve or itself. Next, place a new dot on this curve. Players take turns, and the player who makes the last move wins. Each dot can have at most three curves connected to it.

After just a casual inspection, one might guess that a game could keep sprouting forever. However, we now know that when Sprouts starts with $n$ spots, the game will last at least $2n$ moves and at most $3n - 1$ moves. The first player can always win in games that start with three, four, or five dots.

In 2007, researchers used computer programs to help determine which player is the winner for all games with up to 32 spots. The status of the 33-spot game is still unknown. Sprouts experts Julien Lemoine and Simon Viennot write, "Despite the little number of moves... it is difficult to determine whether the first or the second player wins, provided those players play perfectly. The best published and complete hand-checked proof is due to [Riccardo] Focardi and [Flaminia] Luccio, and shows who the winner is for the 7-spot game." Journalist Ivars Peterson writes, "Games can sprout all sorts of unexpected growth patterns, making formulation of a winning strategy a tricky proposition. No one has yet worked out a complete strategy for perfect play."

**SEE ALSO** Königsberg Bridges (1736), Jordan Curve Theorem (1905), and Checkers Is Solved (2007).

*Game of Sprouts. In this example, only two starting points (circled) are used, and the game is not yet finished. Despite its apparent simplicity, the game is very difficult to analyze as the number of starting points modestly increases.*

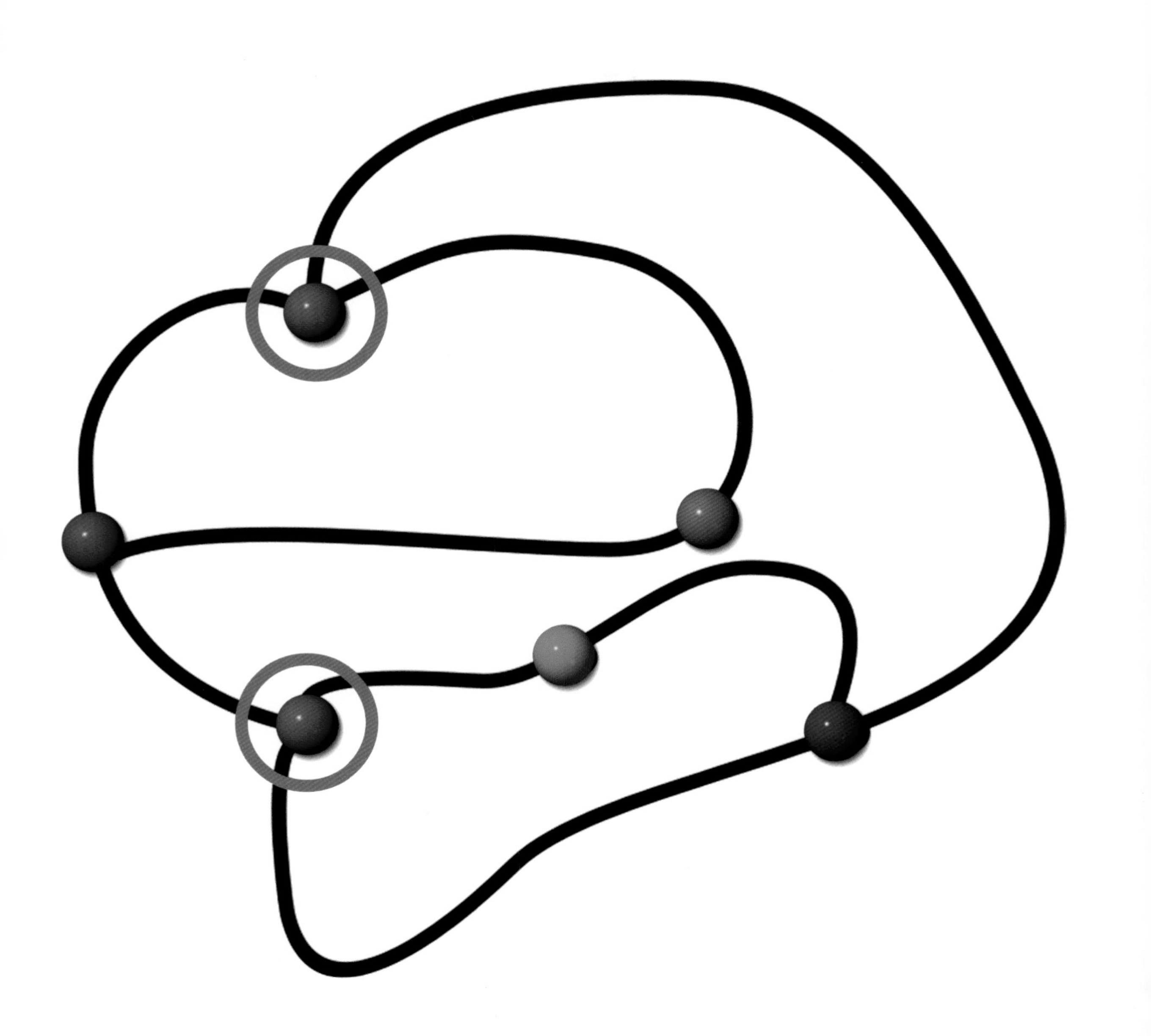

1968

# Catastrophe Theory

**René Thom** (1923–2002)

Catastrophe theory is the mathematical theory of dramatic or abrupt changes. Mathematicians Tim Poston and Ian Stewart give examples: "the roar of an earthquake [or] the critical population density below which certain creatures grow up as grasshoppers, above which as [swarming] locusts....A cell suddenly changes its reproductive rhythm and doubles and redoubles, cancerously. A man has a vision on the road to Tarsus."

Catastrophe theory was developed by French mathematician René Thom in the 1960s. The theory was further promoted in the 1970s by Japanese-born British mathematician Christopher Zeeman, who continued to apply the theory to the behavioral and biological sciences. Thom was awarded the Fields Medal in 1958 for his work in topology, the study of geometrical shapes and their relationships.

Catastrophe theory usually concerns dynamical systems that describe the time dependence of some quantity (like the beating of the heart) and the relationship of these systems to topology. In particular, the theory focuses on certain kinds of "critical points" in which the first derivative of a function, and one or more higher derivatives, are zero. David Darling writes: "Many mathematicians took up the study of catastrophe theory, and it was in tremendous vogue for a while, yet it never achieved the success that its younger cousin chaos theory has because it failed to live up to its promise of useful predictions."

Thom's quest was to better understand how continuous actions (such as smooth and stable behavior in prisons or between countries) could suddenly give way to discontinuous change (prison riots or war). He showed how such phenomena might be described with their own landscapes in the form of abstract mathematical surfaces, with names like the *butterfly* or the *swallowtail*. Salvador Dalí's last painting, *The Swallow's Tail* (1983), was based on a catastrophe surface. Dalí also painted *Topological Abduction of Europe: Homage to René Thom* (1983), which depicted a fractured landscape along with the equation that explained it.

**SEE ALSO** Königsberg Bridges (1736), The Möbius Strip (1858), Fields Medal (1936), Chaos and the Butterfly Effect (1963), Feigenbaum Constant (1975), and Ikeda Attractor (1979).

*Catastrophe theory is the mathematical theory of abrupt changes, such as the swarming behavior of grasshoppers as population density increases. Research has suggested that the sudden swarming behavior is triggered by increased contacts of the insect's hind legs over a few-hour period. Large swarms may consist of billions of insects.*

1969

# Tokarsky's Unilluminable Room

**George Tokarsky** (b. 1946)

Imagine that we are in a dark room with flat walls covered with mirrors. The room has several turns and side passages. If I light a match somewhere in the room, would you be able to see it no matter where you stand in the room, and no matter what the room shape or in which side passage you stand? We can equivalently pose the question in terms of a billiard ball bouncing around a pool table. Must there be a pool shot between any two points on a polygonal pool table?

If we happened to be trapped in an L-shaped room, you'd be able to see the flame no matter where we stood because the light ray could bounce off various walls to get to your eye. But can we imagine a mysterious polygonal room that is so complicated that a point exists that light never reaches? (For our problem, we consider a person and match to be transparent.)

This enigma was first presented in print by mathematician Victor Klee in 1969, although it dates back to the 1950s when mathematician Ernst Straus pondered such problems. It is shocking that no one knew the answer until 1995, when George Tokarsky of the University of Alberta discovered such a room that is not completely illuminable. His published floor plan of the room had 26 sides. Subsequently, Tokarsky found an example with 24 sides, and this strange room is the least-sided unilluminable polygonal room currently known. We do not know if an unilluminable polygonal room with fewer sides is possible.

Other similar problems exist. In 1958, mathematical physicist Roger Penrose and his colleague showed that unlit regions can exist in certain rooms with curved sides. More recently, certain curved rooms have been discovered for which infinitely many matches are needed to illuminate every point. For any finite number of matches, curved rooms exist that cannot be illuminated by these matches.

**SEE ALSO** Projective Geometry (1639) and Art Gallery Theorem (1973).

*In 1995, mathematician George Tokarsky discovered this unilluminable 26-sided polygonal "room." The room contains a location at which a match can be held that leaves another point in the room in the dark.*

1970

# Donald Knuth and Mastermind

**Donald Ervin Knuth** (b. 1938), **Mordecai Meirowitz**

Mastermind is a code-breaking board game invented in 1970 by Mordecai Meirowitz, an Israeli postmaster and telecommunications expert. Mainstream game companies all rejected Meirowitz; thus, he published with the small English game company Invicta Plastics. The game went on to sell more than 50 million copies, making it the most successful new game of the 1970s.

To play the game, a code-maker selects a sequence of four colors, represented by colored pegs that come in 6 different colors. The opponent must guess the code-maker's secret sequence, with as few guesses as possible. Each guess is presented in the form of a sequence of 4 colored pegs. The code-maker reveals how many of those pegs are both the correct color and in the correct position and how many more are the correct color but in the wrong position. For example, the secret code may be *green-white-blue-red*. The guess may be *orange-yellow-blue-white*. Here, the code-maker indicates that the player has 1 peg of the correct color in the correct position and 1 peg of the correct color in the wrong position, but he doesn't mention the specific color names. The game continues with more guesses. A code-maker selects from a possible $6^4$ (or 1,206) possible combinations, assuming there are 6 colors and 4 positions.

Mastermind was significant, partly due to the long stream of research the game triggered. In 1977, American computer scientist Donald Knuth published a strategy that enables a player to always guess the correct code within 5 guesses. This was the first-known algorithm to solve Mastermind, and numerous papers followed. In 1993, Kenji Koyama and Tony W. Lai published a strategy with a maximum of 6 guesses required in the worst case, but with an average number of guesses of only 4.340. In 1996, Zhixiang Chen and colleagues generalized previous results to the case of $n$ colors and $m$ positions. The game has also been studied several times using genetic algorithms, techniques inspired by evolutionary biology.

**SEE ALSO** Tic Tac Toe (1300 B.C.), Go (548 B.C.), Eternity Puzzle (1999), Solving the Game of Awari (2002), and Checkers Is Solved (2007).

*Schematic representation of Mastermind. The normally hidden code at bottom is* green-blue-red-magenta. *The player starts with a guess at the board's top row and converges to a solution in five moves after receiving hints (not shown here) from the opponent.*

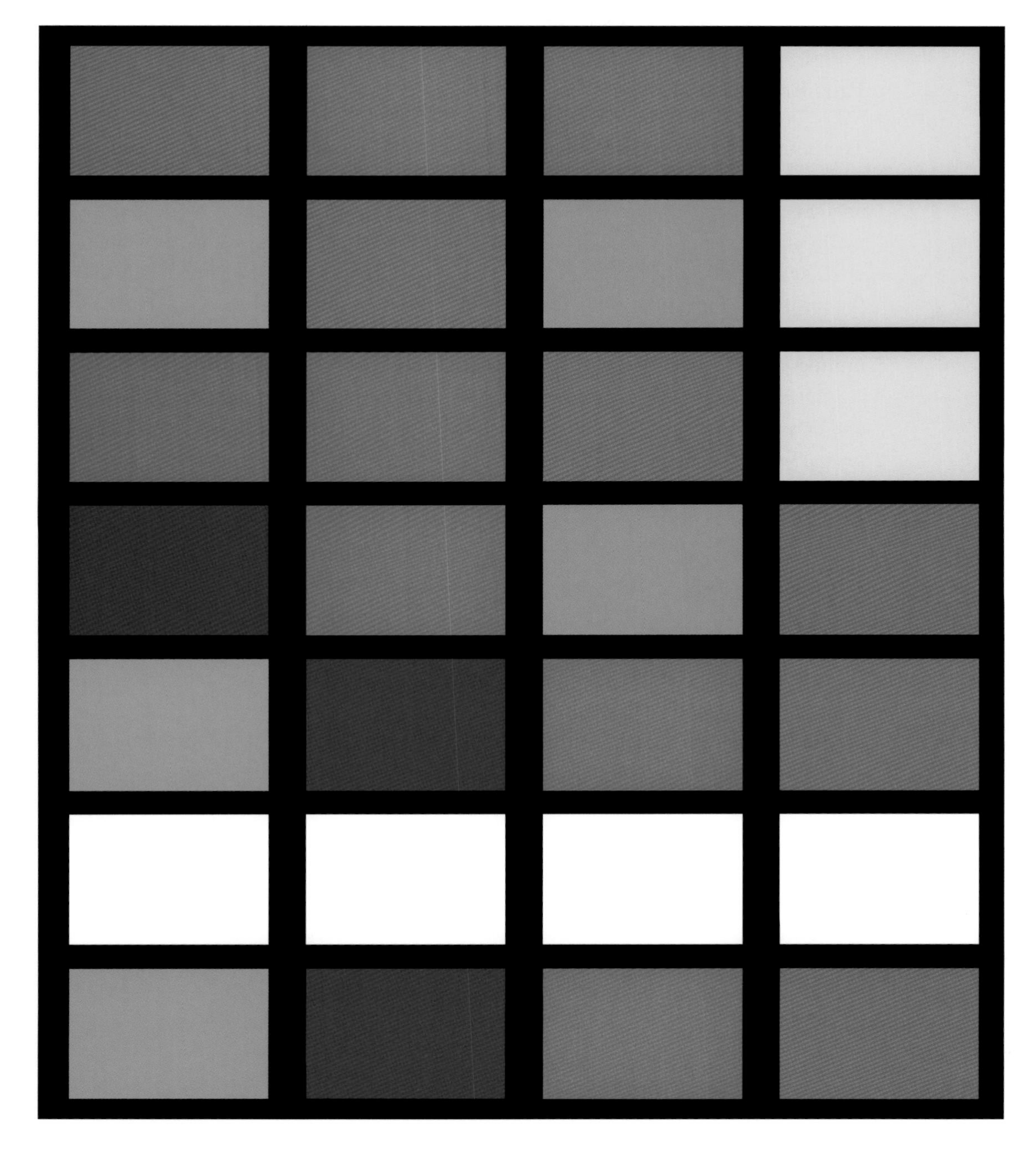

1971

# Erdös and Extreme Collaboration

**Paul Erdös** (1913–1996)

The public often thinks of mathematicians as being sequestered in private rooms, rarely talking to others as they work for days to generate new theorems and solve ancient conjectures. This is true for some, but Hungarian-born Paul Erdös showed mathematicians the value of collaborations and "social mathematics." By the time he died, he had published roughly 1,500 papers—more papers than any mathematician in the history of the world, having worked with 511 different collaborators. His work ranged through a vast landscape of mathematics, including probability theory, combinatorics, number theory, graph theory, classical analysis, approximation theory, and set theory.

During the last year of his life, at age 83, he continued churning out theorems and delivering lectures, defying conventional wisdom that mathematics was a young person's sport. Through all his work, he always shared ideas, caring more that a problem was solved than who solved it. Author Paul Hoffman wrote, "Erdös thought about more problems than any other mathematician in history and could recite the details of some 1,500 papers he had written. Fortified by coffee, Erdös did mathematics 19 hours a day, and when friends urged him to slow down, he always had the same response: 'There'll be plenty of time to rest in the grave.'" After 1971, he took amphetamines almost every day to escape depression and foster mathematical ideas and collaborations. Erdös traveled constantly and lived out of a plastic bag, focusing totally on mathematics at the expense of companionship, sex, and food.

Erdös made an early mark on mathematics at the age of 18, when he discovered an elegant proof of the theorem that, for each integer $n$ greater than 1, there is always a prime number between $n$ and double the number, $2n$. For example, the prime number 3 lies between 2 and 4. Erdös later formulated an elementary proof of the **Prime Number Theorem**, which describes the distribution of prime numbers.

**SEE ALSO** Cicada-Generated Prime Numbers (c. 1 Million B.C.), Sieve of Eratosthenes (240 B.C.), Goldbach Conjecture (1742), Gauss's *Disquisitiones Arithmeticae* (1801), Riemann Hypothesis (1859), Proof of the Prime Number Theorem (1896), Brun's Constant (1919), Gilbreath's Conjecture (1958), and Ulam Spiral (1963).

*Paul Erdös fuelled his superhuman work schedule through constant use of coffee, caffeine tablets, and Benzedrine, and he believed that "a mathematician is a machine for turning coffee into theorems." He often put in 19-hour days, seven days a week.*

CAFE

# HP-35: First Scientific Pocket Calculator

**William Redington Hewlett** (1913–2001) **and team**

In 1972, the Hewlett-Packard (HP) company, with headquarters in Palo Alto, California, introduced the world's first scientific pocket calculator—that is, a handheld calculator with trigonometric and exponential functions. The large numeric range of the HP-35 calculator, with its scientific notation, was from $10^{-100}$ to $10^{+100}$. The HP-35 was introduced at a selling price of US $395. (HP called the device the "35" because it had 35 keys.)

Company co-founder Bill Hewlett began to develop the compact calculator despite market studies that suggested almost no market existed for pocket-size calculators. How wrong they were! In the first few months of sales, orders exceeded the company's expectations with respect to the *entire* market size. In the first year, 100,000 HP-35s were sold, and more than 300,000 were sold by the time it was discontinued in 1975.

When the HP-35 was introduced, slide rules were available for performing high-end scientific calculations. Existing pocket calculators at the time performed addition, subtraction, multiplication, and division. The HP-35 changed everything. The slide rule—which was typically accurate to only three significant figures—"died" and was rarely taught again in many U.S. schools. One wonders what the great mathematicians of yore would have accomplished if they had had access to an HP-35 (along with an endless supply of batteries).

Today, scientific calculators are inexpensive and have significantly changed the mathematics curriculum taught in most countries. Educators no longer teach paper-and-pencil methods for computing values of transcendental functions. In the future, teachers will probably devote even more time to mathematical applications and concepts instead of routine computations.

Author Bob Lewis writes, "Bill Hewlett and Dave Packard founded Silicon Valley in Hewlett's garage. A coin toss made the company Hewlett-Packard instead of Packard Hewlett. . . . Hewlett never showed much interest in being a celebrity. Throughout his life, he remained, at heart, an engineer."

**SEE ALSO** Abacus (c. 1200), Slide Rule (1621), Babbage Mechanical Computer (1822), Differential Analyzer (1927), ENIAC (1946), Curta Calculator (1948), and Mathematica (1988).

*The HP-35 calculator was the world's first scientific pocket calculator, with trigonometric and exponential functions. Bill Hewlett began to develop the compact calculator despite erroneous market studies that suggested almost no market existed for pocket-size calculators.*

3.141592654
OFF
ON
$x^y$
log
ln
$e^x$
$\sqrt{x}$
arc
sin
CLR
1/x
x⇄y
cos
R↓
tan
ENTER↑
STO
CH S
RCL
E EX
CL x
−
+
×
÷
7
8
9
4
5
6
1
2
3
0
•
π
HEWLETT • PACKARD

1973

# Penrose Tiles

**Roger Penrose** (b. 1931)

Penrose tiles refers to two simple geometric shapes that, when put side by side, can cover a plane in a pattern with no gaps or overlaps and that do not repeat periodically. In contrast, the simple hexagonal tile patterns found on some bathroom floors exhibit a simple repeating pattern. Interestingly, Penrose tilings, named after English mathematical physicist Roger Penrose, have five-fold rotational symmetry, the same kind of symmetry exhibited by a five-pointed star. If you rotate the entire tile pattern by 72 degrees, it looks the same as the original. Author Martin Gardner writes, "Although it is possible to construct Penrose patterns with a high degree of symmetry... most patterns, like the universe, are a mystifying mixture of order and unexpected deviations from order. As the patterns expand, they seem to be always striving to repeat themselves but never quite managing it."

Before Penrose's discovery, most scientists believed that crystals based on five-fold symmetry would be impossible to construct, but *quasicrystals* resembling Penrose tile patterns have since been discovered, and they have remarkable properties. For example, metal quasicrystals are poor conductors of heat, and quasicrystals can be used as slippery nonstick coatings.

In the early 1980s, scientists had speculated about the possibility that the atomic structure of some crystals might be based on a nonperiodic lattice—that is, a lattice that does not have periodic repeats. In 1982, Dan Shechtman discovered a nonperiodic structure in the electron micrographs of an aluminum-manganese alloy with an obvious five-fold symmetry reminiscent of a Penrose tiling. At the time, this finding was so startling that some said it was as shocking as finding a five-sided snowflake.

As an interesting aside, in 1997 Penrose filed a copyright lawsuit against a company that had allegedly embossed Penrose tilings on Kleenex quilted toilet paper in England. In 2007, researchers published evidence in *Science* of a Penrose-like tiling in medieval Islamic art, five centuries before its discovery in the West.

**SEE ALSO** Wallpaper Groups (1891), Thue-Morse Sequence (1906), Squaring a Rectangle (1925), Voderberg Tilings (1936), and Outer Billiards (1959).

*Penrose tiling with two geometric shapes that can cover a plane in a pattern with no gaps or overlaps and that does not repeat periodically. (This rendering is by Jos Leys.)*

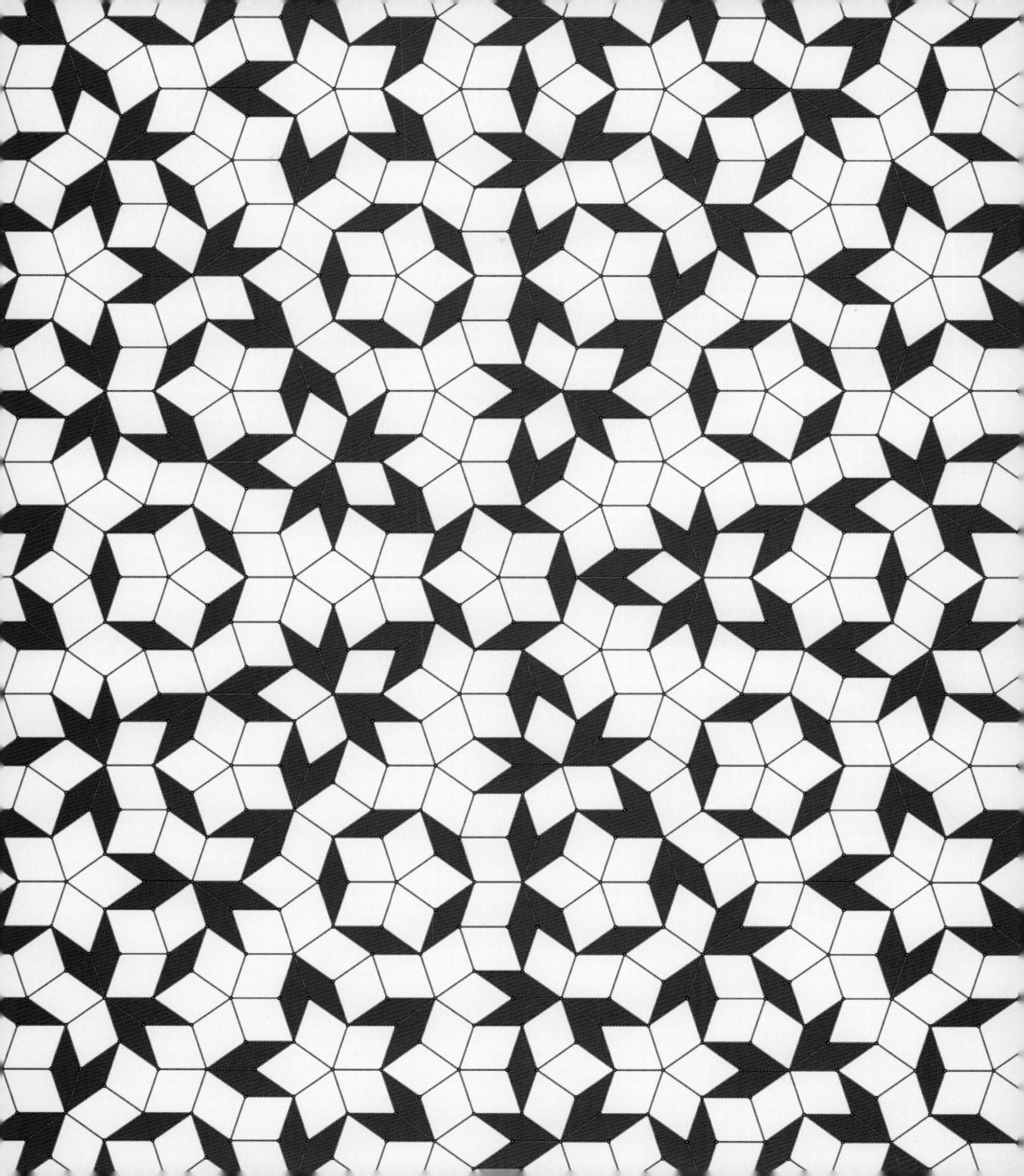

1973

# Art Gallery Theorem

**Václav (Vašek) Chvátal** (b. 1946), **Victor Klee** (1925–2007)

Imagine that you are in an expensive art gallery room represented by a polygon. If we were to place guards at some of the corners (vertices) of the room, what is the minimum number of guards needed so that the entire interior of a polygon can be viewed simultaneously? Assume that the guards can see in all directions at once, but they can't see through walls. Also, the guards are placed in the corners of the gallery so they don't block anyone's view of the art. The problem can be initially explored by drawing polygonal rooms and shading the line of sight for guards placed at several vertices.

Chvátal's art gallery theorem, named after Czechoslovakian-born mathematician and computer scientist Václav Chvátal, states that in an art gallery with $n$ corners, there needs to be at most $\lfloor n/3 \rfloor$ guards at the corners to watch the entire gallery, where the $\lfloor \ \rfloor$ symbols indicate the mathematical floor function that returns the largest integer less than or equal to $n/3$. We assume that the polygon is "simple," which means that the art gallery walls don't self-intersect and that they only meet at their end points.

In 1973, mathematician Victor Klee posed the question about the required number of guards to Chvátal, and Chvátal proved it shortly thereafter. Interestingly, only $\lfloor n/4 \rfloor$ guards are needed to watch a polygonal art gallery having corners that are all right angles. Thus, for this kind of gallery with 10 corners, only 2 rather than 3 guards are needed.

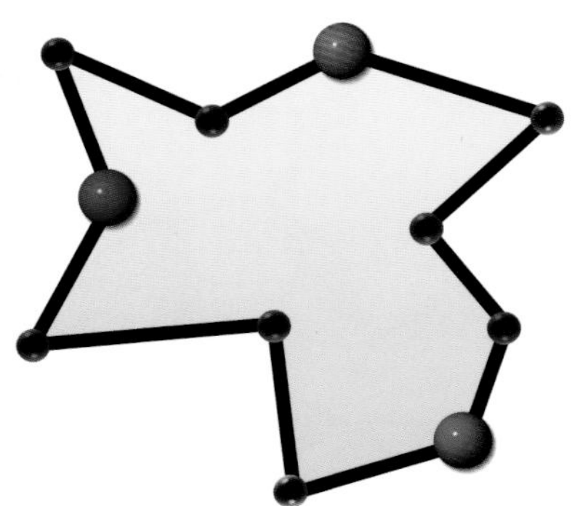

Researchers have since considered the art gallery problem using guards who can move along straight lines rather than remain in fixed positions. The problem has also been contemplated in three dimensions and with walls with holes. Norman Do writes, "When Victor Klee first posed the art gallery problem, he probably had little idea that it would motivate such a wealth of research which still continues over thirty years later. The area is [now] absolutely brimming with interesting problems. . . ."

**SEE ALSO** Projective Geometry (1639) and Tokarsky's Unilluminable Room (1969).

LEFT: *Three guards, positioned at the location of the three large spheres, can simultaneously view the interior of this polygonal room with 11 vertices.* RIGHT: *The art gallery theorem continues to stimulate a wealth of geometrical research using unusual wall arrangements, mobile guards, and higher dimensions.*

1974

# Rubik's Cube

**Ernö Rubik** (b. 1944)

Rubik's Cube was invented by the Hungarian inventor Ernö Rubik in 1974, patented in 1975, and placed on the Hungarian market in 1977. By 1982, as many as 10 million cubes had been sold in Hungary, more than the population of the country. It is estimated that more than 100 million have been sold worldwide.

The cube is a $3 \times 3 \times 3$ array of smaller cubes that are colored in such a way that the six faces of the large cube have six distinct colors. The 26 external sub-cubes are internally hinged so that these six faces can be rotated. The goal of the puzzle is to return a scrambled cube to a state in which each side has a single color. There are

$$43{,}252{,}003{,}274{,}489{,}856{,}000$$

different arrangements of the small cubes, and only one of these arrangements is the initial position where all colors match on each of the six sides. If you had a cube for every one of these "legal" positions, then you could cover the entire surface of the earth (including oceans) about 250 times. A column consisting of all the cube positions would stretch about 250 light years. There are $1.0109 \times 10^{38}$ combinations of the $3 \times 3 \times 3$ Rubik's Cube if you are allowed to remove the colored stickers and place them on different sub-cube faces.

In 2008, Tomas Rokicki proved that all positions of Rubik's cube can be solved in 22 or fewer cube face turns. In 2010, using computer calculations, researchers proved that no configuration requires more than 20 moves to solve.

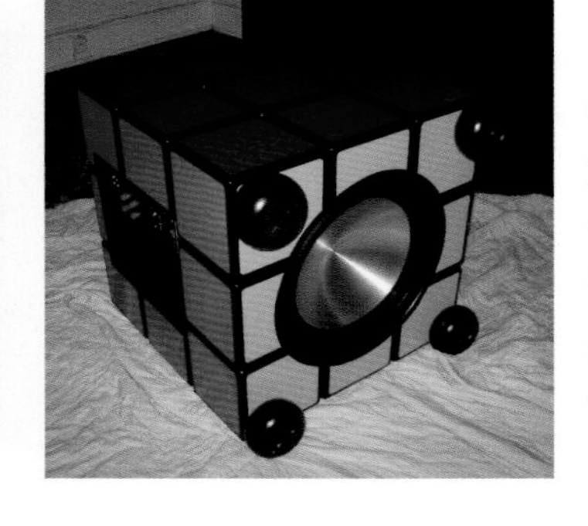

One natural variation that never appeared on toy store shelves is the four-dimensional version of Rubik's Cube—Rubik's tesseract. The total number of positions of Rubik's tesseract is $1.76 \times 10^{120}$. If either the cube or the tesseract changed positions every second since the beginning of the universe, they would still be turning today and not have exhibited every possible configuration.

**SEE ALSO** Group Theory (1832), Fifteen Puzzle (1874), Tower of Hanoi (1883), Tesseract (1888), and Instant Insanity (1966).

LEFT: *Zachary Paisley's handmade speaker enclosure in the form of a Rubik's Cube. This direct-servo subwoofer weighs 150 pounds (68 kilograms). Paisely says that the sounds are "capable of penetrating concrete—almost making it powerful enough to solve itself!"* RIGHT: *In 2008, Hans Andersson built a robot with plastic pieces that can solve Rubik's Cube using a light sensor to detect colors on the cube. The robot does not require a separate connection to a PC to perform calculations and cube manipulations.*

NXT

1974

# Chaitin's Omega

**Gregory John Chaitin** (b. 1947)

A computer program is said to "halt" when it accomplishes its task—for example, when it has computed the thousandth prime number or the first hundred digits of pi. On the other hand, a program will run forever if the task is unending, such as computing every Fibonacci number.

What happens if we feed a random sequence of bits to a **Turing Machine** for its program? (A Turing machine is an abstract symbol-manipulating device that can simulate the logic of a computer.) When this program is started, what is the probability that the machine will halt? The answer is Chaitin's number $\Omega$ (omega). The number varies depending on the machine, but for a given machine, $\Omega$ is a well-defined irrational number with a value between zero and one. For most computers, $\Omega$ is close to a value of 1 because a completely random program is likely to instruct a computer to do something impossible. Argentine-American mathematician Gregory Chaitin has shown that the digit sequence of $\Omega$ is patternless, that $\Omega$ is definable but utterly incalculable, and that it has infinitely many digits. The features of $\Omega$ have vast mathematical implications and place fundamental limits on what we can know.

Quantum theorist Charles Bennett writes, "The most remarkable property of $\Omega$... is the fact that if the first few thousand digits of $\Omega$ were known, they would, at least in principle, suffice to decide most of the interesting open questions in mathematics...." David Darling says that the properties of $\Omega$ show that solvable problems "form a tiny archipelago in a vast ocean of undecidability." According to Marcus Chown, $\Omega$ "reveals that mathematics...is mostly made of gaping holes. Anarchy...is at the heart of the universe."

*Time* magazine explains, "The concept broadens...Gödel's incompleteness theorem, which says there will always be unprovable statements in any system of math, and Turing's halting problem, which says it's impossible to predict...whether a particular computer calculation can ever be finished."

**SEE ALSO** Gödel's Theorem (1931) and Turing Machines (1936).

*The features of $\Omega$ have vast mathematical implications and place fundamental limits on what we can know. The number $\Omega$ has infinitely many digits, and its properties show that solvable problems "form a tiny archipelago in a vast ocean of undecidability."*

1974

# Surreal Numbers

**John Horton Conway** (b. 1937)

Surreal numbers are a superset of the real numbers, invented by prolific mathematician John Conway for the analysis of games, although the name was coined by Donald Knuth in his popular 1974 novelette *Surreal Numbers*, perhaps one of the few times that a major mathematical discovery was published first in a work of fiction. Surreal numbers have numerous bizarre properties. As background, real numbers include both rational numbers, such as ½, and irrational numbers, such as pi, and they may be visualized as points on an infinitely long number line.

Surreal numbers include the real numbers plus much more. Martin Gardner writes in *Mathematical Magic Show*, "Surreal numbers are an astonishing feat of legerdemain. An empty hat rests on a table made of a few axioms of standard set theory. Conway waves two simple rules in the air, then reaches into almost nothing and pulls out an infinitely rich tapestry of numbers that form a real and closed field. Every real number is surrounded by a host of new numbers that lie closer to it than any other 'real' value does. The system is truly 'surreal.' "

A surreal number is a pair of sets $\{X_L, X_R\}$ where the indices indicate the relative position (left and right) of the sets in the pair. Surreal numbers are fascinating because they are built upon an extremely small and simple foundation. In fact, according to Conway and Knuth, surreal numbers follow two rules: 1) Every number corresponds to two sets of previously created numbers, such that no member of the left set is greater than or equal to any member of the right set, and 2) one number is less than or equal to another number if and only if no member of the first number's left set is greater than or equal to the second number, and no member of the second number's right set is less than or equal to the first number.

Surreal numbers include infinity and infinitesimals, numbers smaller than any imaginable real numbers.

**SEE ALSO** Zeno's Paradoxes (c. 445 B.C.), Discovery of Calculus (c. 1665), Transcendental Numbers (1844), and Cantor's Transfinite Numbers (1874).

LEFT: *John H. Conway at the conference on Combinatorial Game Theory at Banff International Research Station in Alberta, Canada, June 2005.* RIGHT: *The cover of Donald Knuth's* Surreal Numbers, *an example of one of the few times that a major mathematical discovery was published first in a work of fiction. Surreal numbers include infinity and infinitesimals, numbers smaller than any imaginable real numbers.*

# SURREAL NUMBERS

D. E. KNUTH

1974

# Perko Knots

**Kenneth A. Perko, Jr.** (1941–2002), **Wolfgang Haken** (b. 1928)

For centuries, mathematicians have searched for ways to distinguish knots. As just one example, the two configurations shown here represent two knots that for more than 75 years were thought to represent two distinct knot types. In 1974, mathematicians discovered that it was possible to simply change the point of view of one knot to demonstrate that both knots were the same. Today, we call these *Perko pair knots* after the New York lawyer and part-time topologist Kenneth Perko, who showed that they were in fact the same knot, while he manipulated loops of rope on his living room floor!

Two knots are considered to be the same if we can manipulate one of them without cutting it so that it looks exactly like the other one with respect to the locations of the over- and under-crossings. Knots are classified by, among other characteristics, the arrangement and number of their crossings and certain characteristics of their mirror images. Stated more precisely, knots are classified using a variety of invariants, of which their symmetries are one and their crossing number is another, and characteristics of the mirror image play an indirect role in the classification. No general, practical algorithm exists to determine if a tangled curve is a knot or if two given knots are interlocked. Obviously, simply looking at a knot projected onto a plane—while keeping the under- and over-crossing apparent—is not an easy way to tell if a loop is a knot or an unknot. (The unknot is equivalent to a closed loop like a simple circle that has no crossings.)

In 1961, mathematician Wolfgang Haken devised an algorithm to determine if a knot projection on a plane (while preserving the under- and over-crossings) is actually an unknot. However, the procedure is so complicated that it has never been implemented. The paper describing the algorithm in the journal *Acta Mathematica* is 130 pages long.

**SEE ALSO** Knots (c. 100,000 B.C.), Perko Knots (1974), Jones Polynomial (1984), and Murphy's Law and Knots (1988).

*The two configurations shown here represent two knots that for more than 75 years were thought to represent two distinct knot types. In 1974, mathematicians discovered that the knots were in fact the same. (This graphics rendering is by Jos Leys.)*

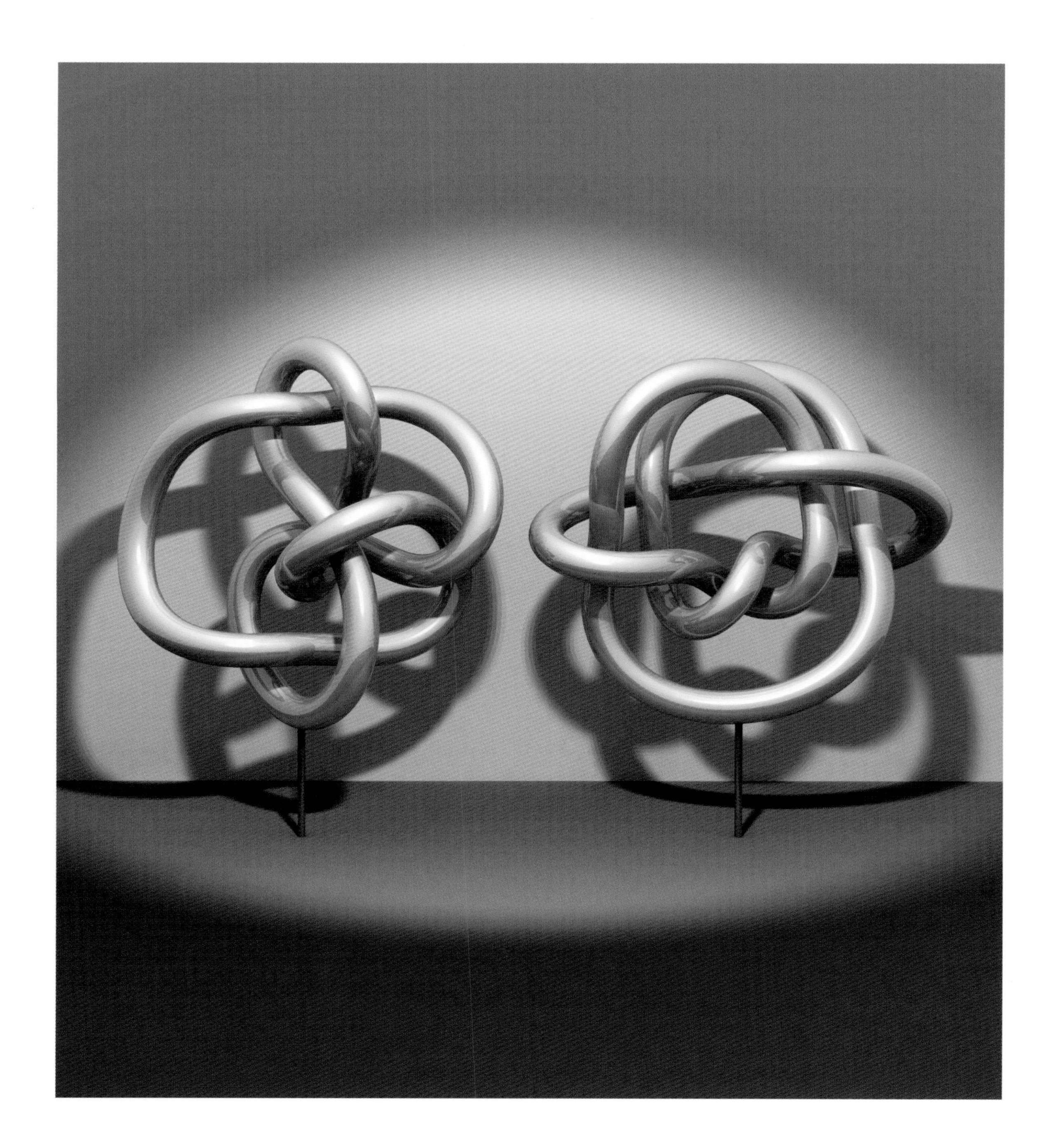

1975

# Fractals

**Benoît B. Mandelbrot** (1924–2010)

Today, computer-generated fractal patterns are everywhere. From squiggly designs on computer art posters to illustrations in the most serious of physics journals, interest continues to grow among scientists and, rather surprisingly, artists and designers. The word *fractal* was coined in 1975 by mathematician Benoît Mandelbrot to describe an intricate-looking set of curves, many of which were never seen before the advent of computers with their ability to quickly perform massive calculations. Fractals often exhibit self-similarity, which suggests that various exact or inexact copies of an object can be found in the original object at smaller size scales. The detail continues for many magnifications—like an endless nesting of Russian dolls within dolls. Some of these shapes exist only in abstract geometric space, but others can be used as models for complex natural objects such as coastlines and blood vessel branching. The dazzling computer-generated images can be intoxicating, motivating students' interest in math more than any other mathematical discovery in the last century.

Physicists are interested in fractals because they can sometimes describe the chaotic behavior of real-world phenomena such as planetary motion, fluid flow, the diffusion of drugs, the behavior of inter-industry relationships, and the vibration of airplane wings. (Chaotic behavior often produces fractal patterns.) Traditionally, when physicists or mathematicians saw complicated results, they often looked for complicated causes. In contrast, many fractal shapes reveal the fantastically complicated behavior of the simplest formulas.

Early explorers of fractal objects include Karl Weierstrass, who in 1872 considered functions that were everywhere continuous but nowhere differentiable, and Helge von Koch, who in 1904 discussed geometric shapes such as the **Koch Snowflake**. In the nineteenth and early twentieth centuries, several mathematicians explored fractals in the complex plane; however, they could not fully appreciate or visualize these objects without the aid of the computer.

**SEE ALSO** Descartes' *La Géométrie* (1637), Pascal's Triangle (1654), Weierstrass Function (1872), Peano Curve (1890), Koch Snowflake (1904), Thue-Morse Sequence (1906), Hausdorff Dimension (1918), Antoine's Necklace (1920), Alexander's Horned Sphere (1924), Menger Sponge (1926), Coastline Paradox (c. 1950), Chaos and the Butterfly Effect (1963), and Mandelbrot Set (1980).

*Fractal structure by Jos Leys. Fractals often exhibit self-similarity, which suggests that various structural themes are repeated at different size scales.*

1975

# Feigenbaum Constant

**Mitchell Jay Feigenbaum** (b. 1944)

Simple formulas can produce amazingly diverse and chaotic behaviors while characterizing phenomena ranging from the rise and fall of animal populations to the behavior of certain electronic circuits. One formula of special interest is the logistic map, which models population growth and was popularized by biologist Robert May in 1976 and based on the earlier work of Belgian mathematician Pierre François Verhulst (1804–1849), who researched models of population changes. The formula may be written as $x_{n+1} = rx_n(1 - x_n)$. Here, $x$ represents the population at time $n$. The variable $x$ is defined relative to the maximum population size of the ecosystem and therefore has values between 0 and 1. Depending on the value of $r$, which controls the rate of growth and starvation, the population may undergo many behaviors. For example, as $r$ is increased, the population may converge to a single value, or bifurcate so that it oscillates between two values, then oscillates between four values, then eight values, and finally becomes chaotic such that slight changes in the initial population yield very different, unpredictable outcomes.

The ratio of the distances between two successive bifurcation intervals approaches the Feigenbaum constant, 4.6692016091…, a number discovered by American mathematical physicist Mitchell Feigenbaum in 1975. Interestingly, although Feigenbaum initially considered this constant for a map similar to the logistic map, he also showed that it applied to all one-dimensional maps of this kind. This means that multitudes of chaotic systems will bifurcate at the same rate, and thus his constant can be used to predict when chaos will be exhibited in systems. This kind of bifurcation behavior has been discovered in many physical systems before they enter the chaotic regime.

Feigenbaum quickly realized that his "universal constant" was important, remarking that "I called my parents that evening and told them that I had discovered something truly remarkable, that, when I had understood it, would make me a famous man."

**SEE ALSO** Chaos and the Butterfly Effect (1963), Catastrophe Theory (1968), and Ikeda Attractor (1979).

*Bifurcation diagram (rotated clockwise by 90°), by Steven Whitney. This figure reveals the incredibly rich behavior of a simple formula as a parameter r is varied. Bifurcation "pitchforks" can be seen as small, thin, light branching curves amidst the chaos.*

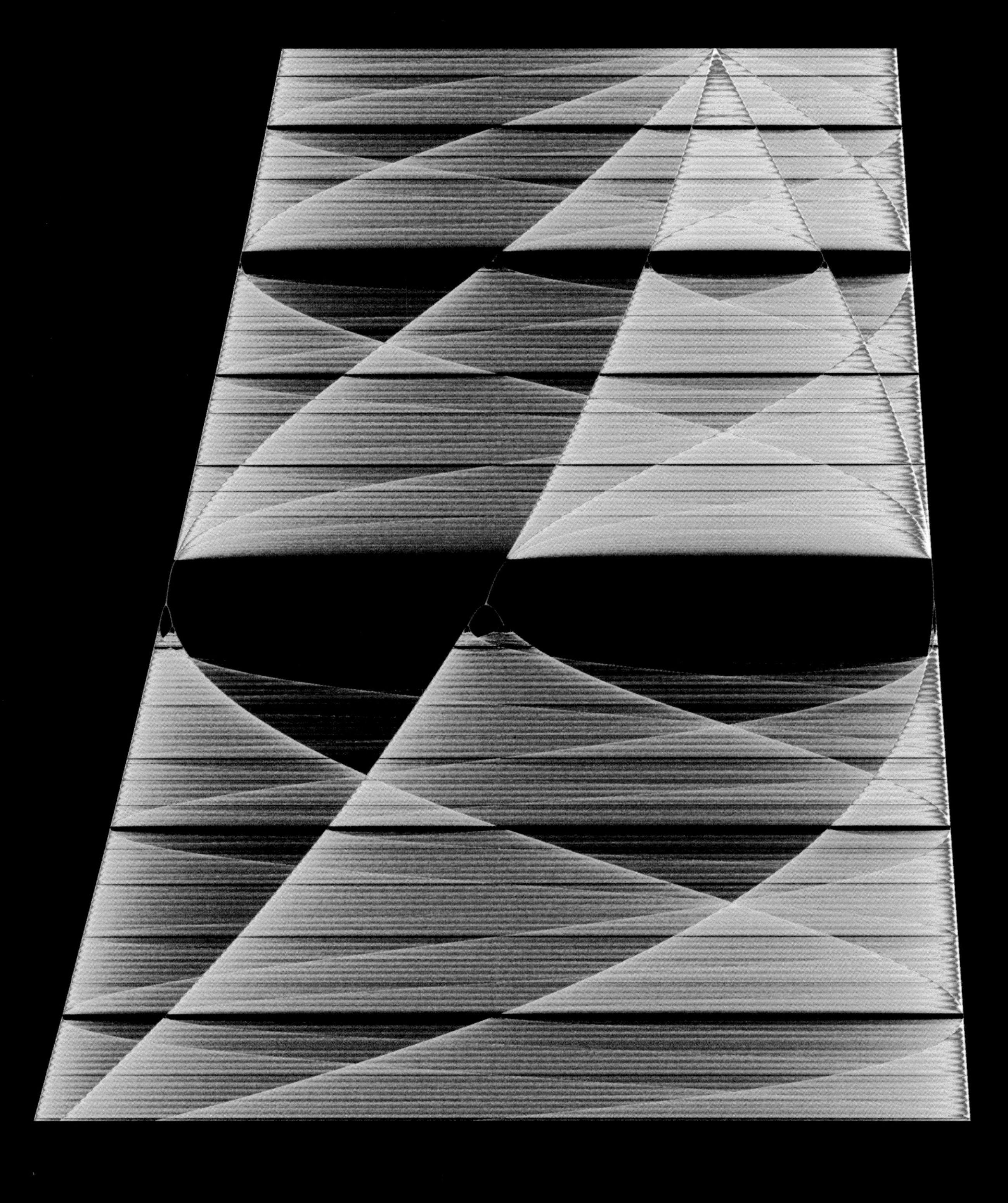

1977

# Public-Key Cryptography

**Ronald Lorin Rivest** (b. 1947), **Adi Shamir** (b. 1952), **Leonard Max Adleman** (b. 1945), **Bailey Whitfield Diffie** (b. 1944), **Martin Edward Hellman** (b. 1945), **Ralph C. Merkle** (b. 1952)

Throughout history, cryptologists have sought to invent a means for sending secret messages without the use of cumbersome code books that contained encryption and decryption keys that could easily fall into enemy hands. For example, the Germans, between 1914 and 1918, lost four code books that were recovered by British intelligence services. The British code-breaking unit, known as Room Forty, deciphered German communications, giving Allied forces a crucial strategic advantage in World War I.

In order to solve the key management problem, in 1976, Whitfield Diffie, Martin Hellman, and Ralph Merkle at Stanford University, California, worked on public-key cryptography, a mathematical method for distributing coded messages through the use of a pair of cryptographic keys: a public key and a private key. The private key is kept secret, while, remarkably, the public key may be widely distributed without any loss of security. The keys are related mathematically, but the private key cannot be derived from the public key by any practical means. A message encrypted with the public key can be decrypted only with the corresponding private key.

To better understand public-key encryption, imagine a mail slot in the front door to a home. Anyone on the street can stuff something into the mail slot; the public key is akin to the house address. However, only the person who possesses the key to the house door can retrieve the mail and read it.

In 1977, MIT scientists Ronald Rivest, Adi Shamir, and Leonard Adleman suggested that large prime numbers could be used to guard the messages. Multiplication of two large prime numbers is easy for a computer, but the reverse process of finding the two original prime numbers given their product can be very difficult. It should be noted that computer scientists had also developed public-key encryption for the British intelligence at an earlier date; however, this work was kept secret for reasons of national security.

**SEE ALSO** Cicada-Generated Prime Numbers (c. 1 Million B.C.), Sieve of Eratosthenes (240 B.C.), *Polygraphiae Libri Sex* (1518), Goldbach Conjecture (1742), Gauss's *Disquisitiones Arithmeticae* (1801), and Proof of the Prime Number Theorem (1896).

*Enigma machine, used to code and decode messages before the age of modern cryptography. The Nazis used Enigma-produced ciphers, which had several weaknesses, such as the fact that the messages could be decoded if a code book was captured.*

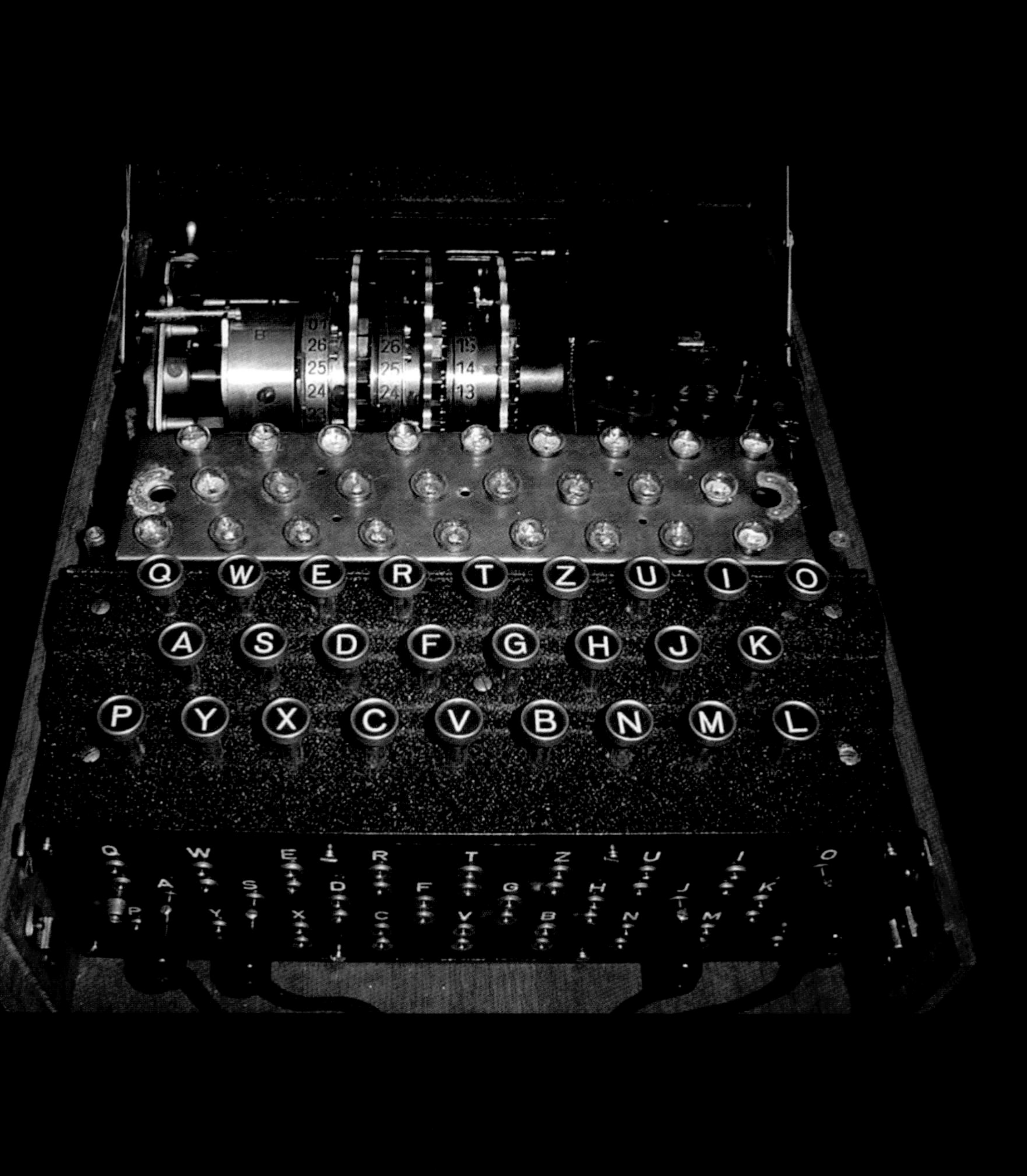
Q W E R T Z U I O
A S D F G H J K
P Y X C V B N M L

1977

# Szilassi Polyhedron

**Lajos Szilassi** (b. 1942)

Polyhedra are three-dimensional solids with flat faces and straight edges. Common examples include the cube and regular tetrahedron, which is a pyramid that is composed of four faces in the shape of equilateral triangles. If the polyhedron is regular, each face has the same size and shape.

The Szilassi polyhedron was discovered in 1977 by Hungarian mathematician Lajos Szilassi. This polyhedron is a heptahedron with seven 6-sided faces, 14 vertexes, 21 edges, and a hole. If we were to smooth the surface of the Szilassi polyhedron to make the edges less apparent, we could see that, from a topological standpoint, the Szilassi polyhedron is equivalent to a doughnut (or torus). The polyhedron has an axis of 180-degree symmetry. Three pairs of faces are congruent—that is, they have the same shape and size. The other unpaired face is a symmetrical hexagon.

Remarkably, the tetrahedron and the Szilassi polyhedron are the only two known polyhedra in which each face shares an edge with each other face. Gardner writes that "until Szilassi's computer program found the structure, it was not known that it could exist."

The Szilassi polyhedron also provides insight into the problem of coloring maps. A traditional map may be colored with a minimum of four colors so that no two adjacent regions are the same color. For a map on the surface of a torus, the number is seven. This means that each face of the Szilassi polyhedron must be a different color to ensure that no two adjacent faces have the same color. For comparison, a tetrahedron demonstrates that four colors are necessary for a map on a surface that is topologically equivalent to a sphere. The properties of the two polyhedra may be summarized like this:

| | | | | |
|---|---|---|---|---|
| Tetrahedron | 4 faces | 4 vertices | 6 edges | 0 holes |
| Szilassi polyhedron | 7 faces | 14 vertices | 21 edges | 1 hole |

**SEE ALSO** Platonic Solids (350 B.C.), Archimedean Semi-Regular Polyhedra (c. 240 B.C.), Euler's Formula for Polyhedra (1751), Four-Color Theorem (1852), Icosian Game (1857), Pick's Theorem (1899), Geodesic Dome (1922), Császár Polyhedron (1949), Spidrons (1979), and Solving of the Holyhedron (1999).

*The Szilassi polyhedron forms the basis for this lamp, created by Hans Schepker.*

1979

# Ikeda Attractor

**Kensuke S. Ikeda** (b. 1949)

A deep reservoir for striking images is the *dynamical system*. Dynamical systems are models comprising the rules that describe the way some quantity undergoes a change through time. For example, the motion of planets about the sun can be modeled as a dynamical system in which the planets move according to Newton's Laws. The figure shown here represents the behavior of mathematical expressions called *differential equations*. One way of understanding the behavior of differential equations involves us imagining a machine that takes in values for variables at an initial time and then generates the new values at some later time. Just as one can track the path of a jet by the smoke path it leaves behind, computer graphics provide a way to follow paths of particles whose motion is determined by simple differential equations. The practical side of dynamical systems is that they can sometimes be used to describe real-world behaviors such as fluid flows, the vibration of bridges, the orbital motion of satellites, the control of robotic arms, and the response of electrical circuits. Often the resulting graphic patterns resemble smoke, swirls, candle flames, and windy mists.

The Ikeda attractor, shown here, is an example of a strange attractor, which has an irregular, unpredictable behavior. An attractor is a set to which a dynamical system evolves, or settles to, after some amount of time. With "tame" attractors, initially close points stay together as they approach the attractor. With strange attractors, initially adjacent points eventually follow widely divergent trajectories. As with leaves in a turbulent stream, it is impossible to predict where the leaves will end up given their initial positions.

In 1979, Japanese theoretical physicist Kensuke Ikeda published "Multiple-Valued Stationary State and Its Instability of the Transmitted Light by a Ring Cavity System," which describes a variation of this attractor. Numerous other famous attractors and related mathematical mappings exist in the mathematical literature, including the Lorenz attractor, logistic map, Arnold's cat map, horseshoe map, Hénon map, and Rössler map.

**SEE ALSO** Harmonograph (1857), Differential Analyzer (1927), Chaos and the Butterfly Effect (1963), and Feigenbaum Constant (1975).

*Dynamical systems are models comprising the rules that describe the way some quantity undergoes a change through time. The Ikeda attractor, shown here, is an example of a strange attractor, which has an irregular, unpredictable behavior.*

1979

# Spidrons

**Dániel Erdély** (b. 1956)

Journalist Ivars Peterson writes of Spidrons, "A field of triangles crumples and twists into a wavy crystalline sea. A crystal ball sprouts spiraling, labyrinthine passages. Faceted bricks stack snugly into a tidy, compact structure. Underlying each of these objects is a remarkable geometric shape made up of a sequence of triangles—a spiral polygon that resembles a seahorse's tail."

In 1979, graphic artist Dániel Erdély created an example of the Spidron system, as a part of his homework for Ernö Rubik's theory of form class at the Budapest University of Art and Design. Erdély had experimented with earlier versions of this work as early as 1975.

To create a Spidron, draw an equilateral triangle, and then draw lines from the three corners of the triangle to a point at its center, creating three identical isosceles triangles. Next, draw a reflection of one of these isosceles triangles so that it juts from the side of the original triangle. Create a new, smaller equilateral triangle, using one of the two short sides of the jutting isosceles triangle as a base. By repeating the procedure, you'll create a spiraling triangulated structure that gets increasingly small. Finally, you can erase the original equilateral triangle, and join two of the triangulated structures along the long side of the largest isosceles triangle to create the seahorse shape.

The Spidron's significance arises from its remarkable spatial properties, including its ability to form various space-filling polyhedra and tiling patterns. If we crawl like an ant along the deeper regions of the seahorse's tail, we find that the area of any equilateral triangle equals the sum of the areas of all the smaller triangles. The infinite collection of smaller triangles could all be crammed into such an equilateral triangle without overlap. When crinkled in just the right manner, Spidrons provide an infinite reservoir for magnificent 3-D sculptures. Possible practical examples of Spidrons include acoustic tiles and shock absorbers for machinery.

**SEE ALSO** Platonic Solids (350 B.C.), Archimedean Semi-Regular Polyhedra (c. 240 B.C.), Archimedes' Spiral (225 B.C.), Logarithmic Spiral (1638), and Voderberg Tilings (1936).

LEFT: *Spidron, a spiraling triangulated structure that grows increasingly small at its two tips.* RIGHT: *Spidrons have the ability to form various tiling patterns and space-filling polyhedra, such as this sculpture, courtesy of Dániel Erdély.*

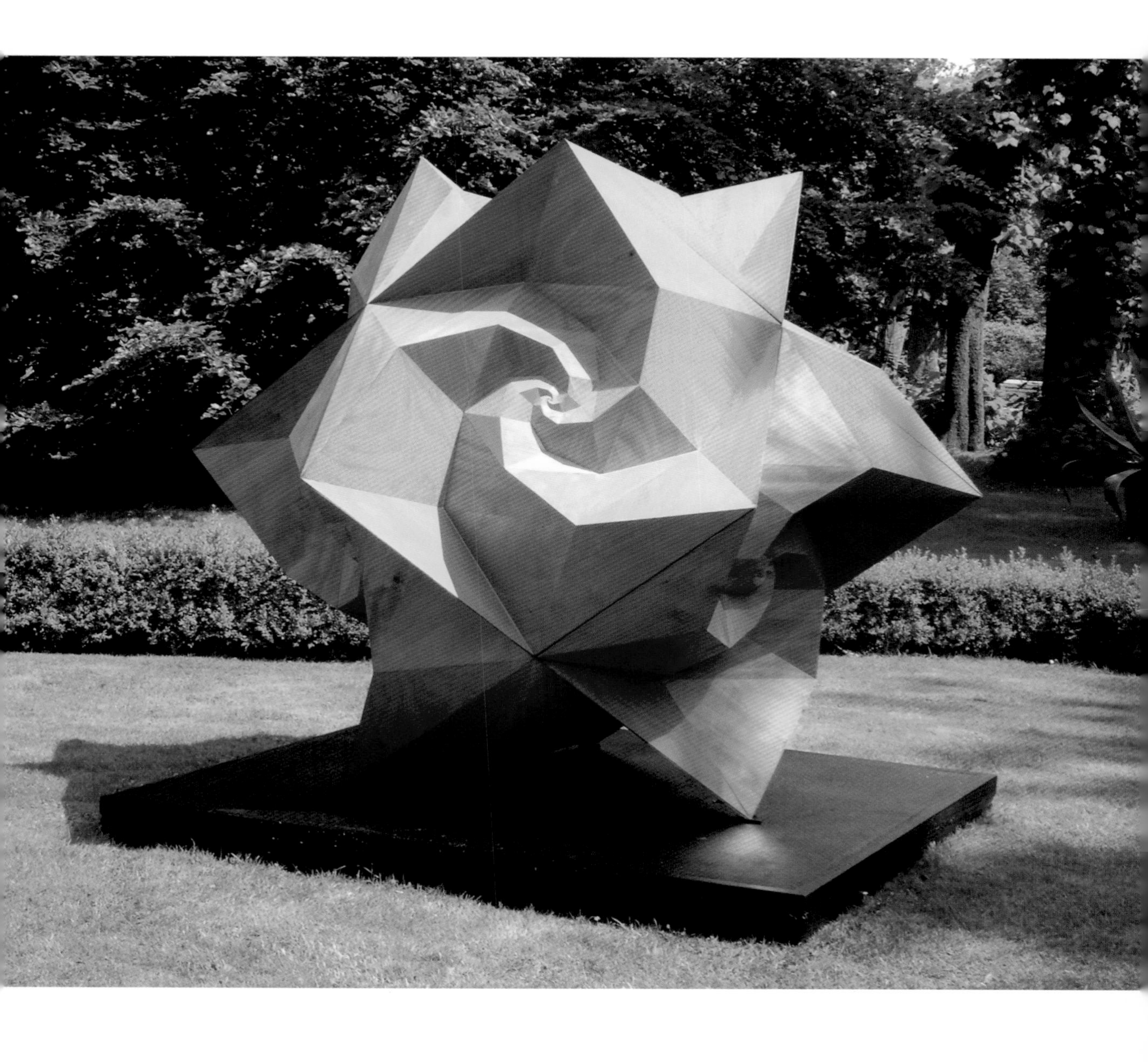

1980

# Mandelbrot Set

**Benoît B. Mandelbrot** (1924–2010)

David Darling writes that the Mandelbrot set, or M-set for short, is the "best known fractal and one of the most... beautiful mathematical objects known." The *Guinness Book of World Records* called it "the most complicated object in mathematics." Arthur C. Clarke emphasizes the degree to which the computer is useful for gaining insight: "In principle, [the Mandelbrot set] could have been discovered as soon as men learned to count. But even if they never grew tired, and never made a mistake, all the human beings who have ever existed would not have sufficed to do the elementary arithmetic required to produce a Mandelbrot set of quite modest magnification."

The Mandelbrot set is a fractal, an object that continues to exhibit similar structural details no matter how much the edge of the object is magnified. Think of the beautiful M-set images as being produced by mathematical feedback loops. In fact, the set is produced by iteration, or repetition, of the very simple formula $z_{n+1} = z_n^2 + c$, for complex values of $z$ and $c$, and for $z_0 = 0$. The set contains all points for which the formula does not produce values that diverge to infinity. The first crude pictures of the M-set were drawn in 1978 by Robert Brooks and Peter Matelski, followed by the landmark paper by Mandelbrot in 1980 on its fractal aspects and the wealth of geometric and algebraic information it conveys.

The M-set structure contains super-thin spiral and crinkly paths, connecting an infinite number of island shapes. Computer magnifications of the M-set will easily yield pictures never seen before by human eyes. The incredible vastness of the M-set led authors Tim Wegner and Mark Peterson to remark: "You may have heard of a company that for a fee will name a star after you and record it in a book. Maybe the same thing will soon be done with the Mandelbrot set!"

**SEE ALSO** Imaginary Numbers (1572) and Fractals (1975).

*The Mandelbrot set is a fractal and continues to exhibit similar structural details no matter how much the edge of the object is magnified. Computer magnifications of the M-set will easily yield pictures never seen before by human eyes. (This rendering is by Jos Leys.)*

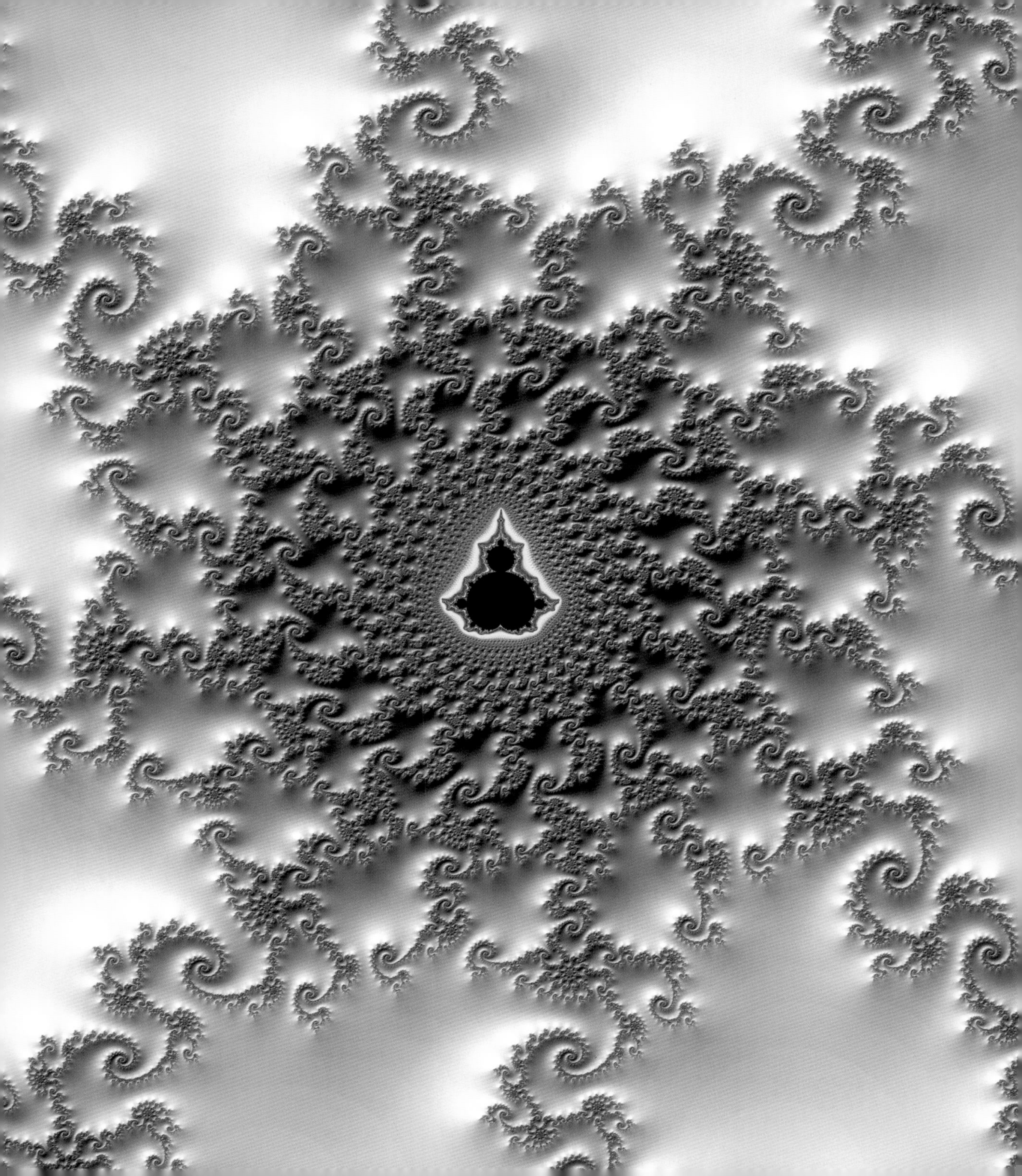

1981

# Monster Group

**Robert L. Griess, Jr.** (b. 1945)

In 1981, American mathematician Robert Griess constructed the Monster—the largest and one of the most mysterious of the so-called sporadic groups, a particular set of groups in the field of **Group Theory**. The quest to comprehend the Monster has helped mathematicians understand some of the basic building blocks of symmetry and how such building blocks, along with some of their exceptional subfamilies, can be used to solve deep problems involving symmetry in mathematics and in mathematical physics. We can think of the Monster group as a mind-boggling snowflake with more than $10^{53}$ symmetries that exists in a space of 196,884 dimensions!

Griess said that he became "addicted" to Monster construction in 1979, the year he was married—and his wife was "very understanding" during his intense pursuit, when he took time off only on Thanksgiving and Christmas Day. In 1982, his 102-page paper on the Monster was finally published. Mathematicians marveled that Griess could construct the Monster without using a computer.

More than a mere curiosity, the structure of the Monster suggests deep connections between symmetry and physics, and it may even have a connection with string theory, which posits that all of the fundamental particles in the universe are made of tiny vibrating loops of energy. Mark Ronan, in his book *Symmetry and the Monster*, writes that the Monster has "arrived before its time—a piece of twenty-second-century mathematics that slipped by chance into the 20th century." In 1983, physicist Freeman Dyson wrote that the Monster may be "built in some unsuspected way into the structure of the universe."

In 1973, Griess and Bernd Fischer predicted the existence of the Monster, and John Conway gave the object its name. In 1998, Richard Borcherds was awarded a Fields Medal for his work on understanding the Monster and its profound connections with other areas of mathematics and physics.

**SEE ALSO** Group Theory (1832), Wall Paper Groups (1891), Fields Medal (1936), and The Quest for Lie Group $E_8$ (2007).

*American mathematician Robert Griess (pictured here) constructed the Monster in 1981. The quest to comprehend the Monster has helped mathematicians understand some of the basic building blocks of symmetry. The Monster group involves a space of 196,884 dimensions!*

1982

# Ball Triangle Picking

## Glen Richard Hall (b. 1954)

In 1982, Glen Hall published his famous research paper "Acute Triangles in the $n$-Ball." This was Hall's first published mathematics paper, and it discusses work he conducted while taking a graduate school class on geometric probability, taught at the University of Minnesota. Imagine picking three points at random in a circle to create a triangle. Hall wondered what the probability would be for obtaining an "acute triangle," not just for triangles inside a circle but also in higher dimensions, such as inside spheres and hyperspheres. These generalizations of a circle are called $n$-dimensional balls. An acute triangle is one in which each of the three angles is less than 90 degrees.

Below are several values for $P_n$, the probability of picking an acute triangle in an $n$-ball if the three points on the triangle are chosen independently and uniformly:

$P_2 = 4/\pi^2 - 1/8 \approx 0.280285$ (circle)
$P_3 = 33/70 \approx 0.471429$ (sphere)
$P_4 = 256/(45\pi^2) + 1/32 \approx 0.607655$ (four-dimensional hypersphere)
$P_5 = 1415/2002 \approx 0.706793$ (five-dimensional hypersphere)
$P_6 = 2048/(315\pi^2) + 31/256 \approx 0.779842$ (six-dimensional hypersphere)

Hall noticed that as the dimension of the sphere increased, the probability of choosing an acute triangle also increased. By the time we reach the ninth dimension, we have a probability of 0.905106 of selecting an acute triangle. The triangle work is notable in that mathematicians had no generalization of triangle picking to higher dimensions until the early 1980s. Hall, in a personal communication to me, notes that he was amazed by the potential alternation of probabilities between rational and irrational solutions according to the dimension of the ball, a dimensional oscillation that mathematicians would probably never have conjectured before this research. Rational numbers are those that can be expressed by a ratio of two integers. Note that mathematician Christian Buchta in 1986 was responsible for providing closed-form evaluations for Hall's integrals.

**SEE ALSO** Viviani's Theorem (1659), Buffon's Needle (1777), Laplace's *Théorie Analytique des Probabilités* (1812), and Morley's Trisector Theorem (1899).

*Select three points at random in a circle to create a triangle. What is the probability of obtaining a triangle in which each of the three angles is less than 90°?*

1984

# Jones Polynomial

**Vaughan Frederick Randal Jones** (b. 1952)

In mathematics, even the most tangled loop in three dimensions can be represented as a projection, or shadow, on a flat surface. When mathematical knots are diagrammed, tiny breaks in the lines often indicate when a strand crosses over or under another strand.

One of the goals of knot theory is to find invariants of knots, where the term *invariant* refers to a mathematical characteristic or value that is the same for equivalent knots so that it can be used to show that two knots are different. In 1984, knot theorists were all abuzz with a startling invention of New Zealand mathematician Vaughan Jones, an invariant, now called the Jones polynomial, that could distinguish more knots than any previous invariant. Jones had made his breakthrough discovery by chance, while working on a physics problem. Mathematician Keith Devlin writes, "Sensing that he had stumbled onto an unexpected, hidden connection, Jones consulted knot theorist Joan Birman, and the rest, as they say, is history. . . ." Jones's research "opened the way to a whole array of new polynomial invariants, and led to a dramatic rise in research in knot theory, some of it spurred on by the growing awareness of exciting new applications in both biology and physics. . . ." Biologists who study DNA strands are interested in knots and how they can help elucidate the functioning of genetic material in cells or even aid in resistance to viral attacks. A systematic procedure, or algorithm, allows mathematicians to express the Jones polynomial for any knot, based on its pattern of crossings.

The use of knot invariants has had a long history. Around 1928, James W. Alexander (1888–1971) introduced the first polynomial associated with knots. Alas, the Alexander polynomial was not useable for detecting the difference between a knot and its mirror image, something that the Jones polynomial could do. Four months after Jones announced his new polynomial, the more general HOMFLY polynomial was announced.

**SEE ALSO** Knots (c. 100,000 B.C.), Perko Knots (1974), and Murphy's Law and Knots (1988).

*Knot with 10 crossings, rendered by Jos Leys. One of the goals of knot theory is to find a mathematical characteristic that is the same for equivalent knots so that it can be used to show that two knots are different.*

1985

# Weeks Manifold

**Jeffrey Renwick Weeks** (b. 1956)

Hyperbolic geometry is a **Non-Euclidean Geometry** in which Euclid's parallel postulate does not hold. In this geometry for two dimensions, for any line and any point not on it, many other lines pass through the point without intersecting the first line. Hyperbolic geometry is sometimes visualized using saddle-shaped surfaces on which the sum of angles of a triangle is less than 180 degrees. Such strange geometries have implications for mathematicians and even cosmologists who ponder possible properties and shapes for our universe.

In 2007, David Gabai of Princeton University, Robert Meyerhoff of Boston College, and Peter Milley of the University of Melbourne in Australia proved that a particular hyperbolic three-dimensional space, or 3-manifold, has least volume. This shape, called the Weeks manifold after its discoverer American mathematician Jeffrey Weeks, holds immense interest to topologists who catalogue shapes of this kind.

In traditional Euclidean geometry, the concept of a "least volume" for a three-dimensional space is meaningless. Shapes and volumes can be scaled to any size. However, the spatial curvature of hyperbolic geometry provides an intrinsic unit for length, area, and volume. In 1985, Weeks had found a small manifold with a volume of approximately 0.94270736. (The Weeks manifold is related to the space around a pair of intertwined loops, known as the Whitehead link.) Until 2007, no one knew for sure if the Weeks manifold was the smallest.

MacArthur Fellow Jeffrey Weeks received his Ph.D. in mathematics from Princeton University in 1985, under the supervision of William Thurston. One of his main passions is the use of topology to bridge the gap between geometry and observational cosmology. He has also developed interactive software to introduce geometry to young students and to let them explore universes that are finite yet have no boundaries.

**SEE ALSO** Euclid's *Elements* (300 B.C.), Non-Euclidean Geometry (1829), Boy's Surface (1901), and Poincaré Conjecture (1904).

*This model of a Weeks manifold contains only one galaxy, but we see images of that galaxy repeating in a crystalline pattern, giving the illusion of an infinite space. The effect is similar to a hall of mirrors, which also gives an illusion of infinite space.*

# Andrica's Conjecture

**Dorin Andrica** (b. 1956)

A prime number is an integer that has exactly two distinct integer divisors: 1 and itself. Examples of prime numbers include 2, 3, 5, 7, 11, 13, 17, 19, 23, 29, 31, and 37. The great Swiss mathematician Leonhard Euler (1707–1783) remarked, "Mathematicians have tried in vain to this day to discover some order in the sequence of prime numbers, and we have reason to believe that it is a mystery into which the mind will never penetrate." Mathematicians have long searched for patterns in the sequence of prime numbers and also in the gaps between them, where the term *gap* refers to the difference between two successive prime numbers. The value of the average gap between primes increases as the natural logarithm of the prime number at either end of the gap. As an example of a known large gap, consider the gap of 879 non-prime numbers after the prime 277,900,416,100,927. In 2009, the largest known prime gap had a length of 337,446.

In 1985, Romanian mathematician Dorin Andrica published "Andrica's Conjecture," which concerns the gaps between prime numbers. In particular, the conjecture states that $\sqrt{p_{n+1}} - \sqrt{p_n} < 1$, where $p_n$ is the $n$th prime number. For example, consider the prime numbers 23 and 29. Applying Andrica's conjecture, we have $\sqrt{29} - \sqrt{23} < 1$. Another way of writing this is $g_n < 2\sqrt{p_n} + 1$, where $g_n$ is the $n$th prime gap, and $g_n = p_{n+1} - p_n$. As of 2008, the conjecture has been shown to hold true for $n$ up to $1.3002 \times 10^{16}$.

If we examine the left side of the inequality in Andrica's conjecture, $A_n = \sqrt{p_{n+1}} - \sqrt{p_n}$, the highest value for $A_n$ ever found is at $n = 4$, where $A_n$ is approximately equal to 0.67087. Andrica's conjecture was stated just at the time when computers were becoming ubiquitous, which encouraged an ongoing flurry of activity in an attempt to understand and find counterexamples that might defeat the conjecture. So far, Andrica's conjecture still stands, although it also remains unproven.

**SEE ALSO** Cicada-Generated Prime Numbers (c. 1 Million B.C.), Sieve of Eratosthenes (240 B.C.), Goldbach Conjecture (1742), Gauss's *Disquisitiones Arithmeticae* (1801), Möbius Function (1831), Riemann Hypothesis (1859), Proof of the Prime Number Theorem (1896), Brun's Constant (1919), Gilbreath's Conjecture (1958), Sierpiński Numbers (1960), Ulam Spiral (1963), and Erdös and Extreme Collaboration (1971).

*The function $A_n$ for the first 100 primes. The highest vertical point in this graph (the bar near the left of the plot) is at 0.67087, and the x-axis range is from 1 to 100.*

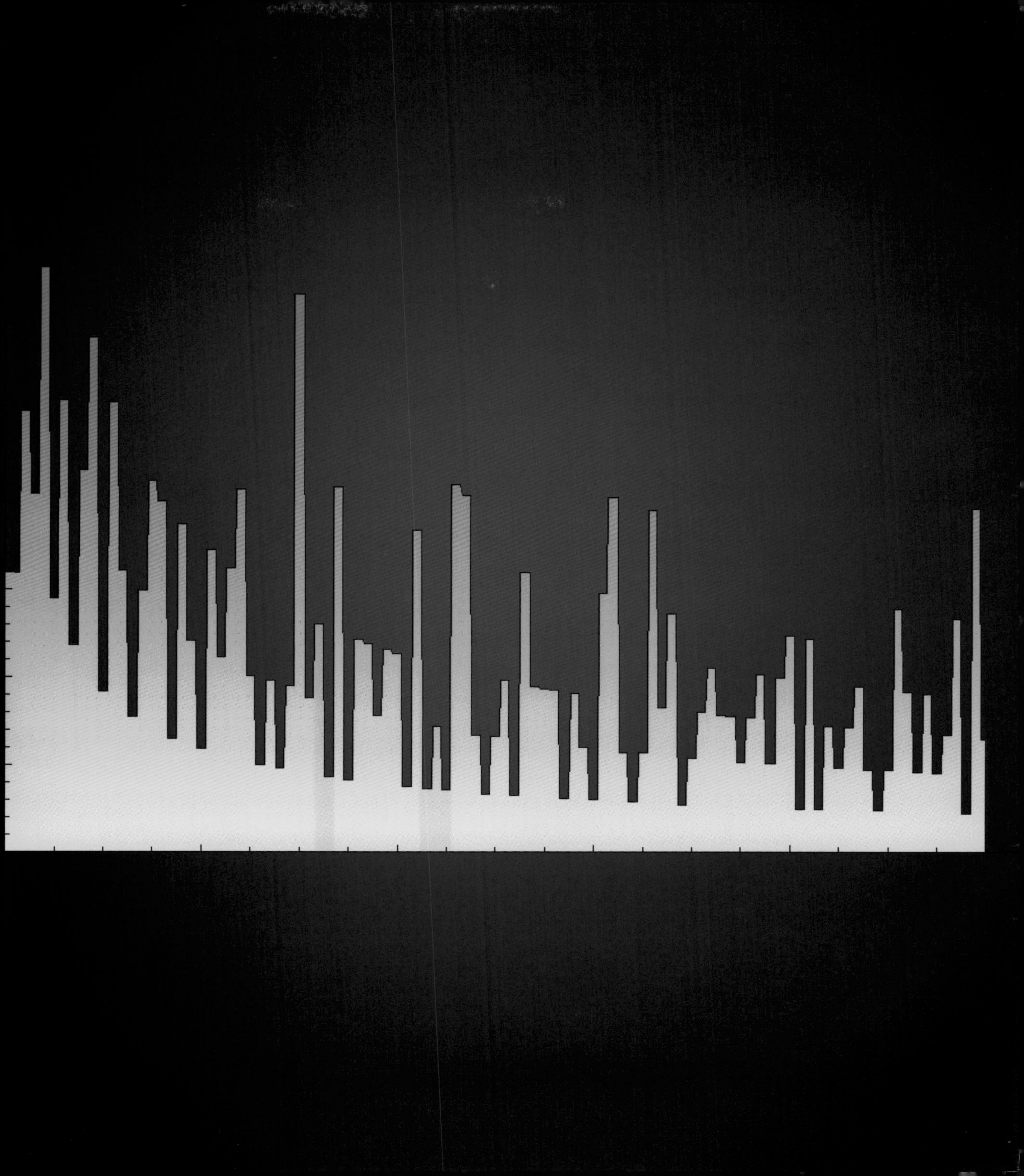

1985

# The *ABC* Conjecture

**David Masser** (b. 1948), **Joseph Oesterlé** (b. 1954)

The *ABC* conjecture is considered to be one of the most important unsolved problems in number theory, the study of the properties of whole numbers. If the conjecture is correct, mathematicians will be able to prove many other famous theorems in just a few lines.

The conjecture was first advanced in 1985 by mathematicians Joseph Oesterlé and David Masser. To understand the conjecture, we define a *square-free number* as an integer that is not divisible by the square of any number. For example, 13 is square-free, but 9 (divisible by $3^2$) is not. The square-free part of an integer $n$, denoted sqp($n$), is the largest square-free number that can be formed by multiplying the prime factors of $n$. Thus, for $n = 15$, the prime factors are 5 and 3, and $3 \times 5 = 15$, a square-free number. So sqp(15) = 15. On the other hand, for $n = 8$, the prime factors are all 2, which means that sqp(8) = 2. Similarly, sqp(18) = 6, which is found by multiplying its factors 3 and 2, and sqp(13) = 13.

Next, consider numbers A and *B* that have no factors in common, and C is their sum. For example, consider A = 3 and *B* = 7 and C = 10. The square-free part of the product *ABC* is 210. Notice that sqp(*ABC*) is greater than C, but this is not always the case. It is possible to prove that the ratio sqp(*ABC*)/*C* can get arbitrarily small for the appropriate selection of A, *B*, and C. However, the *ABC* conjecture states that $[\text{sqp}(ABC)]^n/C$ does reach a minimum value if $n$ is any real number greater than 1.

Dorian Goldfeld writes, "The *ABC* conjecture...is more than utilitarian; to mathematicians it is also a thing of beauty. Seeing so many Diophantine [integer-solution] problems unexpectedly encapsulated into a single equation drives home the feeling that all the subdisciplines of mathematics are aspects of a single underlying unity...."

**SEE ALSO** Cicada-Generated Prime Numbers (c. 1 Million B.C.), Sieve of Eratosthenes (240 B.C.), Goldbach Conjecture (1742), Constructing a Regular Heptadecagon (1796), Gauss's *Disquisitiones Arithmeticae* (1801), Riemann Hypothesis (1859), Proof of the Prime Number Theorem (1896), Brun's Constant (1919), Gilbreath's Conjecture (1958), Ulam Spiral (1963), and Andrica's Conjecture (1985).

*The* ABC *conjecture is considered to be one of the most important unsolved problems in number theory. The conjecture was first advanced in 1985 by mathematicians David Masser (pictured here) and Joseph Oesterlé.*

Q - irred
homogeneous
(abc)
P,Q ∈ ℂ[x] \ ℂ
f,g ∈ ℂ(X) \ ℂ
M = 5.5157...
O(1)

1986

# Audioactive Sequence

**John Horton Conway** (b. 1937)

Consider the following sequence of numbers: 1, 11, 21, 1211, 111221,... To appreciate how the sequence is formed, it helps to speak the entries in each row out loud. Note that the second entry has two "ones," thereby giving the 21 for the third entry. The third entry has one "two" and one "one." Extending this pattern, an entire sequence can be generated. The sequence was extensively studied by mathematician John Conway, who called the process "audioactive."

The sequence grows rather rapidly. For example, row 16 is 132113213221133112132113311211131221121321131211132221123113112221131112311332111213211322211312113211. If you were to carefully study the sequence, you would find a predominance of 1s, with 2s and 3s less common, and no numbers greater than 3. Is it possible to prove that 333 can never occur? You can see from the following representation of Row 11 (in which 3s are represented by ■) that the occurrence of 3 seems erratic, like lost ships on an infinite sea: -■--■-■---■■---■--■■----■-----■--■----■-----■-■-----■---■-■■-----■--■----■--■--.

The number of digits in the *n*th term of this sequence is roughly proportional to Conway's constant: $(1.303577269034269391257099112152551890730702504659 4\ldots)^n$. Mathematicians have found it remarkable that the "bizarre" audioactive construction process yields this constant that turns out to be a unique positive real root of a polynomial equation. Interestingly, the constant applies to *all* starting sequences, with the exception of 22.

Many variants exist. British researcher Roger Hargrave has extended the idea to a variation in which a row takes into account *all* occurrences of each character in a previous row. For example, the sequence starting with 123 is 123, 111213, 411213, 14311213, ... Interestingly, he believes that all his sequences finally oscillate between 23322114 and 32232114. Can you prove this? What are the properties of reverse likeness-sequences? Starting from a particular row, can one work backward and compute the starting string of symbols?

**SEE ALSO** Thue-Morse Sequence (1906), Collatz Conjecture (1937), and The On-Line Encyclopedia of Integer Sequences (1996).

*The strange construction method for the audioactive process yields Conway's constant, 1.3035..., which turns out to be a unique positive real root of a 69-term polynomial equation. This root is at the location of the yellow sphere. Other roots of this polynomial are shown as + symbols.*

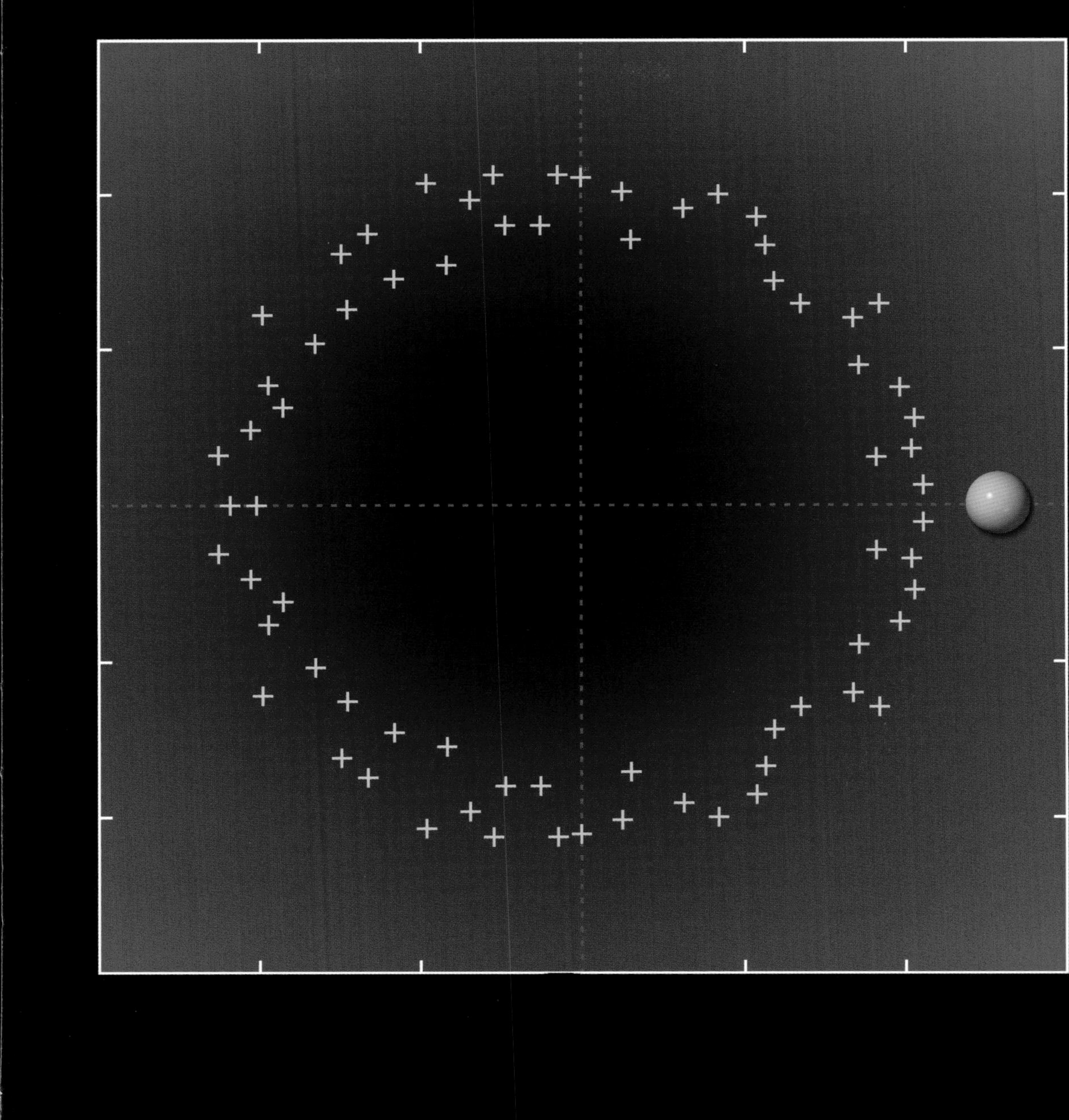

# Mathematica

**Stephen Wolfram** (b. 1959)

A shift has taken place in the way mathematics has been practiced over the last 20 years—a transition from pure theory and proof to the use of computers and experimentation. This shift is due, in part, to computational software packages like Mathematica, sold by Wolfram Research of Champaign, Illinois, and developed by mathematician and theorist Stephen Wolfram. The first version of Mathematica was released in 1988, and today it provides a general computing environment that organizes numerous algorithmic, visualization, and user-interface capabilities. Mathematica is one example of numerous packages available today for experimental mathematics, including Maple, Mathcad, MATLAB, and Maxima.

Since the 1960s, individual software packages have existed for specific numerical, algebraic, graphical, and other tasks, and researchers interested in chaos and fractals have long used computers for their explorations. Mathematica helped unite various features of specialized packages in a convenient fashion. Today, Mathematica is used in engineering, science, finance, education, art, clothing design, and other fields that require visualization and experimentation.

In 1992, the journal *Experimental Mathematics* was launched and helped to show how computation can be used to investigate mathematical structures and identify important properties and patterns. Educator and author David Berlinski writes, "The computer has…changed the very nature of mathematical experience, suggesting for the first time that mathematics, like physics, may yet become an empirical discipline, a place where things are discovered because they are seen."

Mathematicians Jonathan Borwein and David Bailey write, "Perhaps the most important advancement along this line is the development of broad spectrum mathematical software products such as Mathematica and Maple. These days, many mathematicians are highly skilled with these tools and use them as part of their day-to-day research work. As a result, we are starting to see a wave of new mathematical results discovered partly or entirely with the aid of computer-based tools."

**SEE ALSO** Abacus (c. 1200), Slide Rule (1621), Babbage Mechanical Computer (1822), Ritty Model I Cash Register (1879), Differential Analyzer (1927), Curta Calculator (1948), and HP-35: First Scientific Pocket Calculator (1972).

*Mathematica provides a general computing environment that organizes numerous algorithmic, visualization, and user-interface capabilities. This sample 3-D graphic is produced by Mathematica and is courtesy of Michael Trott, an expert in symbolic computation and computer graphics.*

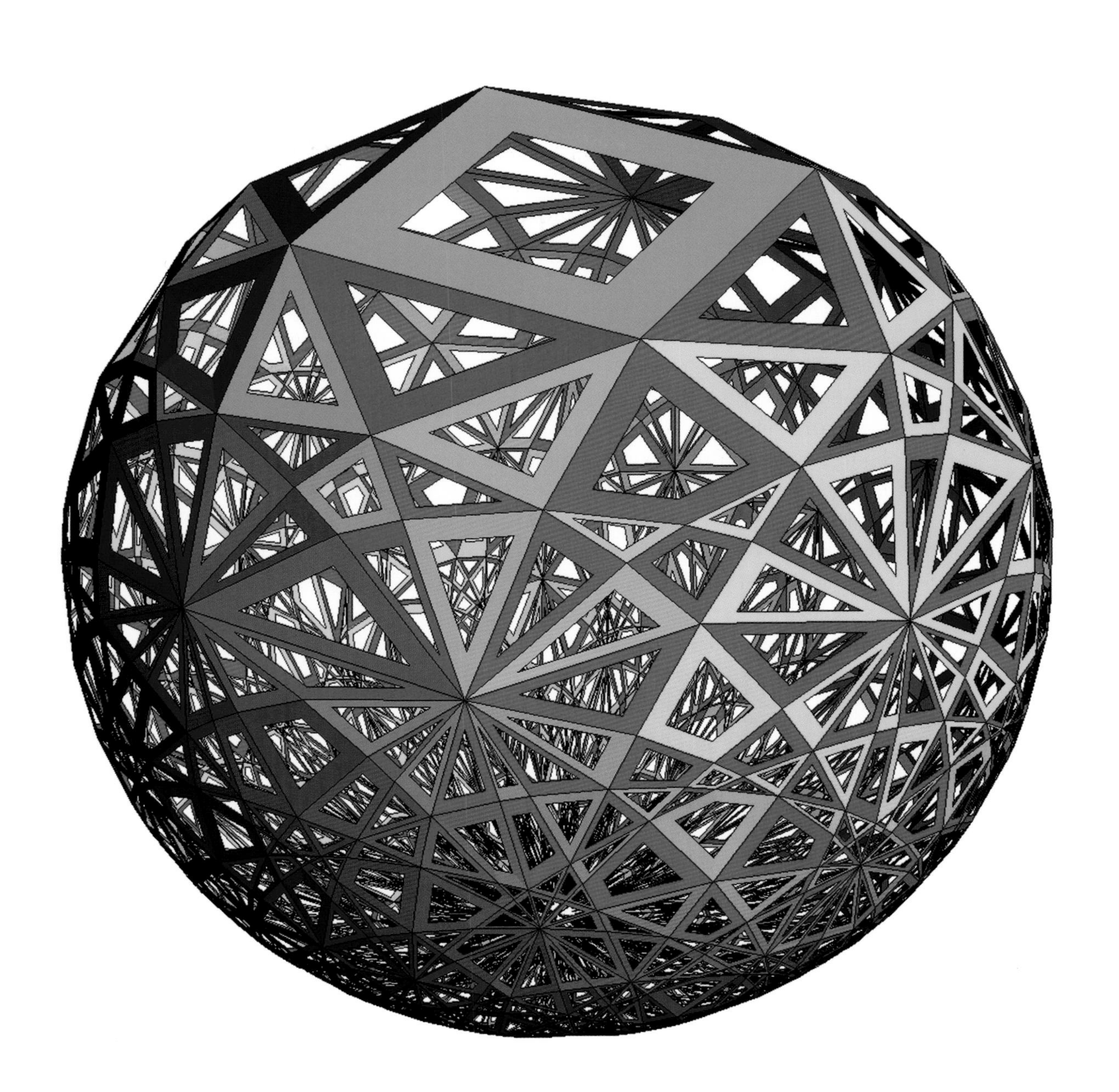

1988

# Murphy's Law and Knots

**De Witt L. Sumners** (b. 1941), **Stuart G. Whittington** (b. 1942)

Since ancient days, frustrated sailors and weavers have observed the apparent tendency of ropes and strings to tangle and knot—a manifestation of Murphy's famous law that states, if something can go wrong, it will go wrong. However, until recently, no rigorous theory existed that explained this maddening phenomenon. Consider just one practical ramification: A single knot in a mountain climber's rope can reduce the amount of stress the rope can withstand without breaking by as much as 50 percent.

In 1988, mathematician De Witt L. Sumners and chemist Stuart G. Whittington clearly elucidated the phenomena by modeling ropes and other string-like objects, such as chemical polymer chains, as self-avoiding random walks. Imagine an ant resting at a point in a cubic grid. It can randomly walk in any of six directions as it traces a path through the lattice (meaning backward or forward in any of three directions). In order to mimic physical objects that cannot occupy the same space at once, the ant's walk is self-avoiding so that no point in space is visited more than once. Based on their research, Sumners and Whittington proved a general result: Nearly all sufficiently long self-avoiding random walks contain a knot.

Not only does their research help to explain why the longer a garden hose in your garage is, the more likely it is to be knotted—or why a knotted rope found at a crime scene may have no forensic significance—this work has vast implications for our understanding of the tangling of DNA and protein backbones. Long ago, protein folding experts believed that forming a knot was beyond the ability of a protein, but today a number of such knots have been found. Some of the knots may stabilize the protein structure. If scientists can accurately predict protein structure, they may be able to better understand diseases and develop new drugs that rely on a 3-D shape of the protein.

**SEE ALSO** Knots (c. 100,000 B.C.), Borromean Rings (834), Lost in Hyperspace (1921), Perko Knots (1974), and Jones Polynomial (1984).

LEFT: *Tangled fishing nets.* RIGHT: *A single knot in a mountain climber's rope can severely reduce the breaking strength of the rope.*

1989

# Butterfly Curve

**Temple H. Fay** (b. 1940)

Parameterizations are sets of equations that express a set of quantities as functions of a number of independent variables. A curve in the plane is often said to be parameterized if the set of coordinates $(x, y)$ on the curve is represented as functions of a variable $t$. For example, in the usual Cartesian coordinates, we have the standard equation of a circle: $x^2 + y^2 = r^2$, where $r$ is the radius of the circle. We can also define a circle in terms of parametric equations: $x = r \cdot \cos(t)$, $y = r \cdot \sin(t)$, where $0 < t \leq 360$ degrees or $0 < t \leq 2\pi$ radians. To create a graph, computer programmers increment the value for $t$ and connect the resultant $(x, y)$ points.

Mathematicians and computer artists often resort to parametric representations because certain geometric forms are very difficult to describe as a single equation, the way one could for a circle. For example, to draw a *conical helix*, try $x = a \cdot z \cdot \sin(t)$, $y = a \cdot z \cdot \cos(t)$, and $z = t/(2\pi c)$, where $a$ and $c$ are constants. Conical helices are used today in certain kinds of antennas.

Many algebraic and transcendental curves express beauty in their symmetry, leaves and lobes, and asymptotic behavior. Butterfly curves, developed by Temple Fay while at the University of Southern Mississippi, are one such class of beautiful, intricate shapes. The equation for the butterfly curve can be expressed in polar coordinates by $\rho = e^{\cos\theta} - 2\cos(4\theta) + \sin^5(\theta/12)$. This formula describes the trajectory of a point as it traces out the butterfly's body. The variable $\rho$ is the radial distance of the point to the origin. The butterfly curve is significant because of the degree to which it has fascinated both students and mathematicians since 1989 when it was first presented, and it has encouraged students to experiment with variants with longer periods of repetition such as $\rho = e^{\cos\theta} - 2.1\cos(6\theta) + \sin^7(\theta/30)$.

**SEE ALSO** Harmonograph (1857).

*Many algebraic and transcendental curves express beauty in their symmetry, lobes, and asymptotic behaviors. This butterfly curve, developed by Temple Fay, can be expressed in polar coordinates by* $\rho = e^{\cos\theta} - 2\cos(4\theta) + \sin^5(\theta/12)$.

# The On-Line Encyclopedia of Integer Sequences

**Neil James Alexander Sloane** (b. 1939)

The *On-Line Encyclopedia of Integer Sequences* (OEIS) is an extremely large, searchable database of integer sequences used by mathematicians, scientists, and laypeople who are intrigued by number sequences in disciplines ranging from game theory, puzzles, and number theory to chemistry, communications, and physics. The astonishing diversity of the OEIS is exemplified by two sample entries: the number of ways to lace a shoe that has *n* pairs of eyelets, and the winning positions of the ancient board game Tchoukaillon solitaire, as a function of the number of stones. The Web site for the OEIS (www.research.att.com/~njas/sequences/) contains more than 150,000 sequences, making it the largest database of its kind.

Each entry includes the first several terms of the sequence, keywords, mathematical motivations, and literature references. Neil Sloane, British-born American mathematician, started collecting integer sequences in 1963, as a graduate student at Cornell University, and his first incarnation of the OEIS was stored on punched cards—and then in the form of a 1973 book called A *Handbook of Integer Sequences*, containing 2,400 sequences, and a 1995 follow-up with 5,487 sequences. The Web version became available in 1996, and it continues to add about 10,000 new entries a year. If it were published as a book today, it would occupy 750 volumes the size of the 1995 book.

The OEIS is a monumental achievement and is frequently used to identify sequences or to determine the current status of a known sequence. However, its most profound use may be as an aid for suggesting new conjectures. For example, mathematician Ralf Stephan recently formulated more than 100 conjectures in many fields simply through a study of the OEIS number sequences. By comparing sequences with the same leading terms (or sequences related by simple transformations), mathematicians may start to consider new conjectures concerning power series expansions, number theory, combinatorics, nonlinear recurrences, binary representations, and other areas of mathematics.

**SEE ALSO** Thue-Morse Sequence (1906), Collatz Conjecture (1937), Audioactive Sequence (1986), and Bed Sheet Problem (2001).

*The OEIS includes a sequence that characterizes the number of ways to lace a shoe that has* n *pairs of eyelets such that each eyelet has at least one direct connection to the opposite side: 1, 2, 20, 396, 14976, 907200. . . . The path must begin and end at the extreme pair of eyelets.*

1999

# Eternity Puzzle

**Christopher Walter Monckton, 3rd Viscount Monckton of Brenchley** (b. 1952)

The extremely difficult jigsaw puzzle known as the Eternity Puzzle became a craze in 1999 and 2000 and has been subject to serious mathematical and computer analysis. The 209 puzzle pieces, all different, are constructed from equilateral triangles and half triangles with the same total areas as six triangles. The task is to fit the pieces together into one large, almost regular dodecagon (12-sided polygon).

Christopher Monckton, inventor of the puzzle, announced a prize of 1 million pounds when the puzzle was released commercially by Ertl Toys in June, 1999. Monckton's initial computer experiments suggested to him that the puzzle would not be solved for several years or perhaps much longer. Indeed, exhaustively searching *all* possibilities would take so long that the fastest computer would have required many millions of years to find a solution using simple-minded searches.

Perhaps to Monckton's disappointment, two British mathematicians, Alex Selby and Oliver Riordan, revealed a correct tiling on May 15, 2000, which they obtained with the help of computers, and they claimed the prize. Interestingly, they discovered that as the number of pieces in an Eternity-like puzzle increased, the difficulty increased up to about 70 pieces. However, beyond 70 pieces, the number of possible correct solutions begins to increase. The official Eternity Puzzle is thought to have at least $10^{95}$ solutions—far more than the number of atoms in our galaxy. Nevertheless, the puzzle is still fiendishly difficult because far more non-solutions exist.

Because Selby and Riordan realized that many solutions should be possible, they decided to deliberately discard Monckton's clues for his own solution in order to consider possibly easier solutions. In 2007, Monckton released the Eternity II Puzzle with 256 square puzzle pieces, whose colored edges must match when the pieces are fit into a $16 \times 16$ grid. The possible number of configurations is estimated to be $1.115 \times 10^{557}$.

**SEE ALSO** Squaring a Rectangle (1925), Voderberg Tilings (1936), and Penrose Tiles (1973).

*An example single piece of the Eternity Puzzle, shown here in the yellow triangulated polygon. Every piece is made up of triangles and "half triangles."*

1999

# Perfect Magic Tesseract

## **John Robert Hendricks** (1929–2007)

The traditional **Magic Square** contains integers arranged in the form of a square grid so that the numbers in each row, column, and diagonal add up to the same total. If the integers are consecutive numbers from 1 to $N^2$, the square is said to be of Nth order.

In a magic tesseract (a four-dimensional cube), the object contains the numbers 1 through $N^4$ arranged in such a way that the sum of the numbers in each of the $N^3$ rows, $N^3$ columns, $N^3$ pillars, $N^3$ *files* (a term used to imply a fourth spatial direction), and in the 8 major *quadragonals* (that pass through the center and join opposite corners) is a constant sum $S = N(1 + N^4)/2$, where N is the order of the tesseract. A total of 22,272 magic tesseracts exist of order 3.

The term *perfect magic tesseract* implies that a magic sum is achieved not only in the rows, columns, pillars, files, and quadragonals, but also in all the diagonals and *triagonals* (space diagonals of the cubes of the tesseract). A perfect magic tesseract requires all cubes to be perfect, and all squares must be perfect (that is, *pandiagonal* so that all of the broken diagonals of the square add up to the magic constant).

Canadian researcher John Hendricks was one of the world's foremost experts in higher-dimensional magic objects, and he proved that a perfect magic tesseract cannot be achieved with any orders below 16 and that a perfect magic tesseract of order 16 exists. This perfect magic tesseract of order 16 contains the numbers 1, 2, 3,…65,536 and has the magic sum of 534,296. In 1999, he and I computed the first-known perfect 16th-order magic tesseract. We can summarize what is known today: The smallest perfect tesseract is of order 16, the smallest perfect cube is of order 8, and the smallest perfect (pandiagonal) magic square is of order 4.

**SEE ALSO** Magic Squares (2200 B.C.), Franklin Magic Square (1769), and Tesseract (1888).

*An order-16 perfect magic tesseract is difficult to visualize, so we display one of John Hendrick's third-order magic tesseracts, showing a sample row (yellow), column (green), pillar (red), file (light blue), and quadragonal (formed by the three magenta numbers) that sum to 123.*

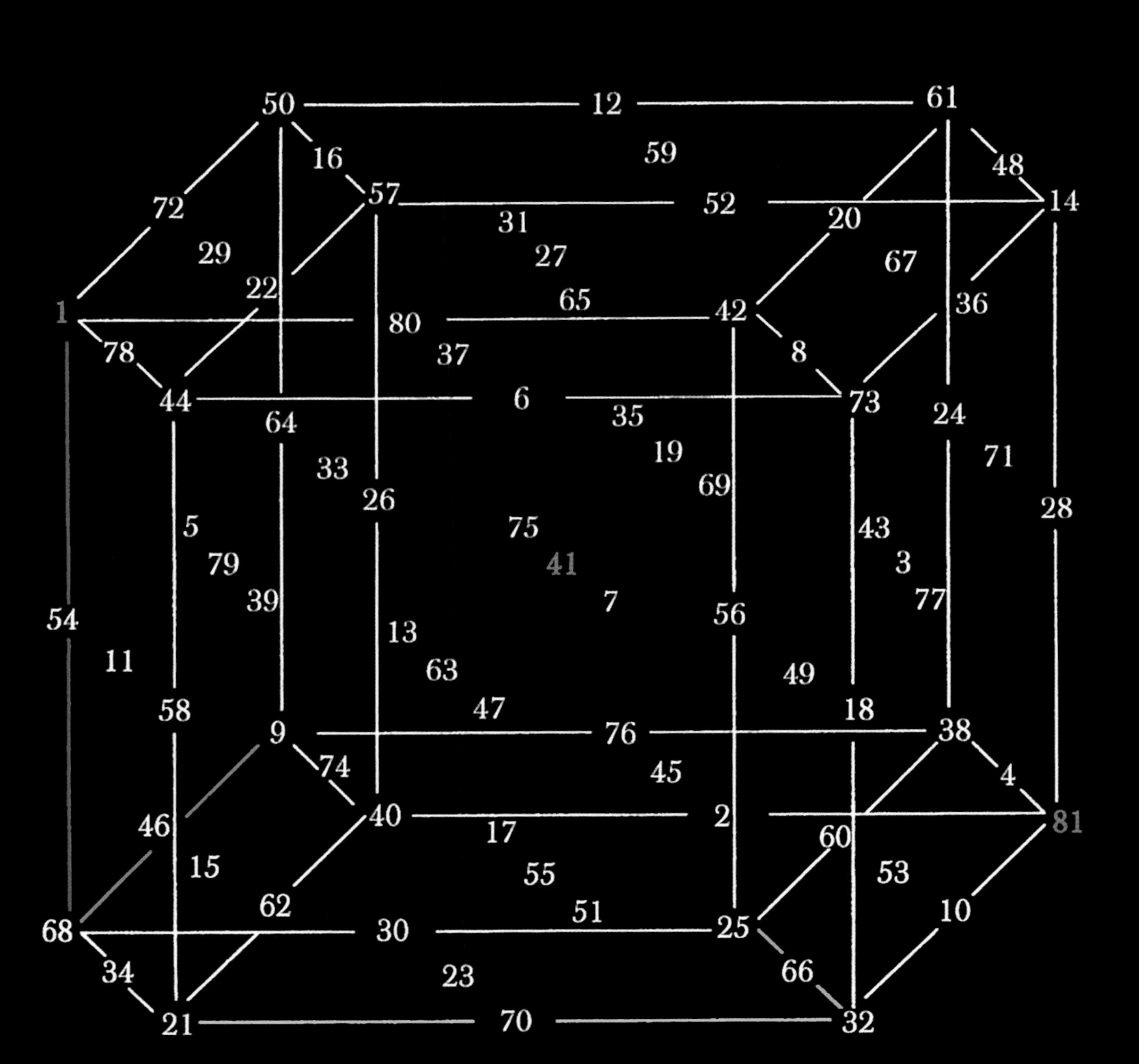

50
12
61
16
59
48
72
57
52
14
31
20
29
27
67
22
65
36
1
80
42
78
37
8
44
6
73
24
64
35
19
71
33
69
26
28
5
75
43
79
41
3
39
7
77
54
56
13
11
63
49
58
47
18
9
76
38
74
45
4
40
2
81
46
17
60
15
53
55
62
51
10
68
30
25
34
23
66
21
70
32

1999

# Parrondo's Paradox

**Juan Manuel Rodríguez Parrondo** (b. 1964)

In the late 1990s, Spanish physicist Juan Parrondo showed how two games guaranteed to make a player lose all his money can be played in alternating sequence to make the player rich. Science writer Sandra Blakeslee writes that Parrondo "discovered what appears to be a new law of nature that may help explain, among other things, how life arose out of a primordial soup, why President Clinton's popularity rose after he was caught in a sex scandal, and why investing in losing stocks can sometimes lead to greater capital gains." The mind-boggling paradox has applications ranging from population dynamics to the assessment of financial risk.

To understand the paradox, imagine you are playing two gambling games that involve biased coins. In game A, each time the coin is tossed, you have a probability $P_1$ of winning that is less than 50 percent, expressed as $P_1 = 0.5 - x$. If you win, you get \$1; otherwise, you lose \$1. In game *B*, you first examine your earnings to see if they are a multiple of 3. If no, you toss another biased coin with the probability of winning $P_2 = (3/4 - x)$. If yes, you toss a third biased coin with the probability of winning a mere $P_3 = (1/10) - x$. In game A or game *B* played separately, for example with $x = 0.005$, you are guaranteed to lose in the long run. However, if you play them alternately (or even if you randomly switch between the games), you'll eventually be rich beyond your wildest dreams! Note that the outcome of game A affects game *B* during this alternation of play.

Parrondo initially devised his paradoxical game in 1996. Biomedical engineer Derek Abbott at the University of Adelaide, Australia, coined the term *Parrondo's paradox*, and, in 1999, he published his work that verified Parrondo's counterintuitive result.

**SEE ALSO** Zeno's Paradoxes (c. 445 B.C.), Aristotle's Wheel Paradox (c. 320 B.C.), Law of Large Numbers (1713), St. Petersburg Paradox (1738), Barber Paradox (1901), Banach-Tarski Paradox (1924), Hilbert's Grand Hotel (1925), Birthday Paradox (1939), Coastline Paradox (c. 1950), and Newcomb's Paradox (1960).

*Physicist Juan Parrondo was inspired by ratchets such as this, the behavior of which can lead to counterintuitive behavior especially when considered for use in microscopic devices. Parrondo extended insights relating to physical devices to games.*

1999

# Solving of the Holyhedron

**John Horton Conway** (b. 1937), **Jade P. Vinson** (b. 1976)

Consider a traditional polyhedral solid constructed from a collection of polygons joined at their edges. A *holyhedron* is a polyhedron with each face containing at least one polygon-shaped hole. The boundaries of the holes share no point with each other or the boundary of the faces. For example, consider a solid cube with its 6 faces. Next, imagine thrusting a pentagonal rod through 1 face, all the way through the cube to the other side to produce (for example) a pentagonal tunnel. At this point in the construction, we have created an object with 11 faces (the 6 original faces and the 5 new faces of the pentagonal tunnel), and only 2 of those 11 faces have holes punched in them. Each time we punch a hole, we are also creating more faces. The immense challenge to finding a holyhedron is to make the holes such that they eventually punch through more than one face to reduce the number of faces that have no holes.

The holyhedron concept was first introduced by Princeton mathematician John H. Conway in the 1990s, who offered a prize of $10,000 to anyone who could find such an object. He also stipulated that his cash reward would be divided by the number of faces in such an object. In 1997, David W. Wilson coined the word *holyhedron* to indicate a hole-filled polyhedron.

Finally, in 1999, American mathematician Jade P. Vinson discovered the world's first holyhedron specimen with a total of 78,585,627 faces (which, obviously, made Vinson's monetary award rather small)! In 2003, computer graphics specialist Don Hatch discovered a holyhedron with 492 faces. The search continues.

**SEE ALSO** Platonic Solids (350 B.C.), Archimedean Semi-Regular Polyhedra (c. 240 B.C.), Euler's Formula for Polyhedra (1751), Prince Rupert's Problem (1816), Icosian Game (1857), Pick's Theorem (1899), Geodesic Dome (1922), Császár Polyhedron (1949), Szilassi Polyhedron (1977), and Spidrons (1979).

LEFT: *An example of thrusting a triangular rod through a cube.* RIGHT: *The holes and tunnels within an Antarctic ice cave are reminiscent of the gorgeous, porous structures of a holyhedron. Of course, a holyhedron must have tunnels bounded by polygons, and each of the holyhedron's flat tunnel walls must contain at least one polygon-shaped hole.*

2001

# Bed Sheet Problem

**Britney Gallivan** (b. 1985)

You have insomnia one night and decide to remove your bed sheet, which is only about 0.4 millimeter thick. You fold it up once, and it becomes 0.8 mm thick. How many times do you fold it if you want to make the bed sheet thickness equal to the distance between the Earth and the moon? The remarkable answer is that if you fold your sheet only 40 times, then you will sleep on the moon! In another version of the problem, you are handed a sheet of paper with a typical thickness of 0.1 millimeter. If you could fold it 51 times, the stack would reach further than the sun!

Alas, it is not physically possible to create many folds in physical objects like these. The prevailing wisdom throughout much of the 1900s was that a sheet of real paper could not be folded in half more than 7 or 8 times, even if the starting paper sheet was large. However, in 2002, high school student Britney Gallivan shocked the world by folding a sheet in half an unexpected 12 times.

In 2001, Gallivan determined equations that characterize the limit on the number of times we can fold a sheet of paper of a given size in a single direction. For the case of a sheet with thickness $t$, we can estimate the initial minimal length $L$ of a paper that is required in order to achieve $n$ folds: $L = [(\pi t)/6] \times (2^n + 4) \times (2^n - 1)$. We may study the behavior of $(2^n + 4) \times (2^n - 1)$. Starting with $n = 0$, we have the integer sequence 0, 1, 4, 14, 50, 186, 714, 2,794, 11,050, 43,946, 175,274, 700,074.…This means that for the eleventh act of folding the paper in half, 700,074 times as much material has been lost to folding, at the curved edges along the folds, as was lost on the first fold.

**SEE ALSO** Zeno's Paradoxes (c. 445 B.C.) and The On-Line Encyclopedia of Integer Sequences (1996).

*In 2001, Britney Gallivan determined equations that characterize the limit on the number of times we can fold a bed sheet or piece of paper of a given size in a single direction.*

2002

# Solving the Game of Awari

**John W. Romein** (b. 1970) **and Henri E. Bal** (b. 1958)

Awari is a 3,500-year-old African board game. Today, Awari is the national game of Ghana, and it is played throughout West Africa and the Caribbean. Classified as a count-and-capture game, Awari is a member of a set of strategy games called Mancala games.

The Awari board consists of two rows of six cup-like hollows, with four markers (beans, seeds, or pebbles) in each hollow. Six cups belong to each player, who takes turns moving the seeds. On a turn, a player chooses one of his six cups, withdraws all seeds from that cup, and drops one seed in each cup counterclockwise from this cup. The second player then takes the seeds from one of the six cups on his side and does the same. When a player drops his last seed into a cup on the opponent's side containing only one or two seeds (making a total of two or three seeds), that player removes all the seeds from this cup, removing them from the game. The same player also takes any seeds in cups immediately before the emptied cup if they now also total two or three. Players take seeds only from their opponent's side of the board. The game ends when one player has no seeds left in the cups on his side. Whoever captures the majority of seeds wins.

Awari has been of immense attraction to researchers in the field of artificial intelligence, in which algorithms are sometimes developed to solve puzzles or play games, but until 2002, no one knew if the game was like Tic Tac Toe in which perfect players always ended a game in a draw. Finally, computer scientists John W. Romein and Henri E. Bal of the Free University in Amsterdam wrote a computer program that calculated the outcome for all 889,063,398,406 positions that can occur in the game, and proved that Awari must end in a draw for perfect players. The massive computation required about 51 hours on a computer cluster with 144 processors.

**SEE ALSO** Tic Tac Toe (c. 1300 B.C.), Go (548 B.C.), Donald Knuth and Mastermind (1970), Eternity Puzzle (1999), and Checkers Is Solved (2007).

*Awari has been of immense attraction to researchers in the field of artificial intelligence. In 2002, computer scientists calculated the outcome for all 889,063,398,406 positions that can occur in the game, and proved that Awari must end in a draw for perfect players.*

2002

# Tetris Is NP-Complete

**Erik D. Demaine** (b. 1981), **Susan Hohenberger** (b. 1978), **and David Liben-Nowell** (b. 1977)

Tetris is a very popular falling-blocks puzzle video game, invented in 1985 by Russian computer engineer Alexey Pajitnov. In 2002, American computer scientists quantified the difficulty of the game and showed that it has similarities with the hardest problems in mathematics that do not have simple solutions but instead require exhaustive analysis to find optimal solutions.

In Tetris, playing pieces start at the top of the game board and move downward. As a piece descends, a player can rotate the piece or slide it sideways. The pieces are shapes called tetrominoes, consisting of four squares stuck together into a group that has a shape like the letter T or other simple pattern. When one piece reaches a resting spot at the bottom, the next piece at the top falls. Whenever a row at the bottom is filled with no gaps, that row is removed, and all higher rows drop down one row. The game ends when a new piece cannot fall because it is blocked. The player's goal is to make the game last as long as possible in order to increase his score.

In 2002, Erik D. Demaine, Susan Hohenberger, and David Liben-Nowell researched a generalized version of the game that used a game board grid that could be any number of squares wide and high. The team discovered that if they tried to maximize the number of rows cleared while playing a given sequence of pieces, then the game was NP-complete. ("NP" stands for "nondeterministically polynomial.") Although this class of problems can be checked to determine if a solution is correct, the solution may actually require an outrageously long time to find. The classic example of an NP-complete problem is the traveling salesman problem, which involves the extremely challenging task of determining the most efficient route for a salesman or delivery man who must visit many different towns. These kinds of problems are difficult because no shortcut or smart algorithm exists for quick solutions.

**SEE ALSO** Tic Tac Toe (c. 1300 B.C.), Go (548 B.C.), Eternity Puzzle (1999), Solving the Game of Awari (2002), and Checkers Is Solved (2007).

*In 2002, computer scientists quantified the difficulty of Tetris and showed that it has similarities with the hardest problems in mathematics that do not have simple solutions but instead require exhaustive analysis to find optimal solutions.*

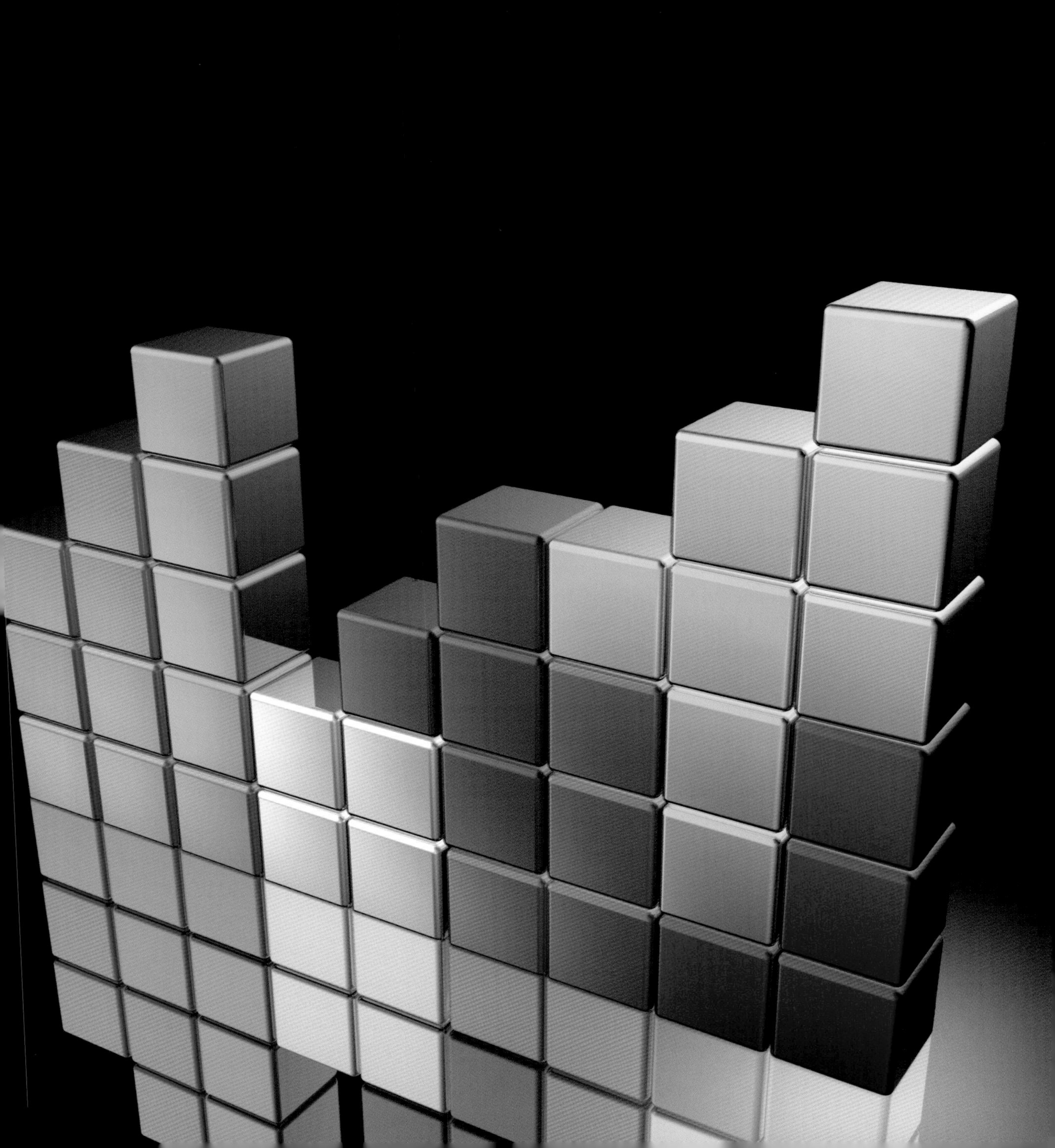

2005

# *NUMB3RS*

## Nicolas Falacci and Cheryl Heuton

*NUMB3RS* is an American television show created by husband-and-wife writing team Nicolas Falacci and Cheryl Heuton. This crime drama concerns a brilliant mathematician, Charlie Epps, who helps the FBI solve crimes using his genius ability in mathematics.

Although it may seem inappropriate to place a TV show in a book along with such famous concepts as Fermat's Last Theorem or the work of Euclid, *NUMB3RS* is significant because it was the first very popular weekly drama that revolved around mathematics, had a team of mathematician advisors, and also received acclaim from mathematicians. The equations seen in the show are real and relevant to the episodes. The mathematical content of the show ranges from cryptanalysis, probability theory, and Fourier analysis to Bayesian analysis and basic geometry.

*NUMB3RS* has also proven to be significant because it has created many learning opportunities for students. For example, mathematics teachers have employed the lessons of *NUMB3RS* in their classrooms, and in 2007, the show and its creators received a National Science Board group Public Service Award for contributions toward increasing scientific and mathematical literacy. Famous mathematicians mentioned in *NUMB3RS* include Archimedes, Paul Erdös, Pierre-Simon Laplace, John von Neumann, Bernhard Riemann, and Stephen Wolfram—the kinds of people discussed throughout this book! Kendrick Frazier writes, "Science, reason, and rational thinking play such a prominent role in the stories that the American Association for the Advancement of Science hosted an entire afternoon symposium at its 2006 annual meeting on the program's role in changing the public's perception of mathematics."

Episodes begin with a spoken tribute about the importance of mathematics: "We all use math everywhere. To tell time, to predict the weather, to handle money....Math is more than formulas and equations. Math is more than numbers. It is logic. It is rationality. It is using your mind to solve the biggest mysteries we know."

**SEE ALSO** Martin Gardner's Mathematical Recreations (1957) and Erdös and Extreme Collaboration (1971).

*Scene from* NUMB3RS, *an American television show featuring a brilliant mathematician who helps the FBI solve crimes using his genius ability in mathematics. The show was the first very popular weekly drama that revolved around mathematics and had a team of mathematician advisors.*

2007

# Checkers Is Solved

**Jonathan Schaeffer** (b. 1957)

In 2007, computer scientist Jonathan Schaeffer and colleagues used computers to finally prove that checkers, when played perfectly, is a no-win game. This means that checkers resembles **Tic Tac Toe**—a game that also can't be won if both players make no wrong moves. Both games end in a draw.

Schaeffer's proof was executed by hundreds of computers over 18 years, making checkers the most complex game ever solved. This also means that it is theoretically possible to build a machine that will never lose to a human.

Checkers, which makes use of an 8 × 8 board, was hugely popular in Europe in the sixteenth century, and early variations of the game have been discovered in the ruins of the ancient city of Ur (c. 3000 B.C.) in modern-day Iraq. The checkers pieces are often in the form of black and red disks that slide diagonally. Players take turns and capture each other's pieces by hopping over them. Of course, given that there are roughly $5 \times 10^{20}$ possible positions, proving that checkers is a guaranteed draw is far harder than proving that Tic Tac Toe can't be won.

The checkers research team considered 39,000 billion arrangements with 10 or fewer pieces on the board and then determined if red or black would win. The team also used a specialized search algorithm to study the start of the game and to see how these moves "funneled" into the 10-checker configurations. The solving of checkers represented a major benchmark in the field of artificial intelligence, which often involves complex problem-solving strategies for computers.

In 1994, Schaeffer's program called Chinook played the world champion, Marion Tinsley, to a series of draws. Tinsley died of cancer eight months later, and some chided Schaeffer for accelerating the death due to the stress Chinook placed on Tinsley!

**SEE ALSO** Tic Tac Toe (c. 1300 B.C.), Go (548 B.C.), Sprouts (1967), and Solving the Game of Awari (2002).

*French artist Louis-Léopold Boilly (1761–1845) painted this scene of a family game of checkers around the year 1803. In 2007, computer scientists proved that checkers, when played perfectly, is a no-win game.*

2007

# The Quest for Lie Group $E_8$

**Marius Sophus Lie** (1842–1899), **Wilhelm Karl Joseph Killing** (1847–1923)

For more than a century, mathematicians have sought to understand a vast, 248-dimensional entity, known to them only as $E_8$. Finally, in 2007, an international team of mathematicians and computer scientists made use of a supercomputer to tame the intricate beast.

As background, consider the *Mysterium Cosmographicum* (*The Sacred Mystery of the Cosmos*) of Johannes Kepler (1571–1630), who was so enthralled with symmetry that he suggested the entire solar system and planetary orbits could be modeled by **Platonic Solids**, such as the cube and dodecahedron, nestled in each other forming layers as if in a gigantic crystalline onion. These kinds of Keplerian symmetries were limited in scope and number; however, symmetries that Kepler could have hardly imagined may indeed rule the universe.

In the late nineteenth century, the Norwegian mathematician Sophus Lie (pronounced "Lee") studied objects with smooth rotational symmetries, like the sphere or doughnut in our ordinary three-dimensional space. In three and higher dimensions, these kinds of symmetries are expressed by Lie groups. The German mathematician Wilhelm Killing suggested the existence of the $E_8$ group in 1887. Simpler Lie groups control the shape of electron orbital and symmetries of subatomic quarks. Larger groups, like $E_8$, may someday hold the key to a unified theory of physics and help scientists understand string theory and gravity.

Fokko du Cloux, a Dutch mathematician and computer scientist who was one of the $E_8$ team members, wrote the software for the supercomputer and pondered the ramifications of $E_8$ while he was dying of amyotrophic lateral sclerosis and breathing with a respirator. He died in November 2006, never living to see the end of the quest for $E_8$.

On January 8, 2007, a supercomputer computed the last entry in the table for $E_8$, which describes the symmetries of a 57-dimensional object that can be imagined as rotating in 248 ways without changing its appearance. The work is significant as an advance in mathematical knowledge and in the use of large-scale computing to solve profound mathematical problems.

**SEE ALSO** Platonic Solids (c. 350 B.C.), Group Theory (1832), Wall Paper Groups (1891), Monster Group (1981), and Mathematical Universe Hypothesis (2007).

*Graph of $E_8$. For more than a century, mathematicians have sought to understand this vast, 248-dimensional entity. In 2007, a supercomputer computed the last entry in the table for $E_8$, which describes the symmetries of a 57-dimensional object.*

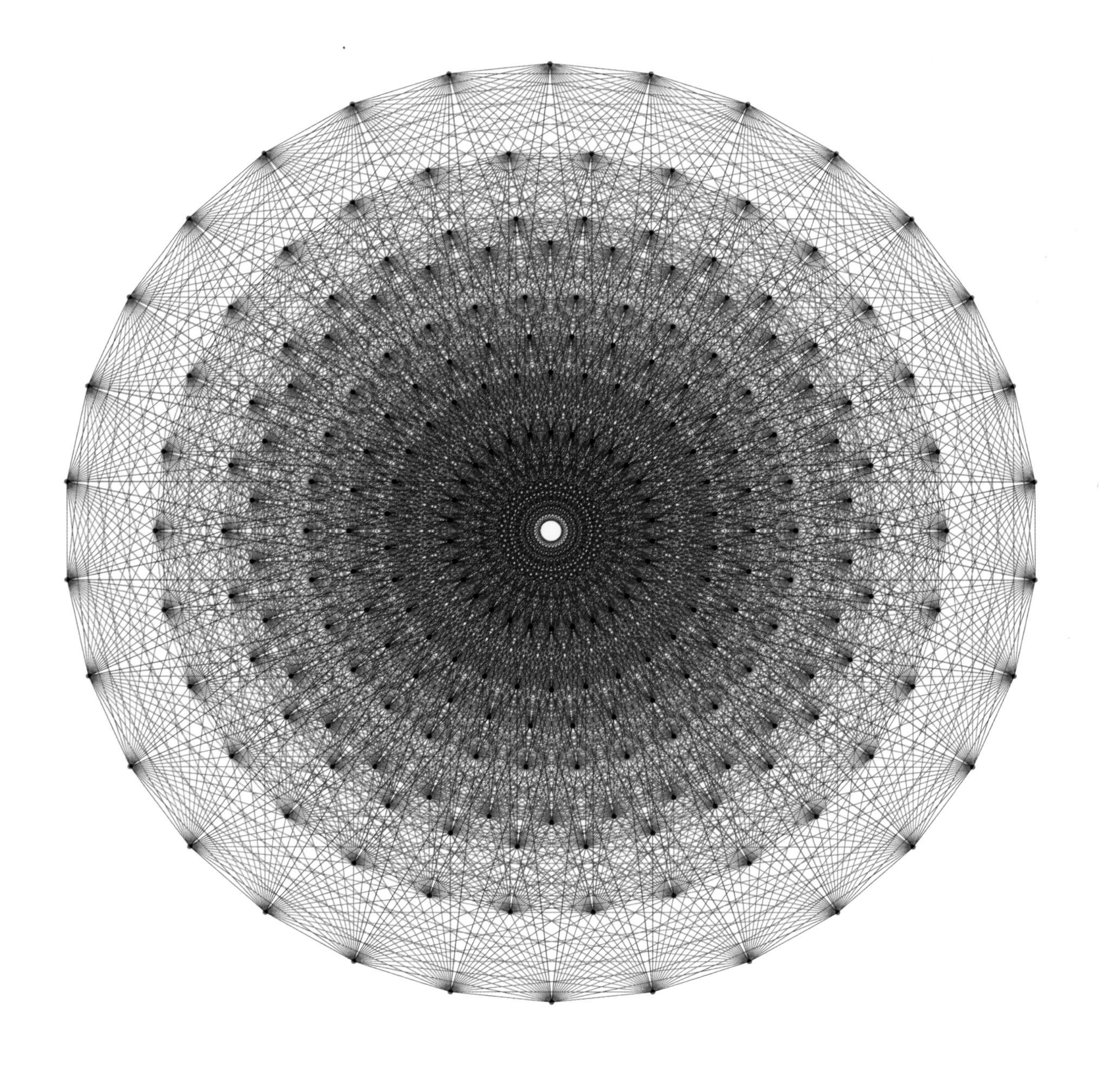

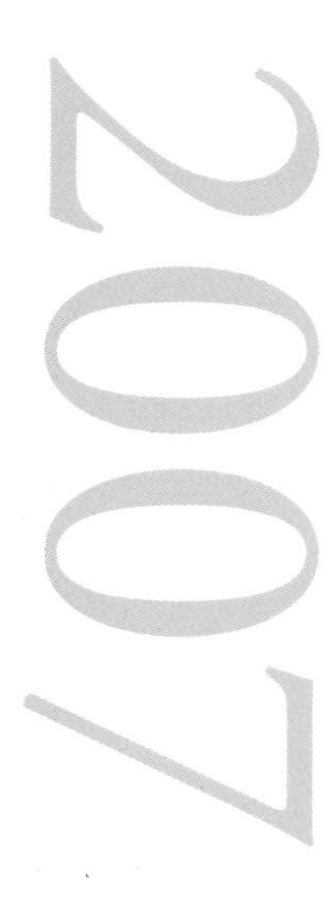


# Mathematical Universe Hypothesis

**Max Tegmark** (b. 1967)

In this book, we have encountered various geometries that have been thought to hold the keys to the universe. Johannes Kepler modeled the solar system with **Platonic Solids** such as the dodecahedron. Large Lie groups, like $E_8$, may someday help us create a unified theory of physics. Even Galileo in the seventeenth century suggested that "nature's great book is written in mathematical symbols." In the 1960s, physicist Eugene Wigner was impressed with the "unreasonable effectiveness of mathematics in the natural sciences."

In 2007, Swedish-American cosmologist Max Tegmark published scientific and popular articles on the Mathematical Universe Hypothesis (MUH) that states that our physical reality is a mathematical structure and that our universe is not just described by mathematics—it *is* mathematics. Tegmark is a professor of physics at the Massachusetts Institute of Technology and scientific director of the Foundational Questions Institute. He notes that when we consider equations like $1 + 1 = 2$, the notations for the numbers are relatively unimportant when compared to the relationships that are being described. He believes that "we don't invent mathematical structures—we discover them, and invent only the notation for describing them."

Tegmark's hypothesis implies that "we all live in a gigantic mathematical object—one that is more elaborate than a dodecahedron, and probably also more complex than objects with intimidating names like Calabi-Yau manifolds, tensor bundles, and Hilbert spaces, which appear in today's most advanced theories. Everything in our world is purely mathematical—including you." If this idea seems counterintuitive, this shouldn't be surprising, because many modern theories, like quantum theory and relativity, can defy intuition. As mathematician Ronald Graham once said, "Our brains have evolved to get us out of the rain, find where the berries are, and keep us from getting killed. Our brains did not evolve to help us grasp really large numbers or to look at things in a hundred thousand dimensions."

**SEE ALSO** Cellular Automata (1952) and The Quest for Lie Group $E_8$ (2007).

*According to the Mathematical Universe Hypothesis, our physical reality is a mathematical structure. Our universe is not just described by mathematics—it is mathematics.*

# Notes and Further Reading

I've compiled the following list that identifies some of the material I used to research and write this book. As many readers are aware, Internet Web sites come and go. Sometimes they change addresses or completely disappear. The Web site addresses listed here provided valuable background information when this book was written.

If I have overlooked an interesting or pivotal moment in mathematics that you feel has never been fully appreciated, please let me know about it. Just visit my web site *pickover.com*, and send me an e-mail explaining the idea and how you feel it influenced the mathematical world. Perhaps future editions of the book will include mathematical marvels such as the Gömböc, the *Suàn Shù Shū*, the Frobenius postage stamp problem, tangrams, and additional ideas of Sophie Germain.

Due to space constraints, many references were removed from the printed edition of this book. However, additional references and notes, as well as fuller citations, can be found at *pickover.com/mathbook.html*.

*Special common publisher abbreviations below*: *AKP*: A. K. Peters, Ltd., Wellesley, MA; *AMS*: American Mathematical Society, Providence, RI; *Dover*: Dover Publications, NY; *CUP*: Cambridge University Press, NY; *Freeman*: W. H. Freeman, NY; *HUP*: Harvard University Press, Cambridge, MA; *MAA*: The Mathematical Association of America, Washington, D.C.; *MIT*: MIT Press, Cambridge, Massachusetts; *Norton*: W. W. Norton & Company, NY; *OUP*: Oxford University Press, NY; *PUP*: Princeton University Press, Princeton, NJ; *RP*: Running Press, Philadelphia, PA; *S&S*: Simon & Schuster, NY; *TMP*: Thunder's Mouth Press, NY; *UBM*: The Universal Book of Mathematics; *UCP*: University of Chicago Press, Chicago, IL; *Wiley*: John Wiley & Sons, Hoboken, NJ; *W&N*: Weidenfeld & Nicholson, London; WS: World Scientific, River Edge, NJ.


**General Reading**

Anderson, M., Victor K., Wilson, R. *Sherlock Holmes in Babylon and Other Tales of Mathematical History*, MAA, 2004.
Boyer, C., Merzbach, U., *A History of Mathematics*, Wiley, 1991.
Darling, D., *The Universal Book of Mathematics*, Wiley, 2004.
Dunham, W., *Journey through Genius*, NY: Penguin, 1991.
Gardner, M., *Martin Gardner's Mathematical Games* (CD-ROM), MAA, 2005.
Gullberg, J., *Mathematics*, Norton, 1997.
Hawking, S., *God Created the Integers*, RP, 2005.
Hodgkin, L., *A History of Mathematics*, OUP, 2005.
O'Connor, J., Robertson, E., "MacTutor History of Math. Archive," *tinyurl.com/5ec5wq*.
Weisstein, E., "MathWorld Wolfram web resource," *mathworld.wolfram.com*.
Wikipedia Encyclopedia, *www.wikipedia.org*.

**Pickover Books**

I have frequently made use of my own books for background information for different entries; however, in order to save space, I do not usually relist them for any of the entries that follow.

Pickover, C., *Keys to Infinity*, Wiley, 1995.
Pickover, C., *Surfing through Hyperspace*, OUP, 1999.
Pickover, C., *Computers, Pattern, Chaos, and Beauty*, Dover, 2001.
Pickover, C., *Wonders of Numbers*, OUP, 2001.
Pickover, C., *The Zen of Magic Squares, Circles, and Stars*, PUP, 2001.
Pickover, C., *The Paradox of God*, NY: Palgrave, 2001.
Pickover, C., *Calculus and Pizza*, Wiley, 2003.
Pickover, C., *A Passion for Mathematics*, Wiley, 2005.
Pickover, C., *The Möbius Strip*, TMP, 2006.
Pickover, C., *From Archimedes to Hawking*, OUP, 2008.
Pickover, C., *The Loom of God*, NY: Sterling, 2009.

**Introduction**

Devlin, K., *tinyurl.com/6kvje4* & *tinyurl.com/5k9wry*.
Dörrie, H., *100 Great Problems of Elementary Mathematics*, Dover, 1965.
Ifrah, G., *The Universal History of Numbers*, Wiley, 1999.
Kaku, M., *Hyperspace*, NY: Anchor, 1995.
Kammerer, P., *Das Gesetz der Serie*, Stuttgart: Deutsche Verlags-Anstalt, 1919.
Klarreich, E., *Sci. News*, **165**:266;2004.
Kruglinski, S., *tinyurl.com/23rosl*.

**Ant Odometer, c. 150 Million B.C.**

Devlin, K., *tinyurl.com/64twpu*.
Wittlinger, M., Wehner, R., Wolf, H., *Science*, **312**:1965;2006.

**Primates Count, c. 30 Million B.C.**

Beran, M., *Animal Cognit.* 7:86;2004.
Kalmus, H., *Nature* **202**:1156;1964.
Matsuzawa, T., *Nature* **315**:57;1985.

**Cicada-Generated Prime Numbers, c. 1 Million B.C.**

Campos, P. et al., *Phys. Review Lett.* **93**:098107-1;2004.
Goles, E., Schulz, O., Markus, M. *Nonlinear Phenom. in Complex Sys.* **3**:208;2000.
Hayes, B., *Am. Scient.* **92**:401;2004.
Peterson, I., *tinyurl.com/66h3hd*.

**Knots, c. 100,000 B.C.**

Bouzouggar, A. et al., *Proc. Natl. Acad. Sci.* **104**:9964;2007.
Meehan, B., *The Book of Kells*, London: Thames & Hudson, 1994.
Sossinsky, A., *Knots*, HUP, 2002.

**Ishango Bone, c. 18,000 B.C.**

Bogoshi, J., Naidoo, K., Webb, J., *Math. Gazette*, **71**:294;1987.
Teresi, D., *Lost Discoveries*, S&S, 2002.

**Quipu, c. 3000 B.C.**

Ascher, M., Ascher, R., *Mathematics of the Incas*, Dover, 1997.
Mann, C., *Science* **309**:1008;2005.

**Dice, c. 3000 B.C.**
Hayes, B., *Am. Scient.* 89:300;2001.

**Magic Squares, c. 2200 B.C.**
Pickover, C., *The Zen of Magic Squares, Circles, and Stars*, PUP, 2001.

**Plimpton 322, c. 1800 B.C.**
Robson, E., *Am. Math. Monthly* **109**:105;2002.

**Rhind Papyrus, c. 1650 B.C.**
Eves, H., *Great Moments in Mathematics (Before 1650)*, MAA, 1983.
Robins, G., Shute, C., *The Rhind Mathematical Papyrus*, Dover, 1990.

**Tic Tac Toe, c. 1300 B.C.**
C. Zaslavasky notes that a 3,300-year-old temple to the memory of Pharaoh Seti I has a Tic-Tac-Toe-like board carved into it. The actual modern rules of Tic Tac Toe may have been first described by C. Babbage around 1820.

Zaslavasky, C. *Tic Tac Toe and Other Three-In-A-Row Games*, NY: Thomas Crowell, 1982.

**Pythagorean Theorem and Triangles, c. 600 B.C.**
Loomis, E., *Pythagorean Proposition*, Washington, D.C.: Natl. Council of Teachers of Math., 1972.
Maor, E., *The Pythagorean Theorem*, PUP, 2007.

**Go, 548 B.C.**
Frankel, K., *Sci. Am.* **296**:32;2007.

**Pythagoras Founds Mathematical Brotherhood, 530 B.C.**
Gorman, P., *Pythagoras*, London: Routledge Kegan & Paul, 1978.
Russell, B., *A History of Western Philosophy*, S&S, 1945.

**Zeno's Paradoxes, c. 445 B.C.**
McLaughlin, W., *Sci. Am.* **271**:84;1994.

**Quadrature of the Lune, c. 440 B.C.**
Dunham, W., *Journey through Genius*, NY: Penguin, 1991.

**Platonic Solids, 350 B.C.**
Platonic solids are convex polyhedrons. A polyhedron is *convex* if for every pair of points that belong to the shape, the shape contains the whole straight line segment connecting the two points. Some astrophysicists have suggested that our entire universe may be in the form of a dodecahedron.

**Aristotle's *Organon*, c. 350 B.C.**
SparkNotes, *tinyurl.com/5qhble*.

**Euclid's Elements, 300 B.C.**
Boyer, C., Merzbach, U., *A History of Mathematics*, Wiley, 1991.

**Archimedes: Sand, Cattle & Stomachion, c. 250 B.C.**
Dörrie cites four scholars who do not believe that the version of the cattle problem that yields this huge solution is due to Archimedes, but he also cites four authors who believe that the problem *should* be attributed to Archimedes.

Dörrie, H., *100 Great Problems of Elementary Mathematics*, Dover, 1965.
Williams, H., German, R., Zarnke, C., *Math. Comput.* **19**:671;1965.

**$\pi$, c. 250 B.C.**
*A New Introduction to Mathematics* (1706) by W. Jones is the first text in which the Greek letter $\pi$ was used for this famous constant. The symbol $\pi$ later became popular after L. Euler started to use it in 1737.

**Archimedean Semi-Regular Polyhedra, c. 240 B.C.**
Semi-regular polyhedra include the 13 Archimedean solids, as well as prisms and antiprisms, if all of their faces are regular polygons.

**Archimedes' Spiral, 225 B.C.**
Gardner, M., *The Unexpected Hanging and Other Mathematical Diversions*, UCP, 1991.

**Ptolemy's Almagest, c. 150**
Grasshoff, G., *The History of Ptolemy's Star Catalogue*, NY: Springer, 1990.
Gullberg, J., *Mathematics*, Norton, 1997.

**Diophantus's Arithmetica, 250**
H. Eves writes, "How and when the new numeral symbols first entered Europe is not settled. They may have been introduced into Spain by the Arabs, who invaded the peninsula in A.D. 711.... The completed system was more widely disseminated by a twelfth-century Latin translation of al-Khwarizmi's treatise..."

Eves, H., *An Introduction to the History of Mathematics*, Boston, MA: Brooks Cole, 1990.
Swift, J., *Amer. Math. Monthly* **63**:163;1956.

**Pappus's Hexagon Theorem, c. 340**
Dehn, M., *Am. Math. Monthly* **50**:357;1943.
Heath, T., *A History of Greek Mathematics*, Oxford: Clarendon, 1921.

**Bakhshali Manuscript, c. 350**
The date of the manuscript is still debated. Many of the earlier scholars have dated it to around A.D. 400. G. Ifrah writes that "the Bakhshali Manuscript cannot have been written earlier than the ninth century [however] it seems likely that the manuscript in present-day form constitutes the commentary or the copy of an anterior mathematical work."

Ifrah, G., *The Universal History of Numbers*, Wiley, 1999.
Teresi, D., *Lost Discoveries*, S&S, 2002.

**Zero, c. 650**
Arsham, H., *tinyurl.com/56zmcv*.

**Alcuin's *Propositiones ad Acuendos Juvenes*, c. 800**
Atkinson, L., *College Math. J.* **36**:354;2005.
Peterson, I., *tinyurl.com/5dyyes*.

**Borromean Rings, 834**
Cromwell, P. et al. *Math. Intelligencer* **20**:53;1998.
Freedman, M., Skora, R., *J. Differential Geom.* **25**:75;1987.
Lindström, B., Zetterström, H., *Am. Math. Monthly* **98**:340;1991.

**Thabit Formula for Amicable Numbers, c. 850**
Gardner, M., *Mathematical Magic Show*, MAA, 1989.

***Chapters in Indian Mathematics*, c. 953**
Morelon, R., *Encyclopedia of the History of Arabic Science*, London: Routledge, 1996.
Saidan, A. S., *Isis*, **57**:475;1966.
Teresi, D., *Lost Discoveries*, S&S, 2002.

**Omar Khayyam's *Treatise*, 1070**
Other individuals also worked on the binomial theorem, including Chinese mathematician Yang Hui (c. 1238–1298), Indian mathematician Pingala, who appears to have lived sometime during the third century B.C., and I. Newton who generalized the formula to other exponents.

**Al-Samawal's *The Dazzling*, c. 1150**
O'Connor, J., Robertson, E., *tinyurl.com/5ctxvh*.
Perlmann, M., *Proc. Am. Acad. Jew. Res.* **32**:15;1964.

**Abacus, c. 1200**
The word *abacus* may come from the Greek *abax* ("calculating table") and/or from *abaq*, the Hebrew word for dust.

Ewalt, D., *tinyurl.com/5psj89*.
Ifrah, G., *The Universal History of Computing*, Wiley, 2002.

**Fibonacci's *Liber Abaci*, 1202**
Today, many authors begin the Fibonacci sequence with a 0, as in 0, 1, 1, 2, 3.... Note that the number of columns in mammalian microtubule is typically a Fibonacci number.

Boyer, C., Merzbach, U., *A History of Mathematics*, Wiley, 1991.

**Wheat on a Chessboard, 1256**
Gullberg, J., *Mathematics*, Norton, 1997.

**Harmonic Series Diverges, c. 1350**
Dunham, W., *College Math. J.* 18:18;1987.

***Treviso Arithmetic*, 1478**
Peterson, I., *tinyurl.com/6a9ngu*.
Smith, D., *Isis* 6:311;1924.
Swetz, F., *Capitalism and Arithmetic*, Chicago: Open Court, 1986.
Swetz, F., *Sci. & Educat.* 1:365;1992.

**Discovery of Series Formula for π, c. 1500**
R. Roy says that although the series formula appears in *Tantrasangraha*, "in the *Aryabhatiyabhasya*, a work on astronomy, Nilakantha attributes the series for sine to mathematician Madhava (1340–1425)."

Roy, R., *Math. Mag.* 63:291;1990.

**Golden Ratio, 1509**
The origin of the term *golden ratio* is disputed but appears to have emerged in the twelfth century. Although the recent history of the golden ratio was triggered by Luca Pacioli's *Divina Proportione* (1509), ancient Greek mathematicians studied the ratio much earlier because it frequently appeared in geometrical studies. Note that the artistic figure for this entry depicts a Fibonacci spiral, based on consecutive terms of a Fibonacci sequence. Because the ratios of consecutive terms in the Fibonacci series approach ø, the two spirals are quite similar in appearance.

***Polygraphiae Libri Sex*, 1518**
Peterson, I., *tinyurl.com/6gvf6k*.

**Cardano's *Ars Magna*, 1545**
Dunham, W., *Journey through Genius*, NY: Penguin, 1991.
Gullberg, J., *Mathematics*, Norton, 1997.
O'Connor, J., Robertson, E., *tinyurl.com/5ue8kh*.

***Sumario Compendioso*, 1556**
Gray, S., Sandifer, C., *Math. Teacher* 94:98;2001.
Smith, D., *Am. Math. Monthly*, 28:10;1921.

**Mercator Projection, 1569**
Short, J., *The World through Maps*, Richmond Hill, Ontario: Firefly Books, 2003.
Thrower, N., *Maps and Civilization*, UCP, 1999.

**Kepler Conjecture, 1611**
Donev, A. et al. *Science*, 303:990;2004.
Hales, T. *Ann. Math.* 162:1065;2005.
Szpiro, G., *Kepler's Conjecture*, Wiley, 2003.

**Logarithms, 1614**
Gibson, G., "Napier and the Invention of Logarithms," in *Handbook of the Napier Tercentenary Celebration*, E. M. Horsburgh, ed., Los Angeles: Tomash Publishers, 1982.
Tallack, P., *The Science Book*, W&N, 2003.

**Slide Rule, 1621**
F. Cajori writes, "It is by no means clear that Delamain [the student] stole the invention from Oughtred; Delamain was probably an independent inventor."

Cajori, F., *William Oughtred*, Chicago: Open Court, 1916.
Oughtred Society, *oughtred.org*.
Stoll, C., *Sci. Am.* 294:81;2006.

**Fermat's Spiral, 1636**
Mahoney, M., *The Mathematical Career of Pierre de Fermat*, PUP, 1994.
Naylor, M., *Math. Mag.* 75:163;2002.

**Fermat's Last Theorem, 1637**
Aczel, A., *Fermat's Last Theorem*, NY: Delta, 1997.
Singh, S., *Fermat's Last Theorem*, NY: Forth Estate, 2002.

**Descartes' *La Géométrie*, 1637**
Boyer, C., Merzbach, U., *A History of Mathematics*, Wiley, 1991.
Grabiner, J., *Math. Mag.* 68:83, 1995.
Gullberg, J., *Mathematics*, Norton, 1997.

**Cardioid, 1637**
Vecchione, G., *Blue Ribbon Science Fair Projects*, NY: Sterling, 2005.

**Logarithmic Spiral, 1638**
Gardner, M., *The Unexpected Hanging and Other Mathematical Diversions*, UCP, 1991.

**Projective Geometry, 1639**
Other prominent people of the fifteenth and early sixteenth centuries who advanced the mathematical theory of perspective were P. Francesca, L. da Vinci, and A. Dürer.

**Torricelli's Trumpet, 1641**
DePillis, J., *777 Mathematical Conversation Starters*, MAA, 2002.

**Pascal's Triangle, 1654**
Gordon, J. et al., *Phys. Rev. Lett.* 56:2280;1986.

**Viviani's Theorem, 1659**
De Villiers, M., *Rethinking Proof with Sketchpad*, Emeryville, CA: Key Curriculum Press, 2003.

**Discovery of Calculus, c. 1665**
In 1671, Newton wrote *On the Methods of Series and Fluxions*. (Fluxion was Newton's term for derivative in the field of calculus.) This work, although circulated in a manuscript to several peers in 1671, did not appear in print until 1736.

**Newton's Method, 1669**
Hamming, R., *Numerical Methods for Scientists and Engineers*, Dover, 1986.

**Tautochrone Problem, 1673**
Darling, D., *UBM*, Wiley, 2004.

**L'Hôpital's *Analysis of the Infinitely Small*, 1696**
Ball, W., *A Short Account of the History of Mathematics*, NY: Dover, 1960
Devlin, K., *tinyurl.com/6rc8ho*.
Kleiner, I., *J. Educat. Studies in Math.* 48:137;2001.

**Euler's Number, *e*, 1727**
The mathematical constant $e$ is special for many reasons, e.g., $f(x) = e^x$ is its own derivative.

Darling, D., *UBM*, Wiley, 2004.
Kasner, E., Newman, J., *Mathematics and the Imagination*, Dover, 2001.
Maor, Eli, *e: The Story of a Number*, PUP, 1998.

**Stirling's Formula, 1730**
The formula $n! \approx ce^{-n}n^{n+1/2}$ was first discovered by Abraham de Moivre (1667–1754), where $c$ is a constant. Stirling showed that $c = \sqrt{2\pi}$.

Ball, K., *Strange Curves, Counting Rabbits, and Other Mathematical Explorations*, PUP, 2003.

**Normal Distribution Curve, 1733**
Galton, F., *Natural Inheritance*, London: Macmillan, 1889.

**Euler-Mascheroni Constant, 1735**
Havil, J., *Gamma*, PUP, 2003.

**Königsberg Bridges, 1736**
Newman, J., *Sci. Am.* 189:66;1953.

**St. Petersburg Paradox, 1738**
Martin, R., *tinyurl.com/2sbcju*.
Bernstein, P., *Against the Gods*, Wiley, 1998.

**Goldbach Conjecture, 1742**
Doxiadis, A., *Uncle Petros and Goldbach's Conjecture*, NY: Bloomsbury, 2000.

**Agnesi's *Instituzioni Analitiche*, 1748**
Mazzotti, M., *The World of Maria Gaetana Agnesi*, Baltimore, MD: Johns Hopkins Univ. Press, 2007.
O'Connor, J., Robertson, E., *tinyurl.com/3h74kl.*
Struik, D., *A Source Book in Mathematics, 1200–1800*, PUP, 1986.
Truesdell, C., *Arch. for Hist. Exact Sci.* **40**:113;1989.

**Euler's Formula for Polyhedra, 1751**
Darling, D., *UBM*, Wiley, 2004.
Wells, D., *Math. Intelligencer* **12**:7;1990 and **10**:30;1988.

**Euler's Polygon Division Problem, 1751**
Dörrie, H., *100 Great Problems of Elementary Mathematics*, Dover, 1965.

**Knight's Tours, 1759**
Dudeney, H., *Amusements in Mathematics*, Dover, 1970.

**Bayes' Theorem, 1761**
Some historians feel that English mathematician Nicholas Saunderson may have discovered Bayes' theorem before Bayes.

**Franklin Magic Square, 1769**
Patel, L., *J. Recr. Math.* 23:175;1991.

**Minimal Surface, 1774**
Darling, D., *UBM*, Wiley, 2004.

**Thirty-Six Officers Problem, 1779**
Bose, R. et al., *Canad. J. Math.* **12**:189;1960.

**Sangaku Geometry, c. 1789**
Boutin, C., *tinyurl.com/6nqdl5.*
Rothman, T., Fukagawa, H., *Sci. Am.* **278**:85;1998.

**Fundamental Theorem of Algebra, 1797**
Dunham, W., *College Math. J.* **22**:282;1991.

**Gauss's *Disquisitiones Arithmeticae*, 1801**
Hawking, S., *God Created the Integers*, RP, 2005.

**Three-Armed Protractor, 1801**
Huddart, W., *Unpathed Waters*, London: Quiller Press, 1989.
U.S. Hydrographic Office, *Bay of Bengal Pilot*, Washington, D.C.: Govt. Printing Office, 1916.

**Fourier Series, 1807**
Jeans, J., *Science and Music*, Dover, 1968.
Ravetz, J., Grattan-Guiness, I., "Fourier," in *Dictionary of Scientific Biography*, Gillispie, C., ed., NY:Scribner, 1970.

**Laplace's *Théorie Analytique des Probabilités*, 1812**
Hawking, S., *God Created the Integers*, RP, 2005.
Richeson, A., *Natl. Math. Mag.* 17:73;1942.

**Prince Rupert's Problem, 1816**
Although John Wallis was the first to write on this problem, I date the entry according to Pieter Nieuwland's actual finding of the maximal cube that will pass through a cube. Some sources appear to suggest that Wallis's writing on the problem did not appear until the second edition of his book, published in 1693.

Guy, R., Nowakowski, R., *Am. Math. Monthly*, **104**:967;1997.

**Bessel Functions, 1817**
Korenev, B., *Bessel Functions and Their Applications*, Boca Raton, FL:CRC Press, 2004

**Babbage Mechanical Computer, 1822**
Norman, J., *From Gutenberg to the Internet*, Novato, CA: Historyofscience.com, 2005.
Swade, D., *Sci. Am.* **268**:86;1993.

**Cauchy's Le Calcul Infinitésimal, 1823**
Hawking, S., *God Created the Integers*, RP, 2005.
Waterhouse, W., *Bull. Amer. Math. Soc.* 7:634;1982.

**Barycentric Calculus, 1827**
Gray, Jeremy, "Möbius's Geometrical Mechanics," in *Möbius and His Band*, Fauvel, J. et al., eds., OUP, 1993.

**Non-Euclidean Geometry, 1829**
Tallack, P., *The Science Book*, W&N, 2003.

**Möbius Function, 1831**
Although August Möbius worked on the sequence in 1831, C. Gauss did initial work on the sequence more than 30 years before Möbius.

**Group Theory, 1832**
It would be incorrect to imply that all of group theory had come to Galois during his final night. I. Peterson writes, "In fact, Galois had been writing papers on the subject since the age of 17, and the new idea of 'group' that he had introduced is found in all of them. Nonetheless, Galois did help create a field that would keep mathematicians busy for hundreds of years, but not in one night!"

Gardner, M., *The Last Recreations*, NY: Springer, 1997.
Peterson, I., *tinyurl.com/6365zo.*

**Quaternions, 1843**
Hamilton published many papers developing the theory of quaternions, including a paper "On Quaternions," published in installments in *Philos. Mag.* between 1844 and 1850.

**Catalan Conjecture, 1844**
Peterson, I., *tinyurl.com/6g5k8n.*

**Boolean Algebra, 1854**
O'Connor, J., Robertson, E., *tinyurl.com/5rv77h.*

**Harmonograph, 1857**
Lissajous curves were actually first studied by N. Bowditch in 1815 and investigated independently by J. Lissajous in 1857. Perhaps Lissajous' patterns should not be considered "harmonographs" given that harmonograph patterns depend on the gradual decay in the oscillations.

**The Möbius Strip, 1858**
Pickover, C., *The Möbius Strip*, TMP, 2006.

**Holditch's Theorem, 1858**
Cooker, M., *Math. Gaz.* 82:183;1998.

**Riemann Hypothesis, 1859**
Derbyshire, J., *Prime Obsession*, NY: Plume, 2004.

**Beltrami's Pseudosphere, 1868**
Darling, D., *UBM*, Wiley, 2004.

**Gros's *Théorie du Baguenodier*, 1872**
Sometime around 1500, Italian mathematician Luca Pacioli was first to mentioned the puzzle in Europe. John Wallis analyzed it in his *Algebra* in 1685.

Darling, D., *UBM*, Wiley, 2004.
Gardner, M., *Knotted Doughnuts and Other Mathematical Entertainments*, Freeman, 1986.
Knuth, D., *The Art of Computer Programming*, Boston: MA, Addison-Wesley, 1998.

**Fifteen Puzzle, 1874**
Slocum, J., Sonneveld, D., *The 15 Puzzle*, Beverley Hills, CA: Slocum Puzzle Foundation, 2006.

**Cantor's Transfinite Numbers, 1874**
Cantor's most important work relating to transfinite numbers spanned the years from about 1874 to 1883. He fully explored his thoughts on transfinite numbers in his best-known work *Beiträge zur Begründung der transfiniten Mengelehre*,1895.

Cantor's first proof demonstrating that the set of all real numbers is uncountable, and that no one-to-one correspondence can exist between the real numbers and natural numbers, was formulated in 1873 and published in: J. *Reine Angew. Math.* 77:258;1874.

Dauben, J., *Georg Cantor*, HUP, 1979.

**Harmonic Analyzer, 1876**

Montgomery, H. C., *J. Acoust. Soc. Am.* **10**:87;1938.
Thomson, W., *Proc. Royal Soc. London* **27**:371;1878.

**Ritty Model I Cash Register, 1879**

"James Ritty," *tinyurl.com/6u2so*.
Cortada, J., *Before the Computer*, PUP, 1993.

**Venn Diagrams, 1880**

Diagrams quite similar to Venn diagrams appear in Leonhard Euler's *Opera Omnia* a century before Venn's work.

Edwards, A., *Cogwheels of the Mind*, Baltimore, MD: Johns Hopkins Univ. Press, 2004.
Grünbaum, B., *Math. Mag.* **48**:12-23;1975.
Hamburger, P., *tinyurl.com/6pp86o*.

**Klein Bottle, 1882**

Stoll, C., *tinyurl.com/92rp*.

**Peano Axioms, 1889**

"Peano Axioms," *tinyurl.com/6ez7a7*.

**Peano Curve, 1890**

Bartholdi, J., *tinyurl.com/5dtkn4*.
Darling, D., *UBM*, Wiley, 2004.
Gardner, M., *Penrose Tiles to Trapdoor Ciphers*, MAA, 1997.
Platzman, L., Bartholdi, J., *J. Assoc. Comput. Mach.* **36**:719;1989.
Vilenkin, N., *In Search of Infinity*, NY: Springer 1995.

**Wallpaper Groups, 1891**

B. Grünbaum notes that the precise number of wallpaper patterns in the Alhambra is ill-defined until we consider whether colors should be taken into account.

Coxeter, H., *Introduction to Geometry*, Wiley, 1969.
Darling, D., *UBM*, Wiley, 2004.
Gardner, M., *New Mathematical Diversions*, MAA, 1995.
Grünbaum, B., *Notices Am. Math. Soc.* **56**:1;2006.

**Sylvester's Line Problem, 1893**

Malkevitch, J., *tinyurl.com/55ecl5*.

**Proof of the Prime Number Theorem, 1896**

Weisstein, E. *tinyurl.com/5puyan*.
Zagier, D., *Math. Intelligencer* **0**:7;1977.

**Pick's Theorem, 1899**

Darling, D., *UBM*, Wiley, 2004.

**Morley's Trisector Theorem, 1899**

John Conway also recently presented a simple proof of Morley's theorem. For more information, see S. Roberts.

Francis, R., *tinyurl.com/6hyguo*.
Morley, F., *My One Contribution to Chess*, NY: B. W. Huebsch, 1945.
Roberts, S., *King of Infinite Space*, NY: Walker, 2006.

**Hilbert's 23 Problems, 1900**

Yandell, B., *Honors Class*, AKP, 2003.

**Boy's Surface, 1901**

Jackson, A., *Notices Am. Math. Soc.* **49**:1246;2002.

**Barber Paradox, 1901**

Joyce, H., *tinyurl.com/63c5co*.
Russell, B., *Mysticism and Logic and Other Essays*, London: G. Allen & Unwin, 1917.

**Poincaré Conjecture, 1904**

Mackenzie, D., *Science*, **314**:1848;2006.
Nasar, S., Gruber, D., *New Yorker*, p. 44, Aug. 28, 2006.
"Poincaré Conjecture," *tinyurl.com/395gbn*.

**Zermelo's Axiom of Choice, 1904**

Darling, D., *UBM*, Wiley, 2004.
Schechter, E., *tinyurl.com/6bk6zy*.

**Brouwer Fixed-Point Theorem, 1909**

Beran, M. *tinyurl.com/595q4d*.
Darling, D., *UBM*, Wiley, 2004.
Davis, M., *The Engines of Logic*, Norton, 2000.

**Normal Number, 1909**

Darling, D., *UBM*, Wiley, 2004.

**Boole's Philosophy and Fun of Algebra, 1909**

Peterson, I., *tinyurl.com/5bnetc*.

***Principia Mathematica*, 1910–1913**

Irvine, A., *tinyurl.com/aothp*.
Modern Library's Top 100 Nonfiction Books, *tinyurl.com/6pghuw*.

**Hairy Ball Theorem, 1912**

Choi, C., *tinyurl.com/5wfk5h*.
DeVries, G., Stellacci, F., et al. *Science* **315**:358;2007.

**Infinite Monkey Theorem, 1913**

Note also that the use of the term *almost surely* in the first sentence of this entry is a mathematical way of saying that the monkey will type a finite text with probability one, assuming we allow an infinite number of trials.

Borel, É., *J. Phys.* **3**:189;1913.
Eddington, A., *The Nature of the Physical World*, NY: Macmillan, 1928.

**Bieberbach Conjecture, 1916**

Mehrtens, H., "Ludwig Bieberbach and Deutsche Mathematik," in *Studies in the History of Math.*, Phillips, E., ed., MAA, 1987. Source for the Bierberbach quote on the Jews.
Sabbagh, K., *tinyurl.com/5969je*.

**Johnson's Theorem, 1916**

Kimberling, C., *tinyurl.com/6a7o96*.
Wells, D., *The Penguin Dictionary of Curious and Interesting Geometry*, NY: Penguin, 1992.

**Brun's Constant, 1919**

Gardner, M., *Sci. Am.* **210**:120;1964.
Granville, A., *Resonance* **3**:71;1998.
Peterson, I., *tinyurl.com/5db4tw*.

**Googol, c. 1920**

Scholars are sometimes divided as to the birthday of Sirotta (e.g., 1911 or 1929) and the date that *googol* was first coined, 1920 or 1938.

Kasner, E., Newman, J., *Mathematics and the Imagination*, Dover, 2001.

**Antoine's Necklace, 1920**

Brechner, B., Mayer, J., *Coll. Math. J.* **19**:306;1988.
Jackson, A., *Notices of the Am. Math. Soc.* **49**:1246;2002.

**Lost in Hyperspace, 1921**

Asimov, D., *The Sciences* **35**:20;1995.

**Alexander's Horned Sphere, 1924**

Gardner, M., *Penrose Tiles to Trapdoor Ciphers*, MAA, 1997.

**Squaring a Rectangle, 1925**

"Zbigniew Moroń," *tinyurl.com/5v3tqw*.

**Hilbert's Grand Hotel, 1925**

Note that there are certain levels or classes of infinity that the Hilbert's Grand Hotel may not be able to accommodate.
G. Cantor showed that there are infinities too big to be counted, which is essentially what happens when we associate each guest with a room number. Note also that the precise date and origin of Hilbert's Hotel is difficult to determine. Hilbert spoke of this hotel during his lectures in the 1920s. His paper "On the Infinite" was delivered in 1925.

Gamow, G., *One, Two, Three…Infinity*, NY: Viking Press, 1947.

**Menger Sponge, 1926**

"Fractal Fragments," *tinyurl.com/5sog2j*.

"The Menger Sponge," *tinyurl.com/58hy6p*.

**Differential Analyzer, 1927**
Bush, V., Gage, F., Stewart, H., *J. Franklin Inst.* **203**:63;1927.
Bush, V., *tinyurl.com/cxzzf*.

**Ramsey Theory, 1928**
Graham, R., Spencer, J., *Sci. Am.* **263**:112;1990.
Hoffman, P., *The Man Who Loved Only Numbers*, NY: Hyperion, 1999.

**Gödel's Theorem, 1931**
Gödel demonstrated the incompleteness of the theory of *Principia Mathematica*, discussed in the entry *Principia Mathematica*.

Hofstadter, D., *Gödel, Escher, Bach*, NY: Basic Books, 1979.
Wang, H., *Reflections on Kurt Gödel*, MIT, 1990.

**Champernowne's Number, 1933**
Belshaw, A., Borwein, P., *tinyurl.com/6mms3d*.
Von Baeyer, H., *Information*, HUP, 2004.

**Bourbaki: Secret Society, 1935**
Aczel, A., *The Artist and the Mathematician*, TMP, 2006.
Mashaal, M., *Bourbaki*, AMS, 2006.

**Voderberg Tilings, 1936**
Grünbaum, B., Shephard G., *Math. Teach.* 88:50;1979.
Grünbaum, B., Shephard, G., *Tilings and Patterns*, Freeman, 1987.
Rice, M., Schattschneider D., *Math. Teach.* **93**:52;1980.

**Birthday Paradox, 1939**
Gardner, M., *Knotted Doughnuts and Other Mathematical Entertainments*, Freeman, 1986.
Peterson, I., *tinyurl.com/53w78*.

**Polygon Circumscribing, c. 1940**
Bouwkamp, C., *Indagationes Math.* 27:40;1965.
Kasner, E., Newman, J., *Mathematics and the Imagination*, Dover, 2001.

**Hex, 1942**
Gale, D., *Am. Math. Monthly* **86**:818;1979.
Gardner, M., *Hexaflexagons and Other Mathematical Diversions*, S&S, 1959.
Nasar, S., *A Beautiful Mind*, NY: Touchstone, 2001.

**Pig Game Strategy, 1945**
Neller, T., Presser, C., *UMAP J.* **25**:25;2004.
Neller, T., Presser, C. *tinyurl.com/6fqyht*.
Peterson, I., *tinyurl.com/5tnteq*.
Scarne, J., *Scarne on Dice*, Harrisburg, PA: Military Service Publishing Co., 1945.

**Von Neumann's Middle-Square Randomizer, 1946**
Hayes, B., *Am. Scient.* **89**:300;2001,

**Gray Code, 1947**
Gardner, M., *Knotted Doughnuts and Other Mathematical Entertainments*, Freeman, 1986.
"What Are Gray Codes?" *tinyurl.com/5txwee*.

**Information Theory, 1948**
Tallack, P., *The Science Book*, W&N, 2003.

**Curta Calculator, 1948**
More accurately, Herzstark was a "half-Jew," as one Nazi referred to him when Herzstark tried to protect his friends from the Gestapo. His father was Jewish and his mother Catholic.

Furr, R., *tinyurl.com/hdl3*.
Ifrah, G., *The Universal History of Computing*, Wiley, 2002.
Saville, G., *tinyurl.com/5da57m*.
Stoll, C., *Sci. Am.* **290**:92;2004.

**Császár Polyhedron, 1949**
Császár, Á., *Acta Sci. Math. Szeged*, **13**:140;1949. This paper has no figures, which may explain why this did not stimulate further work until the 1970s.
Darling, D., *UBM*, Wiley, 2004.
Gardner, M., *Time Travel and other Mathematical Bewilderments*, Freeman:1987.

**Nash Equilibrium, 1950**
Nasar, S., *A Beautiful Mind*, S&S, 1998.
Tallack, P., *The Science Book*, W&N, 2003.

**Coastline Paradox, c. 1950**
Mandelbrot, B., *Science*, **156**:636;1967.
Richardson, L., *Statistics of Deadly Quarrels*, Pacific Grove, CA: Boxwood Press, 1960.

**Prisoner's Dilemma, 1950**
Poundstone, W., *Prisoner's Dilemma*, NY: Doubleday, 1992.

**Cellular Automata, 1952**
Von Neumann, J., *Theory of Self-Reproducing Automata*, Urbana: IL: U. Illinois Press, 1966.
Wolfram, S., *A New Kind of Science*, Champaign, IL: Wolfram Media, 2002.

**Martin Gardner's Mathematical Recreations, 1957**
Berlekamp, E., Conway, J., Guy, T., *Winning Ways for Your Mathematical Plays*, Burlington, MA: Elsevier, 1982.
Gardner, M., *Martin Gardner's Mathematical Games* (CD-ROM), MAA, 2005.
Jackson, A., *Notices Am. Math. Soc.* **52**:602;2005.

**Gilbreath's Conjecture, 1958**
Norman Gilbreath told me that "Erdös believed my conjecture is probably true and it will be 200 years before it is proved."

Guy, R., *Am. Math. Monthly* **95**:697;1988.
Guy, R., *Math. Mag.* **63**:3;1990.
Guy, R., "Gilbreath's Conjecture," in *Unsolved Problems in Number Theory*, NY: Springer, 1994.
Odlyzko, A., *Math. Comput.* **61**:373;1993.

**Turning a Sphere Inside Out, 1958**
Note that although sphere eversion is theoretically possible, a circle cannot be turned inside out.

**Platonic Billiards, 1958**
Cipra, B., *Science* **275**:1070;1997.

**Outer Billiards, 1959**
Cipra, B., *Science*, **317**:39;2007.
Schwartz, R., *tinyurl.com/2mtqzp*.

**Newcomb's Paradox, 1960**
Gardner, M., *The Colossal Book of Mathematics*, Norton, 2001.
Nozick, R., "Newcomb's Problem and Two Principles of Choice," in *Essays in Honor of Carl Hempel*, Rescher, N., ed., Dordrecht: D. Reidel, 1969.

**Sierpiński Numbers, 1960**
Peterson, I., *tinyurl.com/674cu3*.
"Seventeen or Bust," *seventeenorbust.com*.
Zagier, D., *Math. Intelligencer*, **0**:7;1977.

**Chaos and the Butterfly Effect, 1963**
Gleick, J., *Chaos*, NY: Penguin, 1988.
Lorenz, E., *J. Atmos. Sci.* **20**:130;1963.

**Ulam Spiral, 1963**
Gardner, M., *The Sixth Book of Mathematical Games from Scientific American*, UCP, 1984.

**Continuum Hypothesis Undecidability, 1963**
Recent work by mathematician W. Hugh Wooden may suggest the possibility that the continuum hypothesis is false, and, indeed, the hypothesis continues to remain a hot topic of contemporary research.

Cohen, P., *Proc. Natl. Acad. Sci.* **50**:1143;1963.
Gödel, K., *Am. Math. Monthly* **54**:515;1947.
Woodin, W., *Notices of the Am. Math. Soc.* **48**:567;2001.

**Superegg, c. 1965**
Gardner, M., *Mathematical Carnival*, NY: Vintage, 1977.

**Fuzzy Logic, 1965**
Tanaka, K., *An Introduction to Fuzzy Logic for Practical Applications*, NY: Springer, 1996.

**Instant Insanity, 1966**
Armbruster, F., *tinyurl.com/65epdv*.
Peterson, I., *tinyurl.com/6pthxh*.

**Langlands Program, 1967**
Gelbart, S., *Bull. Am. Math. Soc.* **10**:177;1984.
Gelbart, S., "Number Theory and the Langlands Program," Guangzhou, China, *Intl. Instruct. Conf.*, 2007.
Mackenzie, D., *Science* **287**:792;2000.
Mozzochi, C., *The Fermat Diary*, AMS, 2000.

**Sprouts, 1967**
Berlekamp, E., Conway, J., Guy, R., *Winning Ways for Your Mathematical Plays*, Burlington, MA: Elsevier, 1982.
Focardi, R., Luccio, F., *Discrete Appl. Math.* **144**:303;2004.
Gardner, M., *Sci. Am.* **217**:112;1967.
Lemoine, J., Viennot, S., *tinyurl.com/56bfcd*, *tinyurl.com/6kazbt*.
Peterson I., *tinyurl.com/6l3huh*.

**Catastrophe Theory, 1968**
Darling, D., *UBM*, Wiley, 2004.
Thom, R., with response by Zeeman, E., "Catastrophe Theory," in *Dynamical Systems-Warwick 1974*, Manning, A., ed., NY: Springer, 1975.
Zahler, R., Sussman, H., *Nature* **269**:759;1977.

**Tokarsky's Unilluminable Room, 1969**
Darling, D., *UBM*, Wiley, 2004.
Stewart, I., *Sci. Am.* **275**:100;1996.
Stewart, I., *Math Hysteria*, OUP, 2004.

**Donald Knuth and Mastermind, 1970**
Chen, Z. et al., "Finding a Hidden Code by Asking Questions," in *Proc. 2nd Annual Intl. Conf. Comput. Combinat.*, Hong Kong, 1996.
Knuth, D., J. *Recr. Math.* **9**:1;1976.
Koyama, K., Lai, T., J. *Recr. Math.* **25**:251;1993.

**Erdös and Extreme Collaboration, 1971**
Hoffman, P., *The Man Who Loved Only Numbers*, NY: Hyperion, 1999.
Schechter, B., *My Brain Is Open*, S&S, 2000.

**HP-35: First Scientific Pocket Calculator, 1972**
Lewis, B., *tinyurl.com/5t37nr*.

**Penrose Tiles, 1973**
R. Ammann independently discovered these kinds of tilings at approximately the same time as Penrose. B. Grünbaum and G. Shephard wrote "in 1973 and 1974 Roger Penrose discovered 3 sets of aperiodic prototiles." For example, the first set, denoted by P1, consists of 6 tiles based on rhombs, regular pentagons, pentacles, and "half-pentacles," with edges, modified by projections and indentations. The second set of aperiodic tiles, denoted by P2, was discovered by Penrose in 1974 and contains only 2 tiles.

Chorbachi, W., Loeb, A., "Islamic Pentagonal Seal" in *Fivefold Symmetry*, Hargittai, I., ed.,WS, 1992.
Gardner, M., *Penrose Tiles to Trapdoor Ciphers*, Freeman, 1988.
Grünbaum, B., Shephard, G., *Tilings and Patterns*, Freeman, 1987.
Lu, P., Steinhardt, P., *Science* **315**:1106;2007.
Makovicky, E., "800-Year-Old Pentagonal Tiling from Maragha, Iran, and the New Varieties of Aperiodic Tiling it Inspired," in *Fivefold Symmetry*, Hargittai, I., ed., WS, 1992.
Penrose, R., *Bull. of the Inst. Math. Applic.* **10**:266;1974.
Rehmeyer, J., *tinyurl.com/64ppgz*.
Senechal, M., "The Mysterious Mr. Ammann," *Math. Intell.* **26**:10;2004.

**Art Gallery Theorem, 1973**
If the polygon is convex, its entire interior can be viewed from any single vertex.

Chvátal, V., "A Combinatorial Theorem in Plane Geometry," *J. Combinat. Theory* **18**:39;1975.
Do, N., *Austral. Math. Soc. Gaz.* **31**:288;2004.
Fisk, S., *J. Combinat. Theory, Ser. B* **24**:374;1978.
O'Rourke, J., *Art Gallery Theorems & Algorithms*, OUP, 1987.

**Rubik's Cube, 1974**
Longridge, M., *cubeman.org*.
Velleman, D., *Math. Mag.* **65**:27;1992.

**Chaitin's Omega, 1974**
Chaitin, G., *J. ACM.* **22**:329;1975. Omega first appears in this publication. The term also was used in a 1974 IBM Research Division Technical Report.
Chaitin, G., *Meta Math?*, NY: Pantheon, 2005.
Chown, M., *New Sci.* **169**:28;2001.
Gardner, M., *Fractal Music, Hypercards and More*, Freeman, 1991. Contains the writings of C. Bennett.
Lemonick, M., *tinyurl.com/59q796*.

**Surreal Numbers, 1974**
Conway, J., Guy, R., *The Book of Numbers*, NY: Copernicus, 1996.
Gardner, M., *Mathematical Magic Show*, MAA, 1989.
Knuth, D., *Surreal Numbers*, Reading, MA: Addison-Wesley, 1974.

**Fractals, 1975**
Many visually interesting fractals are generated using iterative methods that were first introduced by mathematicians G. Julia and P. Fatou from 1918 to 1920.

Mandelbrot, B., *The Fractal Geometry of Nature*, Freeman, 1982.

**Feigenbaum Constant, 1975**
Feigenbaum, M., "Computer Generated Physics," in *20th Century Physics*, Brown, L. et al., eds., NY: AIP Press, 1995.
May, R., *Nature* **261**:459;1976.

**Public-Key Cryptography, 1977**
Diffie, W., Hellman, M., *IEEE Trans. Info. Theory* **22**:644;1976.
Hellman, M., *Sci. Am.* **241**:146;1979.
Lerner, K., Lerner, B., eds., *Encyclopedia of Espionage Intelligence and Security*, Farmington Hills, MI, Gale Group, 2004.
Rivest, R., Shamir, A., Adleman, L., *Commun. ACM* **21**:120;1978.

**Szilassi Polyhedron, 1977**
Gardner, M., *Fractal Music, Hypercards and More*, Freeman, 1992.
Peterson, I., *tinyurl.com/65p8ku*.
Szilassi, L., *Struct. Topology* **13**:69;1986.

**Ikeda Attractor, 1979**
Ikeda, K., *Optics Commun.* **30**:257;1979.
Strogatz, S., *Nonlinear Dynamics and Chaos*, NY: Perseus, 2001.

**Spidrons, 1979**
Erdély, D., *www.spidron.hu*.
Peterson, I., *Sci. News* **170**:266;2006.

**Mandelbrot Set, 1980**
Clarke, A., *The Ghost from the Grand Banks*, NY: Bantam, 1990.
Darling, D., *UBM*, Wiley, 2004.
Mandelbrot, B., *The Fractal Geometry of Nature*, Freeman, 1982.
Wegner, T., Peterson, M., *Fractal Creations*, Corte Madera, CA: Waite Group Press, 1991.

**Monster Group, 1981**
Conway, J., Sloane, N., "The Monster Group and its 196884-Dimensional Space" and "A Monster Lie Algebra?"

in *Sphere Packings, Lattices, and Groups*, NY: Springer, 1993.
Griess, R., *Invent. Math.* **69**:1;1982.
Griess, R., Meierfrankenfeld, U., and Segev, Y., *Ann. Math.* **130**: 567;1989.
Ronan, M., *Symmetry and the Monster*, OUP, 2006.

**Ball Triangle Picking, 1982**

Buchta, C., *Ill. J. Math.* **30**:653;1986.
Hall, G., *J. Appl. Prob.* **19**:712;1982.
Weisstein, E., *tinyurl.com/5o2sap*.

**Jones Polynomial, 1984**

The HOMFLY polynomial got its name from the last names of its co-discoverers: Hoste, Ocneanu, Millett, Freyd, Lickorish, and Yetter.

Adams, C., *The Knot Book*, AMS, 2004.
Devlin, K., *The Language of Mathematics*, NY: Owl Books, 2000.
Freyd, P. et al., *Bull. AMS* **12**:239;1985.
Jones, V., *Bull. AMS* **12**:103;1985.
Przytycki, J., Traczyk, P., *Proc. AMS* **100**:744;1987.
Witten, E., *Commun. Math. Phys.* **21**:351;1989.

**Weeks Manifold, 1985**

Cipra, B., *Science* **317**:38;2007.
Gabai, D. et al., *tinyurl.com/6mzsso*.
Weeks, J., *Hyperbolic Structures on 3-Manifolds*, Princeton Univ. Ph.D. thesis, Princeton University, 1985.
Weeks, J., *The Shape of Space*, NY: Marcel Dekker, Inc., 2001.

**Andrica's Conjecture, 1985**

Andrica, D., *Revista Matematică*, **2**:107;1985.
Andrica, D., *Studia Univ. Babeş-Bolyai Math* **31**:48;1986.
Guy, R., *Unsolved Problems in Number Theory*, NY: Springer, 1994.

**The *ABC* Conjecture, 1985**

Darling, D., *UBM*, Wiley, 2004.
Goldfeld, D., *Math Horizons*, Sept:26;1996.
Goldfeld, D., *The Sciences*, March:34;1996.
Masser, D., *Proc. Am. Math. Soc.* **130**:3141;2002.
Nitaq, A., *tinyurl.com/6gaf87*.
Oesterlé, J., *Astérisque* **161**:165;1988.
Peterson, I., *tinyurl.com/5mgwvk*.

**Audioactive Sequence, 1986**

Conway, J., *Eureka*, **46**:5;1986.
Conway, J., Guy, R., *The Book of Numbers*, NY: Copernicus, 1996.

**Mathematica, 1988**

Trademarks (Mathematica: Wolfram Research; Maple: Waterloo Maple; Mathcad: Mathsoft; MATLAB: MathWorks).

Berlinski, D., *The Sciences*, Jul./Aug.:37;1997.
Borwein, J., Bailey, D., *Mathematics by Experiment*, AKP, 2003.
"Wolfram Research," *wolfram.com*.

**Murphy's Law and Knots, 1988**

Deibler, R., et al. *BMC Molec. Biol.* **8**:44;2007.
Matthews, R., *Math. Today* **33**:82;1997.
Peterson, I., *tinyurl.com/5r8ccu, tinyurl.com/5nlrms*.
Raymer, D., Smith, D., *Proc. Natl. Acad. Sci.* **104**:16432;2007.
Sumners, D., Whittington, S., *J. Phys. A* **21**:1689;1988.

**Butterfly Curve, 1989**

Fay, T., *Am. Math. Monthly* **96**,442;1989.

**The On-Line Encyclopedia of Integer Sequences, 1996**

Sloane, N., "My Favorite Integer Sequences," in *Sequences and their Applications*, Ding, C., Helleseth, T., Niederreiter H., eds., NY: Springer, 1999.
Sloane, N., *Notices of the AMS* **50**:912;2003.
Stephan, R., *tinyurl.com/6m84ca*.

**Eternity Puzzle, 1999**

Selby, A., *tinyurl.com/5n6dwf*.
Weisstein, E., *tinyurl.com/6lyxdl*.

**Parrondo's Paradox, 1999**

Note that in game *A* or game *B* played separately, you are guaranteed to lose in the long run for any allowed value of $x$ up to 0.1.

Abbott, D., *tinyurl.com/6xwg44*.
Blakeslee, S., *tinyurl.com/6yvd92*.
Harmer, G., Abbott, D., *Nature* **402**:864;1999.
Harmer, G., Abbott, D., *Stat. Sci.* **14**:206;1999.

**Solving of the Holyhedron, 1999**

Hatch, D., *tinyurl.com/5rttaq*.
Vinson, J., *Discr. Comput. Geom.* **24**:85;2000.

**Bed Sheet Problem, 2001**

Historic. Soc. Pomona Vall., *tinyurl.com/5cv4ce*.

**Solving the Game of Awari, 2002**

Peterson, I., *tinyurl.com/65hnet*.
Romein, J., Bal, J., *IEEE Computer* **36**:26;2003.

**Tetris is NP-Complete, 2002**

The scoring formula for most of the Tetris products takes into account the fact that certain row clearing operations are more difficult than others and thus should be awarded more points.

Breukelaar, R., Demaine, E., et al. *Intl. J. Comput. Geom. Appl.* **14**:41;2004.
Demaine, E., Hohenberger, S., Liben-Nowell, D., "Tetris Is Hard, Even to Approximate," *Comput. Combinat., 9th Ann. Intl. Conf.*, 2003.
Peterson, I., *tinyurl.com/5mqt84*.

***NUMB3RS*, 2005**

Frazier, K., *tinyurl.com/6e2f8h*.
Weisstein, E., *tinyurl.com/5n4c99*.

**Checkers Is Solved, 2007**

Cho, A., *Science* **317**:308;2007.
Schaeffer, J., et al. *Science* **317**:1518;2007.

**The Quest for Lie Group $E_8$, 2007**

Also in 2007, physicist A. G. Lisi speculated that $E_8$ explained how the various fundamental particles in physics may result from different aspects of the strange and beautiful symmetries of $E_8$.

Collins, G., *Sci. Am.* **298**:30;2008.
Lisi, A. G., *tinyurl.com/6ozgdh*.
Mackenzie, D., *Science*, **315**:1647;2007.
Merali, Z., *New Scientist*, **196**:8;2007.
American Institute of Math., *aimath.org/E8/*.

**Mathematical Universe Hypothesis, 2007**

Tegmark's theory was partly based on a talk given at the symposium "Multiverse and String Theory" held in 2005 at Stanford University. The seeds of his theory were planted in other papers in the late 1990s but reached full flower in 2007. Other researchers, such as K. Zuse, E. Fredkin, and S. Wolfram have suggested that the physical universe may be running on a cellular automaton.

Collins, G., *Sci. Am.* **298**:30, 2008.
Fredkin, E., *Physica D* **45**:254, 1990.
Tegmark, M., *New Scientist* **195**:39, 2007.
Tegmark, M., *tinyurl.com/6pjjxp*.
Wolfram, S., *A New Kind of Science*, Champaign, IL: Wolfram Media, 2002.
Zuse, K., *Elektronische Datenverarbeitung* **8**:336;1967.

# Index

# Photo Credits

Because several of the ancient and rare documents shown in this book were difficult to acquire in a clean and legible form, I have sometimes taken the liberty to apply image-processing techniques to remove dirt and scratches, enhance faded portions, and occasionally add a slight coloration to a black-and-white document in order to highlight certain details or simply to make an image more compelling to look at. I hope that historical purists will forgive these slight artistic touches and understand that my goal was to create an attractive book—rich in history and detail—that is both aesthetically interesting and alluring even to students and laypeople. My love for the incredible depth and diversity of mathematics, art, and history should be evident through the photographs and drawings exhibited throughout the book.

**Images © Clifford A. Pickover:** pages 49, 55, 75, 97, 103, 106, 112, 113, 133, 147, 148, 151, 157, 159, 169, 174, 179, 185, 189, 197, 199, 201, 203, 272 (left), 277, 291, 295, 297, 301, 321, 333, 337, 339, 345, 353, 361, 383, 385, 399, 425, 427, 437, 441, 443, 450, 469, 470, 477, 493, 497, 499, 502, 517

**Images © Teja Krašek:** pages 59, 65, 183, 245, 373, 408, 415

**Images by Jos Leys (josleys.com):** pages 125, 139, 145, 225, 377, 449, 459, 461, 473, 479

**Images © Paul Nylander, bugman123.com:** pages 51, 117, 155, 193, 217, 255, 257, 303, 335

**Used under license from Shutterstock.com:** p. 21, Image © shaileshnanal, 2009; p. 23, Image © 2265524729, 2009; p. 31, Image © Mikael Damkier, 2009; p. 33, Image © rfx, 2009; p. 43, Image © GJS, 2009; p. 47 and 177, Image © James Steidl, 2009; p. 61, Image © Hisom Silviu, 2009; p. 63, Image © Andreas Guskos, 2009; p. 67, Image © Ovidiu Iordachi, 2009; p. 91, Image © Sebastian Knight, 2009; p. 93, Image © Olga Lyubkina, 2009; p. 99, Image © Tan Kian Khoon, 2009; p. 101, Image © Ella, 2009; p. 123, Image © Jiri Moucka, 2009; p. 127, Image © Maree Stachel-Williamson, 2010; p. 141, Image © Geanina Bechea, 2009; p. 163, Image © Maxx-Studio, 2009; p. 167, Image © MWaits, 2009; p. 213, Image © Yuri Arcurs, 2010; p. 231, Image © Steve Mann, 2009; p. 243 and 351, Image © photobank.kiev.ua, 2009; p. 275, Image © Lee Torrens; p. 279, Image © Holger Mette, 2009; p. 289, Image © Rafael Ramirez Lee, 2009; p. 293, Image © Yare Marketing p. 305, Image © Jeff Davies, 2009; p. 307, Image © Arlene Jean Gee, 2009; p. 313, Image © Andrey Armyagov, 2009; p. 319, Image © Anyka, 2009; p. 327, Image © Vasileika Aleksei, 2009; p. 329, Image © ChipPix, 2009; p. 355, Image © Elena Elisseeva, 2009; p. 379, Image © Jeff Carpenter, 2009; p. 381, Image © Scott Maxwell/ LuMaxArt, 2009; p. 387, Image © Marcus Tuerner, 2009; p. 395, Image © Wayne Johnson, 2009; p. 401, Image © Tischenko Irina, 2009; p. 405, Image © Lou Oates, 2009; p. 412, Image © kotomiti, 2009; p. 419, Image © Zoran Vukmanov Simokov, 2009; p. 435, Image © Jakez, 2009; p. 439, Image © Dmitrijs Mihejevs, 2009; p. 445, Image © Polina Lobanova, 2009; p. 455, Image © Fabrizio Zanier, 2009; p. 490, Image © Ronald Sumners, 2009; p. 491, Image © vladm, 2009; p. 495, Image © Gilmanshin, 2009; p. 501, Image © Robert Kyllo, 2009; p. 503, Image © Armin Rose, 2009; p. 505, Image © Kheng Guan Toh, 2009; p. 507, Image © imageshunter, 2009; p. 509, Image © suravid, 2009

**Other Images:** p. 19 Image Matthias Wittlinger; p. 27 Royal Belgian Institute of Natural Sciences; p. 29 Photo by Marcia and Robert Ascher. In the collection of the Museo National de Anthropologia y Arqueología, Lima, Peru; p. 39. Paul St. Denis and Patrick Grim developed the Fractal Tic Tac Toe image in the Logic Lab of the Philosophy Department at SUNY Stony Brook. An earlier version appeared in St. Denis and Grim, "Fractal Image of Formal Systems," *Journal of Philosophical Logic* 26 (1997), 181-222, with discussion in Ian Stewart, "A Fractal Guide to Tic-Tac-Toe," *Scientific American* 283, no. 2 (2000), 86-88; p. 69 © istockphoto.com/bkindler; p. 71 © istockphoto.com/fanelliphotography; p. 81 © istockphoto.com/-M-I-S-H-A-; p. 95 Mike Simmons, Astronomers Without Borders; p. 109 Photo by Rischgitz/ Getty Images; p. 111 © istockphoto.com/kr7ysztof; p. 129 © Science Museum/Science & Society; p. 131 © istockphoto.com/ihoe; p. 146 By George W. Hart, http://www.georgehart.com; p. 165 Courtesy of The Swiss Post, Stamps & Philately; p. 175 Wikipedia/Matt Britt; p. 187 Courtesy of Dmitry Brant, http://dmitrybrant.com; p. 191 Library of Congress, David Martin, painted 1767; p. 205 Fractal Art by G. Fowler; p. 211 © istockphoto.com/theasis; p. 219 Wikimedia/Carsten Ullrich; p. 223 Brian Mansfield; p. 233 Wikimedia/Leo Fink, using Gaston software; p. 239 Portrait is the frontispiece to Vol. 4 of *The Collected Mathematical Papers of James Joseph Sylvester*, edited by H. F. Baker, Cambridge University Press 1912; p. 241 © istockphoto.com/ nicoolay; p. 247 Ivan Moscovitch; p. 249 Clifford A. Pickover and Teja Krašek; p. 251 Brian C. Mansfield; p. 253 Tibor Majláth; p. 259 U.S. Patent Office; p. 263 © istockphoto.com/MattStauss; p. 267 U.S. Patent Office; p. 269 photographer Clive Streeter © Dorling Kindersley, Courtesy of the Science Museum, London; p. 271 © National Museum of American History; p. 272 (right) and 273 Venn Diagrams created by Edit Hepp and Peter Hamburger; p. 283 Image created by Robert Webb using Stella4D software, http://www.software3d.com; p. 287 Designed and photographed by Carlo H. Sequin, University of California, Berkeley; p. 311 Robert Fathauer; p. 315 Robert Bosch, Oberlin College, DominoArtwork.com; p. 317 Mark Dow, geek art; p. 323 From the book *George Boole: His Life and Work*, by Desmond MacHale (Boole Press); p. 341 Rob Scharein, visit Rob's website at knotplot.com; p. 347 © istockphoto.com/ Pichunter; p. 349 Cameron Browne; p. 357 Image by Paul Bourke and Gayla Chandler, featuring Sydney Renee; p. 359 NASA Headquarters/Greatest Images of NASA; p. 363 Photographed by Oskar Morgenstern. Courtesy of the Archives of the Institute for Advanced Study, Princeton, NJ; p. 365 Peter Borwein; p. 367 © istockphoto.com/ BernardLo; p. 369 Photographed by Stefan Zachow; p. 389 U.S. Army Photo, from K. Kempf, "Historical Monograph: Electronic Computers Within the Ordinance Corps"; p. 393 U.S. Patent Office; p. 397 Wikimedia/Larry McElhiney; p 400 Wikimedia/Elke Wetzig; p. 403 © istockphoto.com/abzee; p. 407 Wikimedia/ © 2005 Richard Ling; p. 409 "Card Colm" Mulcahy, March 2006, Norman, Oklahoma; p. 411 Norman Gilbreath at Cambridge University, England; p. 413 Designed and photographed by Carlo H. Séquin, University of California, Berkeley; p. 417 Richard Evan Schwartz; p. 421 © 2002–2008 Louie Helm; p. 423 Roger A. Johnston, Fractal image created with Apophysis (software available at www.apophysis.org); p. 429 © Philip Gould/ CORBIS; p. 431 © istockphoto.com/ CaceresFAmaru; p. 433 Edward Rothschild; p. 434 © C. J. Mozzochi, Ph.D., Princeton, NJ; p. 447 © 2008 Hewlett-Packard Development Company, L.P. Reproduced with Permission. Photo by Seth Morabito; p. 451 © istockphoto.com/ dlewis33; p. 452 Photo by Zachary Paisley. Rubik's Cube® is a registered trademark of Seven Towns, Ltd.; p. 453 Hans Andersson; p. 456 Thane Plambeck; p. 457 Front cover from Donald Knuth's book, SURREAL NUMBERS, © 1974. Reproduced by permission of Pearson Education, Inc.; p. 463 Steven Whitney at 25yearsofprogramming.com; p. 465 Robert Lord; p. 467 Hans Schepker, Mathematical Artist and Teacher; p. 471 Dániel Erdély © 2005; p. 475 Photography by Robert Griess; p. 481 Image courtesy of Jeff Weeks, www.geometrygames.org; p. 485 David Masser; p. 489 Michael Trott; p. 511 Courtesy of Photofest; p. 515 Wikimedia/ John Stembridge